Q Fever

Volume I
The Disease

Editor

Thomas J. Marrie, M.D.

Professor
Department of Medicine
Assistant Professor
Department of Microbiology
Dalhousie University
Halifax, Nova Scotia, Canada

CRC Press
Boca Raton   **Ann Arbor**   **Boston**

Transferred to Digital Printing 2010

Library of Congress Cataloging-in-Publication Data

Q fever / editor, Thomas J. Marrie.
       p.   cm.
   Includes bibliographical references.
   Contents: Vol. 1. The disease.
   ISBN 0-8493-5984-8 (vol. 1)
   1. Q fever.   I. Marrie, Thomas J.
   [DNLM: 1. Q fever. WC 625 Q11]
RC182.Q35Q25   1990
616.9′225—dc20
DNLM/DLC   89-20971
for Library of Congress   CIP

Direct all inquiries to CRC Press, Inc., 2000 Corporate Blvd., N.W., Boca Raton, Florida, 33431.

© 1990 by CRC Press, Inc.

International Standard Book Number 0-8493-5984-8

Library of Congress Card Number 89-20971
Printed in the United States

This book is dedicated to the people of Maritime Canada who willingly helped in the investigation of Q fever in Nova Scotia, New Brunswick, and Prince Edward Island.

THE EDITOR

Thomas Marrie, M.D., is Professor of Medicine and Head of the Division of Infectious Diseases at Dalhousie University, Halifax, Nova Scotia. Dr. Marrie received his M.D. degree from Dalhousie University in 1970. Following 3 years of Family Practice in Newfoundland, he trained in Internal Medicine at Dalhousie University and in Infectious Diseases at the University of Manitoba, Winnipeg. He obtained his Fellowship of the Royal College of Physicians and Surgeons of Canada in Internal Medicine in 1977 and a certificate of Special Competence in Infectious Diseases in 1983.

Dr. Marrie is a member of the American Society for Microbiology, the American Federation for Clinical Research, the American Society of Rickettsiology, the Canadian Society for Clinical Investigation, and the Canadian Infectious Disease Society.

Dr. Marrie's research on Q fever and pneumonia has been supported by the National Health Research Development Program, Ottawa, the Nova Scotia Heart Foundation, the New Brunswick Heart Foundation, the Nova Scotia Lung Association, and Eli Lilly Canada.

Dr. Marrie has published more than 100 research papers. His current major research interests are the epidemiology of Q fever, pneumonia, and endocarditis.

CONTRIBUTORS

Oswald G. Baca, Ph.D.
Professor
Departments of Biology and Microbiology
School of Medicine
University of New Mexico
Albuquerque, New Mexico

David T. Janigan, M.D.
Professor and Senior Pathologist
Departments of Pathology
Dalhousie University and Victoria General
 Hospital
Halifax, Nova Scotia, Canada

Theodor Khavkin, M.D., Ph.D., Sc.D.
Consultant
Department of Cell Biology
Interferon Sciences, Inc.
New Brunswick, New Jersey

Gerhard H. Lang, D.V.M.
Associate Professor
Department of Veterinary Microbiology
 and Immunology
Ontario Veterinary College
University of Guelph
Guelph, Ontario, Canada

Joanne M. Langley, M.D.
Assistant Professor
Department of Pediatrics
Dalhousie University
Halifax, Nova Scotia, Canada

Barrie P. Marmion, M.D.
Professor
Department of Pathology
University of Adelaide Medica School
Adelaide, South Australia, Australia

T. J. Marrie, M.D.
Professor
Departments of Medicine and Microbiology
Dalhousie University
Halifax, Nova Scotia, Canada

Joseph E. McDade, Ph.D.
Associate Director
Laboratory Science
Center for Infectious Diseases
Centers for Disease Control
Atlanta, Georgia

Richard A. Ormsbee, Ph.D.
Retired
U.S. Public Health Service
Hamilton, Montana

Didier Raoult, Ph.D.
Professor
Department of Microbiology
University Hospital
Marseille, France

Asma Raza, M.D.
Assistant Professor
Department of Pathology
Dalhousie University
Halifax, Nova Scotia, Canada

David M. Waag, Ph.D.
Investigator
Department of Intracellular Pathogens
Bacteriology Division
U. S., Army Medical Research Institute
 of Infectious Disease
Fort Detrick, Federick, Maryland

Michael R. Yeaman, B.S.
Teaching Assistant
Department of Biology
University of New Mexico
Albuquerque, New Mexico

TABLE OF CONTENTS

INTRODUCTION

Q FEVER — THE DISEASE

The objective of this book is to present in detail the various clinical pictures that result from *Coxiella burnetii* infection.

The first chapter is appropriately entitled "Historical Aspects of Q Fever" by Dr. Joseph E. McDade, a noted rickettsiologist. Dr. McDade leads us through the fascinating history of the early events in the elucidation of the epidemiology of Q fever and the discovery of *C. burnetii*, its etiological agent. He illustrates his chapter with pictures of the scientists who were involved in the initial work on Q fever. The picture of the old Cannon abattoir in Brisbane, Australia, where the first cases of Q fever occurred, is also shown.

In Chapter 2, Dr. G. H. Lang from the Ontario Veterinary College in Guelph, Ontario, Canada reviews coxiellosis in animals. While Dr. Lang does give us a listing of the current status of coxiellosis in a variety of animals including livestock and wildlife, he does more than this. He attempts to show that in the past 50 years since the recognition of this zoonosis, we have completed the circle. Initially "abattoir fever" resulted from exposure to *Coxiella*-infected cattle. Subsequently, the efforts focused on the search for an occupation-related vector and wildlife reservoirs. Then, heavy emphasis was placed on sheep and goats as the source of human disease. Finally, the view that dairy cows and, to a lesser extent, milk goats and camels are potential reservoirs from which accessory hosts, including humans, dogs, cats, sheep, and invertebrates become infected, has evolved.

In Chapter 3, the epidemiology of Q fever is reviewed. The aerosol mode of transmission is the most important one. There is suggestive evidence that in some instances ingestion of infected material rarely results in infection in man, although it is undoubtedly important in animals. Rarely, percutaneous transmission and vertical transmission of Q fever have occurred. Person-to-person transmission is distinctly unusual, so there is no need to isolate these patients when they are hospitalized.

The evolution of Q fever in Nova Scotia is used to illustrate many of the features of the epidemiology of this illness. Q fever was first recognized in Nova Scotia in 1979, and since then, a number of features have emerged that may be unique to Nova Scotia. Pneumonia is the predominant manifestation of Q fever in this province, but endocarditis occurs as well. In contrast, hepatitis as the only manifestation of Q fever has not occurred to date. A unique feature is the implication of infected cats in the transmission of this disease. Transmission occurs at the time that the cat gives birth to kittens. Another unusual feature has been the association of Q fever with exposure to infected rabbits. The questions that are raised by the Nova Scotia experience are still unanswered. How did Q fever spread to this province? Why is the rate of pneumonia so high? Why are cats the major reservoir for the infection?

In Chapter 4, Dr. Khavkin describes the interaction of *C. burnetii* with the host. He uses material that he has accumulated over many years of study to illustrate the process whereby *C. burnetii* infects ticks, mice, guinea pigs, cell cultures, and chicken embryos. The sequence of events following infection of the whole animal is given for the guinea pig, while the infected cell culture presents an important model for the study of the basic aspects of *C. burnetii*-cell interactions. The interaction of professional and nonprofessional phagocytes is also described. Dr. Khavkin's excellent photomicrographs illustrate many of these processes and lead in to Dr. David Waag's discussion in Chapter 5 of the immune response to *C. burnetii* infection. The interaction of *C. burnetii* with the host's immune system is a complex one. This microorganism can both enhance and suppress various limbs of the immune system. The inflammatory response may aid the growth and dissemination of *C. burnetii*, for example, the alveolar macrophages sequester *C. burnetii* in acidic vacuoles during translocation to the regional tracheal lymph nodes. Acute phase proteins enhance uptake of *C. burnetii* by cells before the initiation of a specific response. Ceruloplasmin, a glyco-protein involved in most of the plasma copper recycled in the liver, may explain the tissue tropism of *C. burnetii*. Ceruloplasmin receptors are found in the aorta and the cardiac valves.

The role of antibodies in the host defense against *C. burnetii* is equivocal. Opsonizing antibody may alter the course of infection by accelerating initiation of the cellular immune response. It is the latter, however, that is most important in the control of *C. burnetii* infection. The antibody response to phase I and phase II antigens is used to diagnose Q fever serologically by means of a variety of tests including the complement fixation, microimmuno-fluorescence, and enzyme-linked immunosorbent assay.

Much has yet to be learned about the role of cytokines in modulation of the host response to *C. burnetii,* but we do know that interferon inhibits the growth of *C. burnetii* in mouse fibroblasts. Also, in chronic Q fever there is lymphocyte unresponsiveness to *C. burnetii* antigens while these lymphocytes proliferate normally in response to other antigens.

The clinical features of acute Q fever are described in Chapter 6. A study of pneumonia due to *C. burnetii, Mycoplasma pneumoniae,* and *Legionella pneumophila* allows one to compare and contrast the features of "atypical pneumonia" due to these three pathogens. Case histories of individual patients are used to give the reader a feel for the clinical course of many of these patients. Finally, the literature on the less frequent manifestations of Q fever is reviewed under headings for the organs that are infected, such as the thyroid, myocardium, etc.

In Chapter 7, Drs. Janigan and Marrie describe the pathology of Q fever pneumonia. There is little information regarding the pathology of the lung in Q fever pneumonia, but these workers have had the opportunity to examine material from three patients with Q fever pneumonia. In two of these patients *C. burnetii* was the only organism implicated in pneumonia. Both alveolar and interstitial inflammation was noted, and the inflammatory infiltrate consisted of macrophages and lymphocytes. The second patient had a pseudotumor due to *C. burnetii.*

Chapter 8 deals with Q fever hepatitis. This entity has its own chapter because the liver can be involved in both acute and chronic Q fever. Some investigators feel that the liver may be the site where the organism remains in the latent period between acute and chronic disease. It is evident that features of this illness as it occurred in Ontario, Canada may be unique in that these patients presented with fever of unknown origin and granulomatous hepatitis due to *C. burnetii.*

The most serious complication of Q fever infection — endocarditis — is discussed in Chapter 9. In this chapter, data from Dr. Raoult's experience in France and Dr. Marrie's experience in Nova Scotia are combined. The gross, microscopic, and ultrastructural pathology of the valves in Q fever endocarditis are shown. The course of two patients is given to illustrate the diagnostic and therapeutic problems that these patients pose. Finally, recommendations regarding the management of these patients from the authors' experience are given.

In Chapter 10, Dr. Joanne Langley asks the question, "Is *Coxiella. burnetii* a perinatal pathogen?". She uses strict rules of evidence in her review of the literature to try to answer this question and concludes that, while perinatal Q fever does occur, overall it poses a small burden of illness. She also states that at present it is speculative whether or not this organism is associated with early pregnancy loss in humans.

In Chapter 11, Drs. Yeaman and Baca discuss the antibiotic susceptibility of *C. burnetii.* They used a persistently infected L929 cell line to determine the susceptibility of *C. burnetii* to various antibiotics. One of their most important observations is that the Priscilla isolate, which has been implicated in chronic disease in animals and man, appears to be resistant to a high degree to rifampin and compared with the Nine Mile isolate, which commonly causes acute disease, is less sensitive to the quinolones. A combination of rifampin and ciprofloxacin was most effective against the Priscilla isolate. This finding may have great clinical applicability.

In Chapter 12, Dr. Marmion from Adelaide, Australia, in conjuction with Dr. Ormsbee, discusses prevention of *C. burnetii* infection. His work has resulted in the licensure of a safe and effective vaccine for the prevention of this infection.

More than 50 years following its discovery, there are still many "queries" regarding Q, or query, fever. The clinical and epidemiological queries are raised in this volume. The basic science queries are in the companion volume, *"Coxiella burnetii* — The Microorganism".

Chapter 1

HISTORICAL ASPECTS OF Q FEVER

Joseph E. McDade

TABLE OF CONTENTS

I. INTRODUCTION

It is axiomatic that chance is a major element of discovery, but never has a laboratory infection played such a pivotal role in discovery as it did with Q fever. In the 1930s, scientists on one side of the world worked to identify the cause of a new respiratory disease, while scientists in the other hemisphere struggled to identify a novel microorganism they had isolated from ticks. Were it not for patient "X" and his laboratory-acquired infection, their relationship might not have been established for decades.

II. DISCOVERY OF Q FEVER AND ITS ETIOLOGIC AGENT

A. HAMILTON, MONTANA — 1935

By 1935 the Rocky Mountain Laboratory (RML) was well established in Hamilton, MT. The rudiments of the RML had been formed at the turn of the century in response to a public outcry for assistance in combating a local plague. For decades, settlers in the area had been afflicted with a disease known as "spotted fever of the Rockies", with as many as 80% of the adult patients in some areas succumbing to the disease. In 1901, settlers in the Bitterroot Valley officially appealed to the newly created State Board of Health, asking that they find a way to control the disease, and the seeds for RML were sown.

A parade of distinguished scientists made their way into the valley in the succeeding years, including Howard Taylor Rickets, who in 1906 established the fundamental features of the ecology and epidemiology of the disease that we now know as Rocky Mountain spotted fever (RMSF). In 1928 the state of Montana erected the first building of what would eventually become a Public Health Service research facility in hamilton (Figure 1). In 1937 the status of the facility was officially upgraded from "Hamilton Station" to is current designation as "Rocky Mountain Laboratory". By that time the mission of the RML had been expanded to include studies of typhus, tularemia, mosquito-borne encephalitides, and plague.

Gordon Davis (Figure 2A) was among the first generation of scientists who came to the RML. Born in Vermont in 1889, Davis graduated from Oberlin College in 1917. He subsequently held several academic and public health (bacteriologist) positions in Maryland, Arkansas, and Alabama while he continued his education. He received his Sc.D. from Johns Hopkins University in 1930. From 1928 to 1930 Davis had been a Special Member of the West African Yellow Fever Commission in Lagos, Nigeria. After completing that assignment, he took a position as bacteriologist at the RML, where he remained until retiring in 1955. Davis had an exceptionally productive career, with major contributions in bacteriology and entomology. Two species of tick — *Ornithodoros davisi* Fonseca and *Haemaphysalis davisi* Hoogstraal — were named in recognition of his contributions.

In the spring of 1935, the staff members at the "Hamilton Station" were working on several different aspects of RMSF, including the ecology of the disease and vaccine development. Crude extracts of infected ticks were being used as the prototype vaccine (see below). At that time the Civilian Conservation Corps (CCC) maintained a camp near Nine Mile, MT, and they assisted with the collection of ticks, both for vaccine development and for use in other laboratory studies at the Hamilton Station. One day that spring, laboratory attendant Lawrence Humble returned from the Nine Mile Creek area and brought Gordon Davis 200 *Dermacentor andersoni* ticks that had been collected there as part of the CCC effort.[1] Davis had requested the ticks for his studies of the ecology of RMSF and tularemia.

Davis divided the ticks into four groups of 50 each. He placed each group of ticks in a feeding capsule and attached them to the shaved bellies of four guinea pigs. Davis then monitored the rectal temperatures of the guinea pigs for several weeks to determine if the ticks transmitted any infectious agents during feeding. Two of the guinea pigs remained completely afebrile, and one of the guinea pigs died of unknown causes 2 d after the ticks began feeding. The fourth guinea pig, however, developed a fever of 41°C on day 12. The following day, when Davis determined

FIGURE 1. The Rocky Mountain Laboratory, probably taken in the early 1940s.

that the guinea pig was still febrile, he obtained 4 cc of blood from the animal by cardiac puncture and inoculated 2 cc intraperitoneally into each of two normal guinea pigs. Four days later, both of these animals developed a febrile illness. Seven days later the donor guinea pig died. Realizing that he had isolated an infectious organism, Davis continued to pass it in guinea pigs so that he could identify and characterize the agent.[2]

When the first guinea pig became ill, Davis must have suspected that he had isolated *Rickettsia rickettsii*, the etiologic agent of RMSF, because *D. andersoni* was already well established as a vector of RMSF and guinea pigs were known to be quite susceptible to spotted fever rickettsias. The symptoms manifested by these guinea pigs, however, indicated otherwise. In contrast to experimental RMSF, in which guinea pigs develop marked testicular swelling and even scrotal necrosis, the testes and tunica vaginalis of the guinea pigs appeared virtually normal. In addition, no growth was observed when infected tissue suspensions were tested on a variety of bacteriologic media, indicating that the organism was not the agent of tularemia. Although Gordon Davis could not fully appreciate it at the time, he was on the threshold of an important discovery. It would take several years of work with a young, new colleague, however, and a fortuitous laboratory infection before the significance of his finding would become manifest.

B. QUEENSLAND, AUSTRALIA — 1935

Edward Holbrook Derrick (Figure 3A) was a native Australian. Born in Victoria in 1898, he was educated at Melbourne University, receiving his M.D. degree there in 1922. After graduation Derrick spent a short time at the Walter and Eliza Hall Institute in Melbourne. He then devoted a year to the study of pathology at London Hospital. Derrick returned to Australia, eventually becoming the director of the Laboratory of Microbiology and Pathology of the Queensland Health Department at Brisbane, a position he held from 1934 through 1947.

In August 1935, scarcely a year after assuming his post, the new director was contacted by Sir Raphael Cilento, Director-General of Health and Medical Services for Queensland. Sir

A   B

FIGURE 2.   Some of the North American scientists involved in the initial work on Q fever. (A) Gordon Davis, who isolated the Nine Mile strain of *Coxiella burnetii* from ticks. (B) Herald Cox, who was primarily responsible for identifying and characterizing the Nine Mile strain. (C) Ralph Parker, former director of the Rocky Mountain Laboratory, who verified the tick transmissability of the Nine Mile strain. (D) Hideyo Noguchi, a researcher at the Rockefeller Institute, who in 1925, in collaboration with Parker, isolated an organism that is presumed to have been *C. burnetii*. (E) Rolla Dyer, former director of the NIH, whose laboratory infection with Q fever was instrumental in showing that Q fever existed in both Australia and the U.S.

Raphael had just become aware of an outbreak of undiagnosed febrile illness among abattoir workers in Brisbane (Figure 4). To his dismay, Cilento learned that cases of the disease had been occurring periodically since 1933 and that additional cases were still being reported. Cilento briefed Derrick on the outbreak and charged him with investigating the situation and determining its cause.[3]

Derrick decided that his first task would be to determine the characteristics of the illness. As new cases developed, he carefully monitored the clinical features of the disease. Symptoms included fever, lasting from 7 to 24 d; headache; malaise; anorexia; and pain in the limbs. Curiously, all of the patients' blood cultures were negative, and their sera were negative for antibodies to numerous pathogens, including influenza, typhus, relapsing fever, leptospirosis, typhoid, paratyphoid, and several other veterinary and zoonotic diseases.[3] Derrick was not at all disappointed at the uniformly negative results. On the contrary, he was fully aware that he was at the threshold of discovery. Recalling the early stages of his investigation, Derrick later wrote:

> Then the suspicion arose and gradually grew into a conviction that we were here dealing with a type of fever which had not previously been described. It became necessary to give it a name, and "Q" fever* was chosen to denote it until fuller knowledge should allow a better name.[3]

* Macfarlane Burnet later recounted how the name Q fever was derived.

> At about this time, Derrick christened the disease Q-fever, mainly because of the possible industrial implications of such a name as abattoirs fever. Our suggestion from Melbourne, of Queensland rickettsial fever, was regarded as derogatory to that sovereign state and since "X-disease" had already been appropriate in Australia, Derrick decided that Q (for query)-fever was the appropriate name.[4]

FIGURE 2C.

FIGURE 2D.

FIGURE 2E.

A   B

FIGURE 3. Australian scientists who were instrumental in the discovery of Q fever and the isolation of its etiologic agent. (A) Edward Derrick, who investigated the initial outbreak of Q fever and isolated the agent in guinea pigs. (B) F. Macfarlane Burnet, who identified the agent of Q fever as a previously undescribed rickettsia.

Derrick undoubtedly paused at this point in his studies to consider the status of his investigation. Was the disease actually caused by an infectious agent as he presumed, or was it due to environmental factors? Perhaps even more fundamental, was there a common cause for all of the infections? Fortunately, in addition to inoculating blood cultures, Derrick also attempted to isolate the etiologic agent in guinea pigs, and it was this avenue of investigation that eventually led him to the cause of the disease.

Guinea pigs repeatedly developed a febrile illness following inoculation with patient's blood or urine.[3] Furthermore, the disease could be transferred to additional guinea pigs by inoculating them with tissue suspensions prepared from the original group of infected animals. Significantly, convalescent animals from the first group resisted challenge with isolates obtained from other patients, providing the first reliable data to indicate that the outbreak did have a common etiology. Unfortunately, however, Derrick's subsequent attempts to characterize the guinea pig agent were unsuccessful.

Attempts to visualize bacteria in the infected guinea pig tissues proved negative as did his efforts to cultivate an agent on various bacteriologic media. He had correctly surmised that a single infectious species had caused the disease, but his failure to detect bacteria in the infected guinea pig tissues led him to a preliminary, but erroneous, conclusion that the etiologic agent was a virus. In October 1936, unable to characterize the agent further, Derrick pondered his next step. Mindful of his earlier experience at the Walter and Eliza Hall Institute, he sent a saline emulsion of infected guinea pig liver to Macfarlane Burnet in Melbourne for additional study.[3]

C. HAMILTON, MONTANA — 1936

Herald Rea Cox (Figure 2B) grew up in Indiana. After graduating from Indiana State University in 1928, he went to John Hopkins University where he earned his doctoral degree in

FIGURE 4. The old Cannon Hill abattoir in Brisbane, Australia, where the first cases of Q fever occurred. It was one of two abattoirs that were operating in Brisbane at the time. Pregnant animals were occasionally slaughtered at this abattoir, where cases occurred, but not at the other. The significance of this difference was not apparent until many years later when Luoto and Huebner[34] reported the isolation of Q fever rickettsias from the parturient placentas of naturally infected cows.

1931. He stayed on at Hopkins for a year as an instructor in Immunology, and then went to the Rockefeller Institute for Medical Research where he devoted 4 years to research in bacteriology and pathology. In 1936, Cox joined the RML as an associate bacteriologist in the U.S. Public Health Service. Eventually he moved into the house directly across the street from the RML so that he could be close to his work. Ralph Parker (Figure 2C), then the director of the RML, assigned Cox to work with Davis on identifying and further characterizing the "Nine Mile Agent". Fresh from a research environment, Cox was a perfect complement to the public health veteran. Cox and Davis began a series of studies[2,5] that would take them several years to complete.

Characterizing a new infectious agent was considerably more difficult in 1936 than it is today. The tissue culture technique was in its infancy, the electron microscope had just been invented, antibiotics had yet to be developed, and although bacteria were readily cultivable and could easily be visualized with available stains, the staining of rickettsiae was more of an art than a science. Rather crude parameters were frequently used to determine whether an etiologic agent that had been isolated in an experimental animal was a virus or a rickettsia. For example, production of inclusions in inflammatory cells was considered strong evidence of virus infection, whereas the absence of inclusions and the failure of a noncultivable organism to pass through bacterial filters indicated that it was probably a rickettsia. Davis and Cox began to evaluate these and other characteristics of the Nine Mile agent; they soon discovered that the organism that Davis had been passing in guinea pigs had properties of both viruses and rickettsiae.[2,5]

They inoculated guinea pigs with an infected tissue suspension that they had divided into aliquots, with each aliquot treated in different ways. Some guinea pigs received decimal dilutions of unfiltered inoculum, whereas others were inoculated with dilutions of the same suspension after it had been passed through one of two types of Berkefeld filters. All animals became infected, but "... while the infectious agent readily [passed] these filters, it [did] not do so in undiminished quantity".[2]

Infectivity end points with filtered suspensions were 10- to 1000-fold less than those obtained with unfiltered inoculum. Meanwhile, they stained touch preparations of inflammatory cells with Giemsa and examined them for viral inclusions.

> No inclusion bodies were found, but numerous minute, pleomorphic, rickettsia-like organisms were observed ... The organisms most commonly observed free from cells were small lanceolate rods, bipolar rods, diplobacillary forms, and occasionally segmented, filamentous forms ... more slender than typhus rickettsiae ... [they] also closely resemble *Bartonella bacilliformis.*[5]

Cox and Davis summarized their results quite simply: "... the infectious agent is not a filterable virus in the recognized sense of the term".[5]

D. MELBOURNE, AUSTRALIA — 1936

Frank Macfarlane Burnet (Figure 3B) was born in Traralgon, Australia, in 1899. He received his M.D. from Melbourne, Australia in 1922, and 6 years later took his Ph.D. at London University. From 1928 through 1944, the future Nobel laureate worked at the Walter and Eliza Hall Institute in Melbourne, serving as its director from 1934 to 1944. By the 1950s his seminal work in virology and immunology was known throughout the world. Burnet was knighted in 1951.

In 1936, Burnet was a young but seasoned virologist with an imaginative but disciplined mind, the kind of person you turned to when you encountered a problem. In retrospect, it is obvious why Derrick sought his assistance in determining the etiology of the Queensland outbreak.

Upon receipt of the infected guinea pig tissues from Derrick, Burnet and his associate Mavis

Freeman began their investigation.[6] The manuscript describing their initial findings is now more than 50 years old, but it remains a model of logic and clarity.

> We had no difficulty in reproducing in guinea-pigs with this material the characteristic febrile reaction described by Derrick. The virus has been maintained by guinea-pig passage without significant change in the type of reaction for about twenty passages. We have found no characteristic macroscopic or microscopic changes in the organs of infected guinea-pigs. Attempts to cultivate bacteria or leptospira gave negative results, and appropriate microscopic examinations for leptopsira and for Rickettsiae (in smears from the tunica) were equally without result. On the provisional hypothesis that a filtrable virus was responsible for the febrile reaction in guinea-pigs, a series of studies was made along the orthodox lines of virus research to determine the characteristics and range of pathogenicity of the virus responsible.[6]

In addition to guinea pigs, they evaluated other animals as models of infection, including mice and monkeys; they also attempted to cultivate the organism on the chorioallantoic membrane (CAM) of embryonated eggs. Mice and monkeys were susceptible, but sustained growth could not be maintained on the CAM.[6]

Then, they examined tissues of normal and infected mice.

> The first indication that rickettsial organisms were concerned in the condition was obtained from a section of mouse spleen stained with haematoxylin and eosin. Certain oval areas about the size of a nucleus were observed which seemed to be filled with lightly stained material of faint, uniformly granular texture. They suggested cytoplasmic microcolonies of organism like the psittacosis.[6]
>
> Smears were therefore made and stained by Castaneda's method, and with Giemsa stain. In a high proportion of smears from enlarged and infective spleens, bodies which appear to be of rickettsial nature were found, sometimes in enormous numbers.[6]
>
> We have not been able to obtain a completely satisfactory method of staining the organisms. With Castaneda's stain most of them appear blue, but the tint is paler than that taken by psittacosis bodies, and in many clumps a large proportion of the Rickettsiae are pink. We think it probable that in some of the negative smears such pink-staining Rickettsiae may be present, but in the absence of associated blue-staining forms it is impossible to distinguish them with certainty. With Giemsa the bodies stain a reddish-purple colour, and in smears from heavily infected spleens they are easily recognizable.
>
> The organisms take the form of tiny rods less than 1.0 u in length and about 0.3 u across; the shape varies from well-marked rods to coccoid forms.[6]

Like Cox and Davis, Burnet and Freeman also performed filtration experiments, concluding that the agent was "filtrable with difficulty". Later, they obtained a convalescent-phase serum from a Q fever patient and reacted it with an emulsion of infected spleen tissue that had been clarified by centrifugation. The serum agglutinated the suspension, providing additional evidence that the agent had caused Q fever. In other seriological studies, convalescent-phase animal sera, plus a convalescent-phase serum from a typhus patient, failed to do so. After considering all of his own data, plus Derrick's findings that sera from Q fever patients failed to react with typhus rickettsias, Burnet summarized his conclusions.

> Modern work on the rickettsial diseases of human beings indicates that three great divisions may be made primarily on cross immunity reactions in guinea-pigs: (i) classical and endemic typhus, transmitted by lice and rat fleas; (ii) *Tsutsugamushi* and scrub typhus, transmitted by mites, ... (iii) Rocky Mountain spotted fever and the other milder spotted fevers. It is quite obvious that the infection we are dealing with does not correspond with the type form of any of these divisions.[6]

Even though evidence for a new rickettsial agent was mounting, Burnet remained appropriately cautious.

> However, the greatest caution is necessary in interpreting apparent transmissions of human infections to animals, particularly in the early stages of research, before easy identification of the agent by immunological tests *et cetera* can be made. There are many well-known mistakes of this sort in bacteriological literature.[6]

E. HAMILTON, MONTANA — 1937

Ralph Robinson Parker (Fibure 2C) was one of the founders of the RML. Raised and educated in Massachusetts, Parker moved west in 1915 to take his first job as an assistant entomologist with the Montana State Board of Epidemiology. In the 7 years that he spent with that group, he attempted to develop ways to control the wood tick that transmitted RMSF. In 1921, the Public Health Service hired Parker to set up a field station at an abandoned schoolhouse in Hamilton. Shortly thereafter, Parker was joined by Dr. Roscoe Roy Spencer, a Physician who had recently come to Montana from Pensacola, FL where he was the officer in charge of bubonic plague-suppression measures. Spencer and Parker became one of the most productive scientific teams in the early days of infectious diseases. Among their many contributions, they are perhaps best known for developing the first spotted fever vaccine, a phenolized suspension of infected tick tissues. Parker became the director of the RML in 1928, a position that he held until his death in 1949.

In 1937, Parker was still an active, bench-level scientist. He was intrigued by the "virus" that Davis and Cox were studying, and he wanted to verify that ticks were reservoirs of this agent. Although the method of isolation of the Nine Mile agent indicated that ticks were the source of the organism, it was possible that one or more of the original ticks had transiently acquired the agent in a blood meal and, therefore, were not true reservoirs of the organism. Parker began a series of studies [7] to determine if larval ticks could acquire the agent by infectious feeding and pass it transstadially as they molted into nymphs and adults. Perhaps even more importantly, he wanted to know if adult ticks could transmit the organism to their progeny.

> On March 5, 1937, a guinea pig was injected with spleen virus. On the first day of fever it was infested with noninfected *D. andersoni* larvae and 30 engorged larvae were recovered 6 days later. On May 5, the above ticks as nymphs, were placed on 2 guinea pigs for feeding. One of the guinea pigs became typically febrile beginning with the seventh day and was immune to virus injected the twenty-sixth day. On July 28, 1937, 8 adult ticks from the above engorged nymphs were used to infest a guinea pig. The latter was febrile from the fifth to ninth days. The host animal was immune to virus injected on the thirty-fourth day.[7]

And in another series of experiments.

> Groups of larvae from eggs of 3 of the engorged famales were placed on separate host guinea pigs on June 2. All 3 host animals began a 4-day period of fever on the twelfth day after infestation and all remained febrile following an immunity test given on the twenty-sixth day . ... one [nymph] was placed on a guinea pig on July 28 and was removed, fully engorged, on August 2. The host animal became febrile on the twelfth day.[7]

Although his experiments had been conducted with only a limited number of ticks, Parker was convinced that ticks were true reservoirs of the Nine Mile agent.

F. QUEENSLAND, AUSTRALIA — 1937

With the investigation of the clinical aspects of Q fever completed, Derrick turned his attention to the epidemiology of the disease. Derrick had little experience in field investigation at the time, so he wisely enlisted the support of several collaborators, including D. J. W. Smith. The association of Q fever with domestic animals was unmistakable. What was not obvious, however, was the natural reservoir of the Q fever agent or its mechanism of transmission to humans. Burnet, of course, had informed Derrick of his laboratory findings, which showed that the etiologic agent was a rickettsia, and knowing that the disease was caused by a rickettsia suggested that a vector was part of the organism's life cycle. Although tick bite was not commonly reported by Q fever patients, ticks were still considered as possible vectors because they were commonly found on the domestic animals in the abattoir. Even if the ticks did not bite humans, Derrick thought, aerosols of infectious tick feces could have transmitted the infection. Derrick postulated that one or more species of wild animals were the natural reservoirs of the

Q fever agent, domestic animals were secondary reservoirs, and that ticks or some other vector transmitted the organism between animal species. He set out to verify his hypothesis.

> The search for an arthropod vector of Q fever was begun in July, 1937. Ectoparasites collected from a large number and variety of native and domestic animals found within the endemic area were examined for the presence of [the Q fever agent].[8]

Smith and Derrick's work extended well into 1939.

> On six occasions infected parasites were detected; in each instance being ticks collected from bandicoots captured at Moreton Island. No other parasites in the large series examined were found to be infected.[8]

Isolates were obtained from ticks each year of the researchers' 3-year study; all were obtained from *Haemaphysalis humerosa*. In their report, [8] Smith and Derrick summarized their thoughts about the relationship of the isolates to Q fever.

> The homology of the tick strains with strains of human origin was confirmed by their characteristic behaviour during animal passage. The tick *H. humerosa* must be regarded as an important factor in the natural maintenance of a reservoir of infection in the native animal population, but its precise aetiological significance with regard to human infections has yet to be determined.[8]

G. WASHINGTON, D. C. — 1938

Rolla Eugene Dyer (Figure 2D) was the Director of the National Institutes of Health (NIH) from 1942 through 1950. He spent more than 30 years in the Public Health Service, beginning in 1916, 1 year after he had received his M.D. from the University of Texas. In 1938, Dyer was a well-known scientist who had already made major contributions to the study of rickettsial diseases. Dyer had helped develop the fundamental concepts of the ecology and epidemiology of murine typhus and was among the first to discover that RMSF occurred in the eastern U.S. In 1938, Dyer was also an accomplished scientific administrator who had supervisory responsibility for activities at the RML. The staff at RML sent him monthly reports of their work which he reviewed in detail.

The report of April 1938 disturbed him. Herald Cox had reported that he had successfully cultivated rickettsias in large numbers in embryonated eggs. Dyer refused to believe this report, because he and his colleagues at the NIH had repeatedly tried, without success, to cultivate rickettsias in eggs. In May 1938, Dyer went to Hamilton to confront Cox with his conflicting results. Cox later recalled that visit.[9]

> Then in the early part of May 1938, Dr. Parker came into my lab at about 11:00 one morning, and it was obvious that he was very nervous and ill at ease, because he was chain smoking cigarettes and coughing more than usual. He said "Cox, Dr. Dyer, Director of the Division of Infectious Disease at the National Institutes of Health in Bethesda, has just telephoned me from Missoula. Dr. Dyer stated that he will be down to Hamilton by about 1:00 or 1:30 this afternoon, and he doesn't believe what you have written in your April report to the National Institutes of Health, that you are able to cultivate rickettsiae in great numbers in fertile hen's eggs."
> Dr. Dyer came into the lab at about 1:30 p.m. It was the first time we had ever met, and I soon learned that he was one that came immediately to the point in his speech and did not stand for any monkey business. The first thing that Dr. Dyer said to me was, "Cox, I don't believe a damned word in that recent monthly report of yours, in which you state that you are able to cultivate rickettsiae in great numbers in fertile hen's eggs, because Dr. Ida Bengtson and I tried for about 3 years to grow rickettsiae in fertile hen's eggs and we didn't have a bit of luck." I said, "Dr. Dyer, did you ever examine the yolk sac membrane tissue in those eggs to see if any rickettsiae were there?" He said, "No, we didn't." I said, "Well, that was your mistake, because that is where you would find the rickettsiae. Now, let's quit arguing and you sit down and look at these representative slides of spotted fever, epidemic typhus and Nine Mile fever, and then tell me what you think of them." Well, Dr. Dyer sat down and look at the slides for about 10-15 minutes. Then he turned around and said, "Well, I'll be, but you've convinced me. You surely have done what you stated you did." Then he stood up and shook my hand, as if to seal the bargain.

I can only say, that from then on, both Dr. Dyer and Dr. Parker were my firm and steadfast friends for as long as they lived, and I certainly appreciated their friendship.[9]

Dyer stayed on in Hamilton and worked side by side with Cox for several days, handling infected guinea pigs and assisting with the work with embryonated eggs. He returned to Washington on May 21. On May 26 Dyer began to feel ill. Patient "X" had contracted a laboratory infection with Q fever.

Dyer later gave his firsthand account of the episode.

On May 1938, a member of the staff of the National Institutes of Health, "X", spent a few days (May 12—16) at the Public Health Service laboratory in Hamilton, MT, where this infection was being studied in animals, chiefly guinea pigs, and in tissue cultures. Many of the infected guinea pigs were handled by X in company with Y, one of the investigators engaged in the study of this virus at the Hamilton laboratory. X also assisted Y when the latter was making egg culture transfers. No accident is recalled by either X or Y which might explain an infection with this virus. On May 26 and 27, X noticed occasional rather dull to sharp pains in the eyeballs, and in the evening of the 27th felt more than ordinarily tired. He afterwards thought that he might have had a slight fever on the evening of that day. The pains in the eyeballs persisted throughout the 28th, with an increasing feeling of malaise in the later afternoon. Temperature was taken at 8 p.m. and found to be 99.5°F. On the 28th, X felt slightly chilly at times, particularly in the afternoon. These chilly sensations became more pronounced on the 30th, and the patient took to bed during the evening of that day. The chills ceased on June 2 and a drenching sweat was experienced about midnight. This sweat was repeated on each of the following 4 nights, becoming less in intensity each succeeding night. After convalescence was established the return to normal strength was fairly rapid, being gained about 10 days to 2 weeks after defervescence.[10]

Although Dyer suspected that he had contracted a laboratory infection while in Hamilton, he realized that he needed to verify his assumption. Five milliliters of his blood, drawn on the day 6 of his illness, was inoculated into a guinea pig. The animal developed a fever 8 d later. The isolate was then transferred into additional guinea pigs by inoculating them with infected tissues from the original animal. Rickettsiae were identified in the spleen of the latter animals by staining. Then, cross-immunity tests were performed in guinea pigs between the isolate ("X") obtained from Dyer, and murine typhus, epidemic typhus, and RMSF rickettsias, as well as the Nine Mile agent. Although guinea pigs that had recovered from infection with the Nine Mile agent resisted challenge with strain "X", no cross-immunity was observed between strain X and the other rickettsias. Furthermore, Dyer's serum neutralized the infectivity of the isolate.[10] There seemed little doubt that Dyer had been infected with the Nine Mile agent.

Then Dyer began the definitive experiment linking the Nine Mile agent to Australian Q fever. In April 1938, Burnet had sent Dyer the spleens of mice that had been infected with the Q fever agent. Dyer had successfully infected guinea pigs with this material and had passaged the isolate several times in guinea pigs. In the course of these studies, Dyer had handled Q fever-infected guinea pigs, but he had not had contact with infected animals at NIH since April 18, 6 weeks before the onset of his illness. Thus, while it seemed highly unlikely that the Australian isolate had caused Dyer's illness, Dyer's characteristic thoroughness dictated that he should exclude the possibility that he had contracted the infection in his own laboratory. Besides, he was intrigued that the Nine Mile agent, like the Q fever agent, was unrelated to other rickettsias. Dyer challenged five guinea pigs that had convalesced from Q fever infection with strain "X". All five resisted infectious challenge. Unfortunately, because he had lost the Q fever agent in passage, he was unable to perform the reciprocal experiment until later. Nevertheless, his conclusions were inescapable.

The possibility of the infectious agent isolated in Montana and the causative agent of "Q" fever being closely related, as the "one way" cross immunity tests suggest, should not be overlooked. … Epidemiologically, this latter disease has been found in Australia, particularly among workers in abattoirs and among dairy farmers. Such an epidemiological picture is not at variance with the picture of a "tick borne" infection, since it suggests a reservoir in animals and the existence of the infection in their arthropod parasites.[10]

III. NAMING OF AN ORGANISM

Providing an appropriate name for a new microorganism is not always a straightforward matter. Frequently, an organism is discovered simultaneously by two or more investigators, and each can justifiably claim the privilege of choosing its name. The human immunodeficiency virus (HIV) virus is perhaps the most recent example. In addition, many "new" organisms turn out to be unnamed but previously described agents which, because they appear to lack medical significance, remain obscure. Legionella is a good example of the latter. Both of these situations occurred with the discovery of the Australian and Nine Mile agents.

The group at RML was in a quandary when they thought of how best to name the Nine Mile organism. In the spring of 1925, Parker had collected *D. andersoni* ticks in Saw Tooth Canyon, MT as part of his studies of the ecology of spotted fever and tularemia. He sent one lot of ticks to Hideyo Noguchi (Figure 2E) at the Rockefeller Institute. (Noguchi was a bacteriologist who had been involved with some of the early events surrounding the discovery of RMSF and its mechanism of pathogensis)[11] Noguchi isolated a "filter-passing virus", with properties similar to the Nine Mile agent, from the ticks, and he successfully passaged it for several generations in guinea pigs.[12] Unfortunately, because the agent was eventually lost in passage, it was not possible for the RML scientists to compare Noguchi's isolate retrospectively to the Nine Mile agent, and because it was not possible for them to confirm its identify, they certainly could not name it for its original discoverer. On the other hand, it was presumptuous for them to name the organism after either Cox or Davis if Noguchi had, indeed, been the first to isolate and describe the agent.

Cox chose perhaps the only solution available at the time. He gave the organism a name which described its unique property among the rickettsias, i.e., its filterability. After consulting with Professor W. P. Clark, an expert in Greek at the University of Montana, Cox found an appropriate specific epithet.

> Since the outstanding characteristic differentiating this agent from the known pathogenic rickettsiae is its property of filterability, the name *Rickettsia diaporica (diaporica* is derived from the Greek word and means having the property or ability to pass through) is proposed.[13]

On the other side of the world, the situation was perhaps less complicated. Even though Derrick had initially isolated the Q fever agent, Burnet's identification of the organism as a rickettsia and his subsequent characterization of its unique properties were instrumental in describing this new species. Cognizant of Burnet's important contributions, Derrick proposed that it be named for Burnet.

> It has become necessary to name definitely the organism causing "Q" fever, and the name *Rickettsia burneti* is proposed.
> "Q" fever being distinct from other rickettsioses, a new specific name is required; "*burneti*" is chosen after Dr. F. M. Burnet, the discoverer of the organism.[14]

The unique status of *R. burneti* (*R. diaporica*) among the rickettsias became more apparent in succeeding years as its other properties were discovered. Later classification schema elevated it to the status of a subgenus, *Coxiella*. Then, in 1948, Cornelius B. Philip, a long-time entomologist at the RML, proposed that it be considered as the single species of an entirely distinct genus.[15]

> The classification of the rickettsiae has been undergoing revisions as research continues to clarify relationships of the pathogenic forms. At the time of the proposal of the name *Coxiella* as a subgenus of *Rickettsia* for the etiologic agent of Q fever, *R. burneti*, it appeared desirable to use subgenera as a useful systematic category for distinct groups within the genus.

The subgeneric level is recognized in the present bacteriological system in other families. However, it was originally recognized and stated that *Coxiella* possessed certain striking characters that might eventually warrant its full generic recognition. Steinhaus and others have recommended this action, and the writer has been using the name as a full genus in unpublished tables for teaching and other purposes during the War. It is here proposed to validate that usage by elevating *Coxiella* to the status of a full genus, the genotype, of course, remaining the same, i.e., *R. burneti* Derrick, which now becomes *Coxiella burneti** (Derrick).[15]

Coxiella burnetii is occasionally still referred to as *R. burneti* in some articles, but it seems fitting that the names of both Cox and Burnet should always be associated with the organism. Davis actually reported first on Q fever, on June 26, 1936, when he summarized his data at the meeting of the Western Branch of the American Public Health Association in British Columbia. Derrick first published a report on Q fever on August 21, 1937.

Davis and Cox's published report was printed in December 1938. Propriety of first publication seems trivial in retrospect, particularly when both groups were approaching the isolation of *Coxiella* from different directions. Work in Montana and Australia began simultaneously and progressed along similar lines at the same pace, and both Cox and Burnet identified the Q fever agent as a new rickettial species. Half a century after their discoveries, they once more found a common destiny. Cox and Burnet both died in 1986.

IV. EPILOGUE

Although the history of the discovery of Q fever and the isolation and identification of *C. burnetii* is especially interesting, the reports of the other early studies of Q fever are no less enjoyable. Table 1 lists some other milestones in Q fever and selected references to this early work. A careful sifting of these reports should prove particularly rewarding to the serious student of Q fever.

ACKNOWLEDGMENTS

The author wishes to acknowledge the assistance of several people who made contributions to the article. Photos were provided by Leza S. Hamby, Information Officer, (RML); John Pope (Australia); and Lucy Keister, Curator, Prints and Photographs Collection, National Library of Medicine. The author is particularly indebted to Vickie Harden, historian, National Institutes of Health, for reviewing the manuscript and for generously providing supplemental historical information. The special assistance of Dr. B. P. Marmion in locating photos is also gratefully acknowledged.

---

* An additional "i" was added to "*burneti*" in later classification schema.

TABLE 1
Other Selected Milestones in the Early History of Q Fever

Year	Event	Ref.
1940—1946	Several outbreaks of laboratory-acquired Q fever occur among employees at the National Institutes of Health. The highly infectious nature of *Coxiella burnetii* becomes manifest.	16—18
1944—1945	The War Office in London reports that at least eight outbreaks of Q fever occurred among British troops stationed in the Italy, Greece, and Corsica during World War II. Outbreaks are also reported among American troops returning to the United States from duty in the Mediterranean area. The full public health impact of Q fever is realized.	19—22
1946—1947	Outbreaks of Q fever are observed at meat packing houses in Amarillo, TX and Chicago IL. Shepard documents aerosol transmission of Q fever in Chicago by epidemiologic studies.	23—28
1946	Topping and colleagues compare the antigenicity of six strains of Q fever rickettsiae and note their lack of homogeneity, but antigenic phase variation is still unknown.	29
1948	Additional reservoirs of the Q fever are detected: cases of Q fever associated with sheep and goats are documented in northern California by Lennette. *C. burnetii* is isolated from the air of premises housing infected animals.	30—31
1948	"*Rickettsia burneti* is" isolated from raw milk, demonstrating another potential source of infection.	32
1948	Lennette reports that Q fever patients respond to treatment with aureomycin.	33
1950	Luoto and Huebner report the isolation of *C. burnetii* from the placentas of parturient cows, thereby identifying a major source of human infection.	34
1956	Antigenic-phase variation of Q fever rickettsias is described by Stoker and Fiset.	35

REFERENCES

1. Spotted Fever Laboratory Monthly Report, August 1935, pp. 9—10, Rocky Mountain Laboratory Monthly Reports, Montana State Archives, Helena, MT.
2. **Davis, G. E. and Cox, H. R.,** A filter-passing infectious agent isolated from ticks. I. Isolation from *Dermacentor andersoni,* reactions in animals, and filtration experiments, *Public Health Rep.,* 53, 2259, 1938.
3. **Derrick, E. H.,** "Q" fever, a new fever entity: clinical features, diagnosis and laboratory investigation, *Med. J. Aust.,* 2, 281, 1937.
4. **Burnet, M.,** *Walter and Eliza Hall Institute, 1915—1965,* Melbourne University Press, Melbourne, 1971, 106.
5. **Cox, H.,** A filter-passing infectious agent isolated from ticks. III. Description of organism and cultivation experiments, *Public Health Rep.,* 53, 2270, 1938.
6. **Burnet, F. M. and Freeman, M.,** Experimental studies on the virus of "Q" fever, *Med. J. Aust.,* 2, 299, 1937.
7. **Parker, R. R.,** A filter-passing infectious agent isolated from ticks. II. Transmission by *Dermacentor andersoni, Public Health Rep.,* 53, 2267, 1938.
8. **Smith, D. J. W. and Derrick, E. H.,** Studies in the epidemiology of Q fever. I. The isolation of six strains of *Rickettsia burneti* from the tick, *Haemaphysalis humerosa, Aust. J. Exp. Biol. Med. Sci.,* 18, 1, 1940.
9. **Cox, H. R.,** Reminiscences, in *Rickettsiae and Rickettsial Diseases,* Burgdorfer, W. and Anacker, R. L., Eds., Academic Press, New York, 1981, 11.
10. **Dyer, R. E.,** A filter-passing infectious agent isolated from ticks. IV. Human infection, *Public Health Rep.,* 53, 2277, 1938.
11. **Harden, V. A.,** Koch's postulates and the etiology of rickettsial disease, *J. Hist. Med. Allied Sci.,* 42, 277, 1987.
12. **Noguchi, H.,** A filter-passing virus obtained from *Dermacentor andersoni, J. Exp. Med.,* 44, 1, 1926.
13. **Cox, H. R.,** Studies of a filter-passing infectious agent isolated from ticks. V. Further attempts to cultivate in cell-free media. Suggested classification, *Public Health Rep.,* 54, 1822, 1939.
14. **Derrick, E. H.,** *Rickettsia burneti:* the cause of "Q" fever, *Med. J. Aust.,* 1, 14, 1939.
15. **Philip, C. B.,** Comments on the name of the Q fever organism, *Public Health Rep.,* 63, 58, 1948.
16. **Hornibrook, J. W.,** An institutional outbreak of pneumonitis. I. Epidemiological and clinical studies, *Public Health Rep.,* 55, 1936, 1940.
17. **Huebner, R. J.,** Report of an outbreak of Q fever at the National Institute of Health. II. Epidemiological features, *Am. J. Public Health,* 37, 431, 1947.
18. **Spicknall, C. G., Huebner, R. J., Finger, J. A., and Blocker, W. P.,** Report of an outbreak of Q fever at the National Institute of Health. I. Clinical features, *Ann. Intern. Med.,* 27, 28, 1947.
19. **Robbins, F. C. and Ragan, C. A.,** Q fever in the Mediterranean area: report of its occurrence in allied troops. I. Clinical features of the disease, *Am. J. Hyg.,* 44, 6, 1946.
20. **Robbins, F. C., Gauld, R. L., and Warner, F. B.,** Q fever in the Mediterranean area: report of its occurrence in Allied troops. II. Epidemiology, *Am. J. Hyg.,* 44, 23, 1946.
21. **Feinstein, M., Yesner, R., and Marks, J. L.,** Epidemics of Q fever among troops returning from Italy in the spring of 1945. I. Clinical aspects of the epidemic at Camp Patrick Henry, Virginia, *Am. J. Hyg.,* 44, 72, 1946.
22. Commission on Acute Respiratory Diseases, Epidemics of Q fever among troops returning from Italy in the spring of 1945. II. Epidemiological studies, *Am. J. Hyg.,* 44, 88, 1946.
23. **Irons, J. V., Topping, N. H., Shepard, C. C., and Cox, H. R.,** Outbreak of Q fever in the United States, *Public Health Rep.,* 61, 784, 1946.
24. **Topping, N. H., Shepard, C. C., and Irons, J. V.,** Q fever in the United States. I. Epidemiologic studies of an outbreak among stock handlers and slaughterhouse workers, *JAMA,* 133, 813, 1947.
25. **Irons, J. V. and Hooper, J. M.,** Q fever in the United States. II. Clinical data on an outbreak among stock handlers and slaughterhouse workers, *JAMA,* 133, 815, 1947.
26. **Irons, J. V., Murphy, J. N., Jr., and Wolfe, D. M.,** Q fever in the United States. III. Serologic observations in an outbreak among stock handlers and slaughterhouse workers, *JAMA,* 133, 819, 1947.
27. **Cox, H. R., Tesar, W. C., and Irons, J. V.,** Q fever in the United States. IV. Isolation and identification of rickettsias in an outbreak among stock handlers and slaughterhouse workers, *JAMA,* 133, 820, 1947.
28. **Shepard, C. C.,** An outbreak of Q fever in a Chicago packing house, *Am. J. Hyg.,* 46, 185, 1947.
29. **Topping, N. H., Shepard, C. C., and Huebner, R. J.,** Q fever. An immunological comparison of strains, *Am. J. Hyg.,* 44, 173, 1946.
30. **Lennette, E. H., Clark, W. H., and Dean, B. H.,** Sheep and goats in the epidemiology of Q fever in Northern California, *Am. J. Trop. Med.,* 29, 527, 1949.
31. **Lennette, E. H. and Welsh, H. H.,** Q fever in California. X. Recovery of *Coxiella burneti* from the air of premises harboring infected goats, *Am. J. Hyg.,* 54, 44, 1951.
32. **Huebner, R. J., Jellison, W. L., Beck, M. D., Parker, R. R., and Shepard, C. C.,** Q fever studies in Southern California. I. Recovery of *Rickettsia burneti* from raw milk, *Public Health Rep.,* 63, 214, 1948.
33. **Lennette, E. H., Meiklejohn, G., and Thelen, H. M.,** Treatment of Q fever in man with aureomycin, *Ann. N.Y. Acad. Sci.,* 51, 331, 1948.

34. **Luoto, L. and Huebner, R. J.,** Q fever studies in Southern California. IX. Isolation of Q fever organisms from parturient placentas of naturally infected dairy cows, *Public Health Rep.*, 65, 541, 1950.
35. **Stoker, M. G. P. and Fiset, P.,** Phase variation of the Nine Mile and other strains of *Rickettsia burneti, Can. J. Microbiol.*, 2, 310, 1956.

Chapter 2

COXIELLOSIS* (Q FEVER) IN ANIMALS

Gerhard H. Lang

TABLE OF CONTENTS

---

* The term "coxiellosis" has been used as early as 1954;[213] it is defined as "infection by *Coxiella burnetii*" and is more meaningful than Q fever for the *Coxiella* infection in animals since it denotes an etiologic concept.

I. INTRODUCTION

It is now over 50 years since the disease Q fever and its agent were discovered, yet effective control of the infection in man and animals has not been achieved despite numerous epidemiological and epizootiological investigations. Q fever is recognized as a worldwide problem,[1,2] and since the active exploratory decades following World War II, changes have occurred in the prevalence of the infection in several geographic areas: in traditional endemic areas, such as the Balkans and southern Europe, the prevalence of human Q fever has declined; in other countries, such as the Netherlands[56] and Canada,[25,44] both for a long time believed to be little affected by coxiellosis, Q fever in humans has been found on an endemic scale and its source thought to be enzootically infected dairy cattle.

Most of our present concepts of this zoonosis were established during the post-World War II years up to 1960. In the following decades, public and animal health officials, with few exceptions, lost practically all interest in the problem. Today there are few animal health officials who know the approximate prevalence of coxiellosis in the livestock under their jurisdiction. The reasons for this unsatisfactory state must be sought in the elusive clinical expression of coxiellosis and inadequate or impractical methods available for its diagnosis to veterinarians. Technical improvements, in particular serodiagnosis, by fluorescent antibody (FA) and enzyme-linked immunosorbent assay (ELISA) methods are slowly replacing the traditional complement fixation (CF) and capillary or microagglutination (CA or MA) techniques.

Coxiella burnetii can proliferate in a very large number of animal species, of which the natural fauna of an area forms the wildlife reservoir and the domestic animals the livestock reservoir of *C. burnetii*. The spread of coxiellae is usually mediated in the former by bloodsucking ectoparasites, mostly ticks, and, thus, wildlife coxiellosis can be qualified an arthropod-borne infection, while livestock coxiellosis assumes the character of a contagious disease by means of fomites that can be carried over distances with the wind and infect new hosts by inhalation. Coxiellosis in domestic animals is almost exclusively the direct or indirect source of human Q fever and must be at the beginning of any practical attempt to control the disease in humans.

Early stages of the study of animal coxiellosis have been periodically reviewed,[1-8] and the reader is referred to these reviews for more information.

II. COXIELLOSIS IN LIVESTOCK

Animals seem rarely to develop illness from a coxiella infection. Although the most common portal of infection is believed to be, as in humans, the respiratory system, coxiellosis does not cause respiratory pathology in any animal species. Also, chronic infection by *C. burnetii* of adult animals does not have a cardiac or hepatic localization. The most vulnerable site of coxiella localization is the female reproductive system, both uterus and mammary gland, where damage is usually limited despite massive proliferation of coxiella, but from where chronic shedding is the mechanism for spread of *C. burnetii* into the environment.

Domestic ruminants are most frequently the source of human infection; some epidemiologists consider mostly small ruminants, while others believe cattle, are the main reservoirs of *C. burnetii* responsible for human and livestock infection. This chapter will discuss which are the critical hosts for the perpetuation of *C. burnetii* in the environment and how it may be possible to manipulate this host-parasite relationship to prevent Q fever in humans.

A. COXIELLOSIS IN SHEEP

The disease Q fever, or abattoir fever as it was at first described in 1934 in Australia, was traced to infected cattle,[9] The first Australian Q fever cases traced to sheep were reported 14 years later. [10] In January 1958, an outbreak of Q fever occurred on a sheep station near Tambo, where shearing of about 25,000 sheep was carried out by 34 shearers. At least 15 cases of Q fever were diagnosed in the shearers and 17 among 22 males permanently employed on the station. When the sera of 174 sheep were examined, 11% were found to possess antibodies against *C. burnetii*. However, it seemed unlikely that the infecting organism had come from the interior of the sheep. Sheep discharge coxiellae only during parturition or abortion, and no such event had taken place on the station for 6 months. However, the sheep were heavily infested with *Amblyomma triguttatum*, and the suspicion arose that the ticks might have contaminated the wool with coxiella-laden feces. The ticks had never been seen in such numbers on sheep as in January 1958. Coxiella antibodies were also found in 3 of the station's 24 cows, and in 1 of 5 kangaroos. That Q fever could be caught from the fleece of sheep received confirmation when 10 cases occurred at a fellmongery in Brisbane from March to May 1958. This fellmongery handled only sheepskin which were collected from many places in Queensland.[9] Q fever cases showed a sharp increase in 1958. More than half the cases arose near Brisbane. An occupational analysis of the 1958—1959 cases showed that 255 were associated with the meat industry, 77 with the sheep industry, and 26 with dairy or mixed farms. In 52 cases, the occupations were various or not stated. The great increase among the meatworkers was due to a changeover to export of boneless beef. This created a market for animals of poor condition which previously had not been slaughtered. These included old dairy cows, a high proportion of which were pregnant. The increase in work demanded the hiring of new hands, most of whom were young, new to the industry, and without immunity. From January to June 1958, 58 out of a staff of 110 developed Q fever; in due course the development of immunity in many employees brought the outbreak under control. In 1980, McKelvie,[11] in a report of a 10-year survey among meatworkers in Queensland, concluded again that cattle formed the perennial source of infection for slaughterhouse personnel, while sheep provoked occasional limited outbreaks of short duration among the workers. Q fever became a particularly pressing problem in Australia in the late 1970s and early 1980s, when abattoirs started to process feral goats, many of which were pregnant. Large outbreaks occurred in abattoirs in South Australia, Victoria, and New South Wales, particularly where operations had changed from processing pigs to slaughtering goats and where, presumably, workers had not acquired immunity through contact with infected ruminants.[12]

A similar situation was reported from Uruguay, where a population of 3 million inhabitants lives among 10,323,000 cattle, 25,560,000 sheep and 12,000 goats,[217] yet all of 1358 Q fever cases diagnosed in Uruguay during 1975 to 1985 occurred in workers at meat-processing plants, and most of the cases studied were traced to cattle.[13]

Serosurveys by ELISA for coxiella antibodies in livestock in Ontario, Canada, between 1985 and 1988 showed that coxiellosis had apparently increased in dairy cows from 2.7% of tested herds in 1960 to 67% of the herds in 1985;[14] dairy goats had a herd prevalence of 20%,[15] while over 90% of sheep flocks tested negative in the coxiella ELISA. The approximate population census for the three species were 810,000 dairy cows and heifers on 13,135 farms, about 7000 dairy goats in 70 herds, and 320,000 sheep in over 900 flocks. While coxiellosis is now established enzootically in dairy cattle in Ontario, the sheep population is still free of the infection under identical environmental and epidemiological conditions, a clear pointer that the two species react differently to the infection by *C. burnetii*.

These observations do not agree with the view that sheep are the principal livestock reservoirs of *C. burnetii* in countries where Q fever is endemic.[16-18,35,90] This opinion originated during the Q fever epidemics which occurred after World War II in Greece, Italy, southern Germany, and North Africa. Data from the Federal Republic of Germany on the frequencies of animal sources of human Q fever during the years 1967 to 1969 placed sheep and goats (30%) ahead of cattle (10%), game animals (5.2%), donkey (0.9%), other animals (15.2%), and undetermined sources (38.7%).[19] The prejudice against sheep is also apparent in a very recent review of Q fever in Italy,[20] which takes only sheep into consideration as source of the human disease and ignores dairy cattle entirely. The spate of Q fever outbreaks among the personnel of research institutions where pregnant ewes were used for experimentation further implicates sheep as the major source of human Q fever.[21-27]

Sheep, however, generally raised for wool and meat, are not the livestock reservoir of *C. burnetii*; dairy animals are responsible for the spread, implantation, and maintenance of *C. burnetii* in many parts of the world. Generalized misconceptions of coxiellosis in ruminants account for this unjust condemnation of the ewe, such as (1) the infectious process in bovines and ovines is essentially comparable or (2) if new infection is avoided, the herd, as a rule, can be expected to rid itself of coxiellosis.[16]

Epidemiological and pathological evidence indicates differences in coxiellosis of sheep and cows. Ovine coxiellosis seldom becomes chronic, while coxiellosis in cows is very frequently endemic, a feature already reflected in the longer-lasting circulating antibody in the latter.[16] Prolonged excretion, over many months or even years, of coxiella can be expected from lactating cows, but not from sheep; experimental evidence for this view is provided by several studies. During an ecological study on the Hopeland Field Station in northern California from 1964 to 1967, Enright et al.[28] were able to isolate *C. burnetii* by mouse inoculation from 13 (25% of examined) sheep placentas, but not from 66 samples from sheep colostrum or milk. Brooks et al. in 1986[29] carried out *C. burnetii* vaccination trials in ewes and reported that in the unvaccinated and heavily challenged control ewes it was possible to isolate coxiellae from five of six placentas examined, but not from postcolostral milk samples. Schaal and Goetz[30] studied 30 culled seropositive sheep from a flock believed to have caused a Q fever epidemic among a rural community in Germany. Among the 161 tissue samples and secretions tested for *C. burnetii* by animal inoculation were 20 samples of uterine and 30 samples of udder tissue (7 lactating). *C. burnetii* was isolated from only one specimen of amniotic fluid. The authors expressed surprise about the failure of isolation from the udder tissues, since previous studies by Schaal[31] and others[32,33] with infected cows showed prolonged shedding in milk over many months and years, and isolation from the mammary tissue was positive in the majority of studied cases. Babudieri[3] reviewed the earlier studies from Italy and the U.S. on sheep coxiellosis. Experimental infection of sheep produced at the most a short-lasting fever. *C. burnetii* could be isolated from the blood for the first 8 to 10 d, and irregularly from feces, urine, and milk during the first week after infection. Coxiella persisted for at least 43 d in liver, spleen, and kidneys, but could not be isolated after 8 months. Transmission studies by Rosati[218] and by Caporale[219] on more than 69 sheep showed that (1) sheep proved infected, when placed with healthy sheep 2 months after lambing and kept with them for 5 months, were no longer able to transmit the infection and (2) coxiella was no longer present in blood, placenta, or body secretions at the second lambing, and healthy contact sheep did not contract the infection. On the other hand, gravid sheep from an infected flock transmitted the infection at their first lambing after infection to healthy pregnant sheep, but not to nonpregnant sheep.

Three sheep flocks in which a wave of abortions was diagnosed as coxiellosis by microscopy of the fetal membranes, were studied by von Kruederer and Schreyer.[34] In flock A 25% of the 333 ewes miscarried, in flock B 33% of 378 ewes, and in flock C 3% of the ewes. Serology by CF revealed 9.9% seropositives in flock A, 13.2% in flock B, and 13.5% in flock C. The CF titers ranged from 1:20 to 1:80, the majority peaked at 1:40. Fifty seroreactors of flock B were serotested at 4-week intervals; of 6 ewes with an initial titer of 1:80, the 4-week titers were 1:40 in 1 sheep, 1:20 in 3 sheep, and negative 2 sheep. Of the 1:40 reactors (n = 26), the titers read 1:40 in 3 sheep, 1:20 in 8 sheep, and negative in 15 sheep. The remainder of the reactors had similar declines. At the 12-, 16-, and 20-week bleedings all ewes were seronegative again.

The consensus of many authors reporting CF titers in coxiella-infected sheep is that the peak ranges from 1:80 to 1:512.[34] Siegrist and Hess are of the opinion that animals of the ovine species are the least apt (of domestic ruminants) in responding to the infection with the production of humoral antibodies.[90] This is also reported by Jellison et al.[44] A flock of 128 ewes, separated 2 months earlier from their lambs, were tested by CF at the National Institutes of Health; only 13 gave reactions usually regarded as significantly positive, i.e., 3+ at 1:8. Eighty-two sheep gave partial reactions less than 3+ at 1:8; 33 sheep were clearly negative. None of the 128 samples was positive at a dilution of greater than 1:16. Yet, coxiella was isolated by guinea pig injection of 2 of 13 milk pools from this flock, although the number of milk shedders was unclear because of extensive pooling of milk used for injection. Such a large proportion of weak reactions, wrote the authors, has not been experienced in similar surveys of bovine herds, although more than 20,000 bovine samples have been tested.

In comparison, coxiella ELISA titers of four ewes that had aborted from coxiellosis and were tested by ELISA shortly thereafter in the author's laboratory registered between 1:204,800 to 1:819,200.[52] These four sera showed also a very strong prozone; in fact, one serum was negative at the 1:100 dilution. The prozone effect is one plausible explanation for the phenomenon of the seronegative coxiella-shedding ewe.[36,37] Such prozones are also observed with CF and have led to erroneous diagnosis in human chronic Q fever.[150]

Lambings subsequent to coxiella abortion are usually carried to term.[35] Flocks affected by abortion also suffer from an increase of weak newborn lambs.[37]

In the epidemiological context, the evidence presented points to ovine coxiellosis being a transient infection tending to a spontaneous cure — like that in dogs, cats, and man — rather than an enzootic reservoir where coxiella circulates among the host population from generation to generation.

An explanation is, however, necessary for the recurrent Q fever problems created by sheep in southern Germany, Switzerland, southeastern France, and in institutional flocks. In many of these environments, sheep are in contact with other animals, especially true reservoir hosts like cows and goats, and the infection is reintroduced in the flocks, starting another wave of abortions.[38] The practical consequence of this observation is that a regional Q fever control program can probably ignore the ovine species while concentrating on the livestock reservoirs, the dairy cow, and dairy goat.

B. COXIELLOSIS IN GOATS

The propensity of goats for transmitting *C. burnetii* to humans was made known by Caminopetros in Greece[39,40] right after the "Balkangrippe" was identified as Q fever. Confirmation was provided shortly thereafter on a trans-Pacific cargo ship which transported dairy goats; the crew drank raw goats' milk and promptly came down with "goat-boat-fever".[41] Caprine coxiellosis was reported from France in goats in the Vienne Departement in 1979, where in October and November a number of goat herds were plagued by miscarriages.[42] There are an estimated 100,000 goats in California, 24% of which were found to be coxiella seropositive in 1978. A trend exists of replacing the family cow by goats as a source of milk.[43] A 2-year survey in 1982 showed 57% (840/1475) of Californian goats seropositive for *C. burnetii*. The goats originated from 32 herds in 15 counties of California. All but 4 herds were found to be infected. Antibody prevalence among goats was more than twice that among sheep (24%). Dairy cows in California were seropositive at 82%; 26 of 32 goat herds had an antibody prevalence above 30%, with 6 herds having more than 90% seropositive animals. Prevalence of 58% of commercial herds seropositive vs. 59% of noncommercial herds indicated no significant difference in the type of management of the herds.

Two seropositive goats were autopsied, and samples of various tissues were inoculated into mice. One goat with an antibody titer of 1:16 by MA test had *C. burnetii* isolated from the mammary gland, spleen, kidney, uterus, liver, lung, and heart. Nine of 29 (28%) samples of goat whey were positive for *C. burnetii*, indicating that goats shed this microorganism. The placenta from a goat with a titer of 1:512 who had delivered a normal kid elicited a mouse postinoculation coxiella serotiter of over 1:256.[26]

In the state of Michigan, Q fever has been a reportable disease since the 1960s. However, the first two reported cases of human Q fever were received by the Michigan Department of Public

Health in the spring of 1984. The patients lived in adjacent rural counties. The first patient most likely was infected from exposure to his goats; the second had multiple exposures on numerous goat farms in the area. A serological survey on 25 farms detected 19 infected herds; there was no herd with all positive goats. Forty percent (49/123) of the goats tested by CF were reactors. Residents on goat farms were almost three times as likely to be coxiella seroreactors than the reference population (36/86 vs. 7/47). The distribution of titers for residents of goat farms (1:8 to 1:1024) was significantly higher than that of the reference population (1:8 to 1:16). The proportion of reactors on goat farms was highest among 10- to 19-year-old residents. According to the researchers, Q fever seems not to be a new disease in Michigan because of the widespread prevalence of antibodies in both humans and goats.[45]

The island of Cyprus has an appreciable goat population which was one of the sources of the Q fever problem faced by British troops in 1974 and 1975, while islanders seem to be immune to the infection.[46,47]

Goats share with sheep a predisposition for abortion during coxiellosis epizootics,[42,49-51] while they share with dairy cows a predisposition for remaining chronically infected. Abortions occur late during gestation, about 15 d before term, affecting 10 to 25% or more goats of a herd, both primiparous and multiparous. The does deliver the afterbirth generally well and have milk. Kids may be born alive, but die during the following days.[42] The author has seen several commercial dairy goat herds in Ontario that struggled for years with coxiellosis. The average lifespan of a dairy goat is about 5 years, and this rapid population turnover is definitely in favor of the survival of the rickettsial parasite. For this reason coxiellosis control programs of dairy goat herds will often have to consider immune prophylaxis in their strategic plans.

C. COXIELLOSIS IN CATTLE

Bovine coxiellosis is spreading in many parts of the world. As a recent Swiss report[38] stated, the Q fever situation in the 1980s cannot be assumed to be that found in the 1960s. While in the 1960s herd infection rates of 0.23 to 1.6% were reported in Switzerland, 1980 surveys on bulk milk found herd infection rates of 26.8 to 58.3%. Also, coxiella abortions of cows were increasingly observed.[53]

In the Federal Republic of Germany a steep rise in the prevalence of infection in cattle has been found in recent years. Up to 30% of all herds investigated and 80% of those with "infertility problems" have been found to contain shedders of the agent. The progression involves also northern parts of the country, and ticks appear to play no role in this process.[8,54,55,213-215]

In the Netherlands, Q fever was for a long time considered a rare disease; two cases were reported in 1979, but 19 were reported in 1981 and 25 in the years thereafter.[56] The source of the apparently increasing human disease was traced to the Dutch dairy cattle.[57] According to the researchers, coxiellosis was not a new arrival in the country, but it was overlooked until diagnostic serology was shifted from the CF test to the more sensitive indirect immunofluorescence test. With regard to the animal source of human Q fever, the Dutch researchers concluded that cattle may constitute the most common reservoir; sheep and goats may do so as well but to a much lesser extent. The only tick in the area, *Ixodes ricinus,* is not a parasite of domestic ruminants and is thus unlikely to play a role in the epidemiology of livestock coxiellosis.[58]

In France numerous abortions of cattle attributed to *C. burnetii* were noted in 1976 and 1977 simultaneously in several départements (Puy-de-Dôme, Sarthe, Ille-et-Villaine, Maine-et-Loire, Somme, Isère, Nord), which previously were only seen in sheep and goats.[59-61] Subsequent publications cite the départements Haute-Savoie,[62,63] Côte-du-Nord, Finistère, Loire-Atlantique, Mayenne, Deux-Sèvres, Vendée, Marne, Meurthe-et-Moselle,[64] as Q fever-infected.

In Ireland, the first case of Q fever was reported in 1968,[65] and estimates are that *Coxiella burnetii* arrived on the island between 1957 and 1962.[66]

Coxiella burnetii has even made inroads into Scandinavia, hitherto cited as the bastion of coxiella-free existence: serologic evidence of Q fever infection was obtained from people rearing sheep on Gotland and west-central Sweden.[67]

In Russia, the Q fever situation was described as not significantly changed in 1980 to 1985 from that of the 1950s and 1960s. The article also stated, however, that in the last few years recurrent Q fever outbreaks have become an important cause of morbidity in man in some regions of the U.S.S.R., especially the south.[18]

Bovine coxiellosis is enzootic throughout the North American continent. In 1960, Luoto[69] warned in his "Report on the nationwide occurrence of Q fever infections in cattle":

> Since 1947, Q fever has been recognized as a public health problem in certain areas of the United States particularly in the Western States. Special investigations have shown that human cases, some of which are severe and protracted, commonly occur in endemic areas in Texas, California and Idaho. Recent reports indicating that Q fever occurs in other States emphasize the need for further investigation of this infection of animals and man. A systematic study of the infection in livestock and of associated human disease is required to define the problem. Dairy cattle, which are a major reservoir of infection and thus an abundant potential source of human disease, develop only asymptomatic infections. After the causative agent *C. burnetii* is introduced into a herd, many animals develop chronic infections and transmit the agent to other additions of the herd; thus the herd usually remains permanently infected. Sheep and goats are also sources of the disease but are of lesser importance because their more limited distribution results in fewer human contacts. Although infection cycles may occur among rodents and arthropods in nature, *Coxiella burnetii* maintains an independent and more important airborne infection cycle among domestic livestock. This airborne transmission, along with the hardiness of the agent and its ability to persist in the environment, suggests a propensity for spreading and becoming a widespread public health problem ... Data now available demonstrate conclusively that Q fever occurs among dairy cattle in all parts of the United States. Bovine infections have ... been found recently in all States where a concerted search has been made. Earlier studies indicated that bovine infections were frequent in seven States, namely California, Wisconsin, Ohio, Iowa, Texas, Arizona and Idaho. The current survey in 26 States confirm and expand earlier findings in some areas and prove the occurrence of considerable bovine infection in 19 additional States — Oregon, Washington, Montana, Wyoming, Utah, South Dakota, Minnesota, Nebraska, Illinois, Michigan, Georgia, North Carolina, Maryland, Pennsylvania, New Jersey, New York, Connecticut, Massachusetts and Hawaii. Other, more limited data suggest that bovine infection occurs in nine other States, namely Louisiana, Mississippi, Virginia, Colorado, New Mexico, North Dakota, Kansas, Missouri, and Kentucky. It is likely that infections occur in the remaining 15 unstudied States, most of which are adjacent to or surrounded by infected areas.

Luoto then cites examples of the progressive expansion of bovine coxiellosis: dairy herd infections in south central Idaho increased from 1 to 17% from 1951 to 1958. A great increase occurred in an Eastern State where evidence of bovine infection was not detected in 1949, a little was present in 1952, but 47% of 248 herds tested were infected by 1959. A number of other examples follow. Luoto concludes:

> The demonstration of widespread bovine infection indicates that Q   fever is endemic throughout the United States and that a nationwide problem already exists. ... Universal bovine infection, similar to that in Southern California where 98 percent of the herds are infected, may develop in other parts of the country. ... The continuing growth of human and animal populations will result in crowding conditions even more conducive to spread of infection. Continued surveillance will indicate the development and scope of the animal disease problem. ... Regardless of any future implications, the presence of *Coxiella burnetii*, a known pathogen, in animals and their products or environment presents situations which must be faced by responsible agricultural, industrial and public health groups. Only through coordinated studies by many groups will data become available for evaluation of the problem. Public relations problems arise. Recognition and reporting of human infection should be promoted. Educational, diagnostic, and epidemiological services must be provided, along with possible regulatory and control measures.

In 1979, investigators at the Centers for Disease Control (CDC) give the following assessment of the situation:[70]

[Q fever] has been recognized as a widely distributed zoonosis that has the potential for causing both sporadic and epidemic disease. Because only certain states require reporting of Q fever and because illness is most frequently a nonspecific, mild-to-moderate, self-limited respiratory ailment, little information is available on the number and distribution of cases that occur each year. Serologic surveys of dairy cattle have indicated that the percentage of herds harboring infected cows has steadily increased over the past two decades. ... Although Q fever is not on the list of nationally notifiable diseases, cases in humans must be reported in 26 states and cases in animals in five states. [From 1948—1977] 1164 cases have been optionally reported to the CDC by 31 states ... 67% of all cases were reported from California. The average was 39 cases per year with peaks of 105 (1953) and 106 (1954). No distinct trend toward a change in the annual number of cases is obvious from the available data. A survey of all 50 states brought the total to 1247 Q fever cases ... fifteen states reported that surveys of animals had been conducted to assess dairy or livestock seropositivity for *Coxiella burnetii*. ... While 36 states have the laboratory capability to examine serum for antibodies to *C. burnetii*, no state examines serum routinely. In fact, only California examines serum for any reason other than a specific request for Q fever serology. Only 27 states considered Q fever to be a problem of any significance. Of these 24 thought that it was of little significance, while California, Colorado, and Pennsylvania believed that it was of moderate significance.

In Californian dairy cows, the infection rate rose sevenfold in 23 years.[71] An important change occurred in Canada, where extension of coxiellosis was reported from Ontario[72] and the Atlantic provinces.[73] The enzootic existence of livestock coxiellosis in Quebec was already made known in a report from 1960.[74]

In South America, Colombia reported extension of coxiellosis in dairy cows at an alarming rate.[75] Serological investigations in 1961 of dairy and beef cattle from different regions in Colombia showed an incidence of 0.46% (1715 reactors); a 1977 study by the CF test revealed a prevalence of 83% in dairy cows and 77% in beef cattle. Antibody prevalence in humans in Colombia in 1977 was 24%, but no clinical Q fever was registered in the population; the likelihood of misdiagnosis of Q fever cases as influenza, atypical pneumonia, or infectious mononucleosis is mentioned, in view of the lack of awareness of physicians that the agent is so widely spread in the livestock.

A controversial point is whether bovine coxiellosis is harmless to the cow's health, as seems to be the consensus in Great Britain and North America,[2,46,50,76,152] or whether reproductive function is interfered with by coxiella, as many Europeans on the continent now contend.[59,60,62,64,86,91,93,94]

When the importance of cattle as a source of human Q fever was recognized,[78] controlled laboratory studies of the experimental infection of cattle with *C. burnetii* were undertaken.[79-85] Australian investigators[78] inoculated two calves subcutaneously with infected guinea pig tissue; both calves responded with a mild febrile reaction on the third day after injection. One calf was sacrificed on the fourth day, and *C. burnetii* was recovered from its spleen and liver. The other calf yielded the infective agent from a blood sample on the fourth day, and antibodies were found from days 11 to 29. The febrile reaction and development of agglutinating antibodies were also reported from a similar study in Morocco.[79]

American studies of experimental Q fever in cattle were begun at the Rocky Mountain Laboratory in 1947.[80] Unsuccessful attempts to infect immature females by intranasal, intravenous, peroral, and vaginal administration of *C. burnetii* were followed by studies of udder infection of lactating cows. Infection was attempted by inoculation of massive doses of infected yolk sac cultures of the Nine Mile strain through the teat canal, under the skin of the udder, and by intraglandular injection. Subcutaneous injection did not lead to coxiella excretion through milk; intramammary gland injection produced shedding in the milk for over 200 d, while infection through the teat canal produced infection only of the inoculated quarter with elimination of infectious coxiella for 17 and 48 d. Isolation attempts from blood, feces, urine, and nasal washings from each cow were negative. Daily rectal temperature measurements taken for over 2 months did not reveal a significant rise. Complement fixation (CF) antibody titers appeared during the second week after infection and persisted as long as the animal was shedding coxiella in the milk; the cow that cleared the infection from the udder in 48 d showed a drop of the CF titer below the diagnostic level from the 60th day after infection.

In a subsequent experiment,[81] a lactating cow was infected with a coxiella strain isolated from milk (California strain). A heavy inoculum of yolk sac culture was introduced into the uterine cervix. Pools of first-milk samples of the four quarters were tested for infectiousness at 2-d intervals for the first 20 d, then twice a week. The samples from days 6, 10, and 18 proved infective; then all following samples until the 160th day were infective. However, when individual quarters were checked, only one was shedding the coxiella during the experiment. Urine, tested for 80 d, was infectious during the first 8 d only. Blood was negative throughout the experiment. Rectal temperature remained normal for 6 weeks. CF titers were at high levels through the day 191 postinfection.

A study of the pathology of coxiellosis included four cows which were infected with the California strain through the teat canal. This produced an acute mastitis which is uncharacteristic of the natural form of bovine coxiellosis. Coxiellemia was found on two or more of the first 5 d; milk was infectious from day 1 until the day of sacrifice of the cows, at titers of 10^5 and 10^4 on days 2 and 7, thereafter only at 10^1 guinea pig infectious doses. The cows were sacrificed on the days 5, 11, 23 and 63 of the experiment. Since there was coxiellemia during the early days, all tissues samples of the first cow sacrificed yielded coxiella, as did lung, liver, spleen, infected quarter, and its satellite lymph node of the second cow.

Peroral infection was attempted on a 3-week-old and a 3-month-old calf, each given milk from a proven coxiella-excreting cow, mixed with an equal volume of normal milk; the daily ration was about 6 l. The older calf yielded coxiella in blood samples taken on days 4, 6, and 12 after the feeding program had started, and a fecal sample collected on day 8. On day 13, the calf became febrile and was sacrificed on the following day. Of the tissues collected at autopsy, only the prescapular lymph node and the abomasum produced Q fever in guinea pigs. The younger calf yielded infectious coxiella in blood samples from days 4 and 13, and in feces for 2 weeks, but no symptoms were observed for more than 30 d of observation. Q fever antibodies were not demonstrable in its serum by CF. Asymptomatic infection was also observed following instillation of coxiella in the conjunctival sac, by bite from an infected tick, and by cohabitation and milking of uninfected and infected milking cows.

These and other experiments[83-85] and field observations led to the presently prevailing opinion that bovine coxiellosis is a harmless asymptomatic infection without sanitary or economic consequences for the industry.[2,46,76,152,153] The dissenting opinion was formulated in Switzerland and France; Swiss studies on bovine abortion very early named *C. burnetii* as a causative agent.[87-90] In France, researchers acted on an observation from earlier American experiments. Stoenner[82] had reported in 1951 that intradermal inoculation of two bull calves with the Nine Mile strain of *C. burnetii* resulted in clinical illness. At the site of the inoculation there was edema, induration, and central necrosis, and a general febrile syndrome followed. Coxiellemia was found for about 5 days postinoculation and the animals seroconverted. A pregnant lactating cow, inoculated by the same procedure with the California strain, also responded with marked clinical disease: high fever, anorexia, depression. A local skin lesion appeared at the site of the inoculation, and coxiellemia persisted through day 7 postinoculation. *C. burnetii* was shed in the milk of all quarters for 15 d, and the cow aborted on day 7 postinfection; the agent was recovered from the placenta. Urine was infectious particularly during the fourth week after infection, but no coxiella was detected in the feces. A French team under Plommet[86] took up this infection mode with the aim of reproducing clinical disease in 12 heifers aged 8 to 11 months. *C. burnetii* strain C_9, a human isolate from the Pasteur Institute, was inoculated intradermally at 30, 300, and 3000 guinea pig minimal lethal doses to the three groups of heifers. Forty-eight hours after the injection, all animals had a marked hyperthermia (40 to 41.1° C), a respiratory rate of 76 to 100 per minute, and a panting respiratory sound at auscultation without rales. The skin reaction previously described was also noted. Fever and tachypnea were most marked on day 2, while labored, panting breathing was prominent on day 3, which was also the crisis point. Thereafter, fever declined, and the temperature became normal between days 5 and 7. A small watery nasal discharge was noted, but there were no signs of bronchitis or of cardiac troubles. During the clinical phase the animals had a poor general appearance and anorexia. Convalescence was spontaneous, rapid, and apparently without complications. However, $6^{1}/_{2}$ months after infection one heifer died from heart failure without any premonitory signs. At necropsy, the heart was slightly dilated, flaccid, with areas of degeneration and sclerosis of the myocardium, mi-crothrombosis, and localized inflammatory lymphocytic infiltrations in the perivascular spaces. Electrocardiographs of the 11 survivors revealed a flattening of the tracings in heifers 2 and 12, a feature attributed to pericardial adherences or pericardial exudation. At slaughter about 17 months after infection, only heifer 1 showed grossly generalized sclerosis of the left lung, but on histopathology heifers 6 and 12 showed acute inflammatory lesions of the vascular connective tissue spaces of the myocardium, which was attributed to vascular damage.

The 11 heifers were artificially inseminated after synchronization of their heat, and 4 delivered a healthy calf; two heifers were slaughtered during pregnancy and were counted among the heifers with normal gestation. Two heifers aborted at 87 and 124 d of gestation, and 3 heifers remained sterile, although one of these may have had an early abortion. In comparison with the normal control population at the research station, fertility of the coxiellosis group was 8/11 (73%) vs. 88/98 (90%) in the controls; abortion was 3/8 (37%) vs. 1/53 (1.9%), and normal gestation was 6/11 (55%) vs. 43/53 (81%) in the control population.

These investigations remained almost unnoticed outside France. In France, Durand and Strohl[59] were among the first to search actively for reproductive troubles in livestock caused by *C. burnetii*. Twelve cases of bovine abortion, all isolated in time and location except 2 which occurred on the same premises, were observed during a 12-month period in 1976 to 1977; confirmation was by CF serology and isolation of the agent in guinea pigs from membranes of aborted fetuses. Further diagnoses of bovine abortion by coxiella were made in many departe-ments of France, beginning at the end of 1976 and especially in 1977.[59] Clinical observations were reported by Coche.[60] Coxiella-infected herds suffered frequently from metritis[64] and sterility and abortions. The latter occurred towards the end of gestation (from the sixth month onward); the earlier the abortion, the more pronounced were the placental lesions. Fetuses were generally dead, but when born alive, usually after the eigth month, were of poor viability. Metritis and infertility were frequent problems in the affected herds. Acute metritis followed abortion or calving, often beginning with retention of the afterbirth. Manual removal of the placenta was usually difficult. The cotyledons were necrotic, the membrane sometimes thickened and stiff. In other cases it was edematous or gelatinous, with abundant mucus of purplish or brownish coloration or thick, yellowish mucus. The endometritis could last for several weeks.

In neighboring Germany similar reports of reproductive problems in cattle attributed to *C. burnetii* came from veterinary services of the southern and middle parts of the Federal Republic;[55,91-93] in Central Europe coxiella abortion was reported from Hungary.[94] In Canada, coxiella abortion in cattle so far has been reported only from Ontario.[95]

Attempts to treat bovine coxiellosis with antibiotics have failed. Corboz and Zurgilgen[199] administered by deep intramuscular injection 20 mg/kg of bodyweight of Terramycin-LA (Pfizer) to 13 dairy cows positive in the CA test in milk and blood serum; the injection schedule was the same dosage on days 1, 4, and 7, and from days 1 to 10 the cows were given one injection of 200 mg oxytetracycline (Mastalone, Pfizer) daily in each teat. One positive cow remained untreated as a control.

Coxiella isolation was attempted by intraperitoneal injection of mice with the sediment (15,000 g/20 min at 4°C) of 4 ml of milk or colostrum, resuspended in 1 ml 0.15 *M* NaCl. The mice were bled by heart puncture 4 weeks after injection. The serum was examined by CF. Five of the 13 experimental cows excreted *C. burnetii* in milk or colostrum before treatment, three cows were found excreting at 1 month and 4 cows at 2 months after treatment. Coxiella was found in the placenta of one cow which calved 3 weeks following the treatment. The untreated control cow excreted the agent during the entire experimental period. CF blood titers of excreting cows were from 1:10 to 1:640.

D. COXIELLOSIS IN CAMELS

The dromedary (*Camelus dromedarius*) of western Asia and northern Africa is not only bred and used for riding and working purposes, but also for the production of milk consumed by humans. Camels in Pakistan reared for dairy purposes may yield up to 40 l of milk a day. In this respect, the camel takes the place of the cow not only in agricultural economics, but also in the epidemiology of Q fever in these oriental societies. The latter aspect was especially brought into focus in French military personnel serving in Algeria, Morocco, and Tunis during the 1950s and 1960s, who faced there an appreciable Q fever risk.[154] Coxiellosis was serologically diagnosed in camels of Nigeria, Chad, Niger, East Africa,[200] and Egypt.[103] Giroud and Capponi[154] name the dromedary as a possible reservoir in Africa — if not constant, at least occasional — a statement which shows that information of coxiellosis in this species is very rudimentary.

E. COXIELLOSIS IN SWINE

The natural susceptibility of swine to coxiella infection was demonstrated by the presence of antibodies to *C. burnetii* in their sera.[96,97] During a Q fever epidemic in a Montevideo, Uruguary slaughterhouse, 21.2% of 391 healthy slaughter pigs tested by MA reacted positively, but no lesions were observed in the reactors. A repeated testing in the following year of 88 swine sera was negative.[97,98] Marmion and Stoker[99] reported that pigs inoculated with *C. burnetii* develop CF and CA antibodies, but coxiella was not found in the placenta of a sow which was infected during pregnancy.

The risk of butchers contracting Q fever by slaughtering pigs seems to be remote, as seroconversion of such exposed meatworkers is of very low frequency,[11,100] and Q fever cases traced to swine have not been reported.

F. COXIELLOSIS IN HORSES AND MULES

Several surveys indicate the natural susceptibility of equines to infection by *C. burnetii*. In California, the sero-prevalence of antibodies indicative of coxiella infection was reported for 121 horses at 26% with the MA test to phase II antigen.[101] A similar study by the indirect immunofluorescence test of 123 horses from Atlantic Canada revealed 10.5 and 1.6% reactors to phase II and phase I antigens, respectively.[102] In Egypt, ten donkeys of 159 tested (6.2%) reacted in the CF test to coxiella antigen.[103]

A Q fever survey from 1967 to 1969 in the Federal Republic of Germany mentions one case of human Q fever attributed to contact with a donkey.[19]

Blinov[104] reported that horses experimentally infected with *C. burnetii* by the subcutaneous route responded with acute catarrhal gastroenteritis, catarrhal rhinitis, conjunctivitis, and pulmonary signs.

G. COXIELLOSIS IN DOGS AND CATS

Companion animals, when infected by *C. burnetii*, can became a health risk to their owners.[105]

1. Dogs

The role of dogs in the transmission chain of *C. burnetii* was suspected and proved during the early post-World War II years. Their experimental susceptibility was described by Blanc et al.[79] Dogs may become infected by tick bite[105,106] by feeding on infected ruminant placentas and infected milk or by aerosol inhalation.[106,107] The bitch can shed coxiellae in her milk for at least 30 d; urine excretion can persist for 70 d, and isolation from blood, although irregular, was reported 60 and 100 d after infection.[3]

Seroepidemiological data on canine coxiellosis showed 32% antibody prevalence in sheep dogs in Sicily,[107] 24.5% in central Italy,[108] 15.9% of 163 Egyptian dogs,[103] 31.4% of 388 dogs in Switzerland,[38] 13% in dogs in West Germany,[110] and 48% of 724 dogs in northern California, of which strays reacted even at 66%.[101]

The transmission of *C. burnetii* from dogs to humans was reported from Germany,[105,111] France,[153] and the U.S. A family outbreak of Q fever in West Germany was traced to two Bouviers de Flandre owned by the family; the dogs were frequently let out on a pasture where sheep were grazing.[105] Giroud[154] reports a case of serious Q fever in an animal caretaker, who fed offal to his dog; the dog developed coxiellosis and infected his master. In Idaho, a dog obtained from a sheep research station caused human Q fever 4 weeks after having left the research station.[111]

2. Cats

Natural infection of domestic cats in southern California was demonstrated[112] in a serosurvey using both the CA test and the MA test. Sera from 207 cats of the pounds of San Bernardino and Riverside counties yielded 19.8% positives by CA and 18.4% by MA; in the latter test, the titers ranged from 1:2 to 1:16, with 34 of the 38 positives not higher than 1:4.

Human disease acquired from contact with cats was reported from eastern Canada. In Halifax, Nova Scotia, ten members of a family and two friends developed a febrile illness within 2 weeks, which was diagnosed serologically as Q fever. Two weeks earlier a cat had given birth to several kittens in the house where the 12 persons were playing poker, and all had handled the cat or its litter. A serum sample from the cat collected about 6 weeks after parturition was positive at 1:64 by CF test.[113] More than 12 other incidents of this kind were observed in the area, each involving a pregnant female cat; 20% of all cases of Q fever pneumonia observed by Marrie in Nova Scotia have occurred following exposure to the products of feline conception.[73] There is nothing on record indicating that nonpregnant cats can transmit Q fever to humans.

Experimental infection of cats was studied by Gillespie and Baker.[114] Farm cats experimentally infected subcutaneously with *C. burnetii* responded with pyrexia from day 2 to day 5, lethargy, and anorexia; the agent was isolated from their blood irregularly but during a period of 1 to 4 weeks after infection and from their urine from 4 to 8 weeks. Infection by the oral route remained subclinical, although the agent was isolated from blood and urine, as it was following subcutaneous infection. Contact transmission among cats was seen in $^1/_4$ exposed cats by isolation from blood, but without clinical manifestations of the infection. CF antibodies were found in all cats infected subcutaneously (titers 1:16 to 1:128), but agglutinating antibodies were found only in 3 out of 4 (titers 1:16 to 1:32). Orally infected cats were seropositive in CF only 1 of 3, and 2 of 3 by agglutination. The one infected contact cat had both CF and agglutinating titers of 1:8 and 1:16, respectively.

In dogs and cats, the ELISA test is a better test for antibodies than the traditional CF and agglutination tests.[109] The role of dogs and cats in the epidemiology of human Q fever requires closer scrutiny because of the large number of urban cases and also cases without a traceable animal source.

H. COXIELLOSIS IN POULTRY

Birds are part of the host spectrum of *C. burnetii*.[115-120] Infected domestic poultry can transmit coxiellosis to humans through consumption of raw eggs from infected hens,[116] or through aerosolized fomites.[118] Serological evidence of natural infection was found in hens, ducks, geese, turkeys, and pigeons.[3] *C. burnetii* has been isolated from the kidneys of a pigeon[3] as well as from spleen, kidneys, and feces of experimentally infected hens.[118] Reisolation from infected pigeons was possible for up to 40 d and from sparrows up to 25 d.[115] Experimentally infected hens shed coxiellae in their feces from days 7 to 14.[118]

The risk to humans in contact with infected poultry was assessed in a study in India.[121] Sixteen of 25 poultry farms in Uttar Pradesh and Rajastan were found infected, with reactor rates of the chickens ranging from 5 to 50%. At a farm with 6% reacting chickens the personnel seroreacted at 27.52%, but clinical Q fever was not mentioned in the report. One Q fever case in Ontario was observed in a poultry farmer,[122] but the report did not indicate whether the birds were the source of the infection and if so, how this was experimentally verified.

III. COXIELLOSIS IN LABORATORY ANIMALS

The adult male guinea pig and the adult white mouse are the traditional laboratory hosts for the isolation of rickettsias.

A. GUINEA PIG

Guinea pigs from random-bred colonies may show variable susceptibility to infection with *C. burnetii*: in particular, poorly expressed clinical signs or low antibody responses; in extreme cases even no reaction at all following injection with virulent material from clinical cases. Mature animals (550 to 650 g) tend to respond more uniformly, but animals within the weight range 450 to 750 g can be used. To minimize error from poorly responding animals, it is recommended to inoculate intraperitoneally at least three animals with inocula from 2 to 4 ml. Isolation of *C. burnetii*, is almost always indicated by hyperthermy (40°C and higher) after 5 to 12 d of incubation. The duration of the febrile response depends on the virulence of the organism and the amount of the injected infectious material, but animals usually recover without sequelae; *C. burnetii* infection is not accompanied by visible scrotal reaction as with other rickettsias. Specific antibodies appear during convalescence which can be measured by serological tests (CF, MA, ELISA). Q fever antibodies in guinea pig serum is sufficient evidence to identify the agent as *C. burnetii*, since no immunologic cross-reactions with other organisms are known. Serum collected 2 weeks after infection contains moderate levels of only phase II CF antibodies; after 50 d, significant levels of both phase I and phase II antibodies are found.[123] Coxiella may persist in guinea pigs and be excreted in the urine over many months.[124]

Animal passages are carried out with spleen suspension or blood collected during the early febrile phase. In infection experiments by intraperitoneal injection with the California AD strain of *C. burnetii* up to 150 ID_{50} seldom caused fever in guinea pigs, but with increasing dosage all guinea pigs regularly showed fever. Death was rare unless inocula of $10^{7.5}$ ID_{50} were given, and the LD_{50} level was at $10^{8.0}$ ID_{50}.[125]

Pathology — Like humans, experimentally infected guinea pigs develop specific pathomorphological changes. Spleen and testes show especially well-marked histopathology.[126,128-132] In the lung the dominant feature is nonpurulent inflammation diffusely extending in the vascularized septa and best observed at the crisis of the disease. Inguinal and mesenteric lymph nodes may be up to fourfold enlarged; kidneys and lungs may be congested. Typical are large (20 to 45 nm) reticuloendothelial cells containing one or several vacuoles (phagolysosomes) filled with coxiellae. Inflammatory infiltrates consist mainly of lymphocytes and plasma cells, but few leukocytes, except when complicated by bacterial superinfection. Hepatitis is mainly seen in the portobiliary spaces, but vacuolized cells (phagolysosomes) do occur between the hepatic trabeculae. Degenerative myocarditis is frequently found in guinea pigs dying during convalescence. Vascular thrombosis is often seen in lungs and spleen. Lesions evanesce during convalescence without leaving scars.

Guinea pigs sacrificed on the first day of fever after intraperitoneal inoculation regularly showed pulmonary congestion.[133] Thus, a pulmonary route of infection is not necessary for the eventual development of lung lesions.[125]

B. MICE

Burnet and Freeman[137] described numerous rickettsias in the liver and spleen of mice experimentally infected with the agent of Q fever and stated that the only feature calling for comment was the character and distribution of rickettsias in the spleen. In the liver they found diffuse infiltrations with cells of vascular endothelial origin, variable numbers of small necrotic foci, and rickettsias in Kupffer cells. The histopathology of Q fever in mice was studied in more detail by Perrin and Bengtson.[138] Mice infected by intranasal and intraperitoneal inoculations showed consistently nodular and patchy granulomatous lesions of large mononuclear cells in spleen, liver, kidneys, and adrenals, appearing on the fourth day in spleen and liver, and the sixth day in kidney and adrenals. Pneumonic reactions were only seen in intranasally inoculated mice. They were characterized by early proliferative changes and exudate with mostly large and small mononuclear cells. Pneumonia was also seen in a number of control mice given normal yolk sac suspension intranasally, but in these the exudative changes predominated over the proliferative ones, and polymorphnuclears were mostly seen in the exudate. Lesions were absent in heart, thyroid, esophagus, stomach, intestines, pancreas, testicles, brain, meninges, spinal chord, and spinal ganglia. Intranasal administration of coxiella preparations to white mice led to focal purulent bronchopneumonia. Lung impression smears stained by Giemsa or Victoria blue showed coxiella mainly intracellularly in alveolar epithelial cells and histiocytes, but not in leukocytes.[129] Intraperitoneal injection of yolk sac or spleen suspensions into mice produced granulomatous or nodular lesions throughout spleen, liver, kidney, adrenal, and retroperitoneal and mediastinal lymph nodes, mainly composed of mononuclear cells. Similar nodules and areas of aplasia and degeneration were observed in the bone marrow.[138]

C. HAMSTER

Hamsters are particularly indicated for animal passages of *C. burnetii*. They are easily infected, and after intraperitoneal infection organisms are very numerous in the spleen. They have the advantage that contamination with latent viruses is rare.[154] Five coxiella strains isolated from rodents (*Peromyscus maniculatus, Dipodomys ordii, Dipodomys microps*) were reported to be 30 to 300 times more infectious in hamsters than in guinea pigs and mice; they also stimulated in hamster higher antibody titers.[135]

D. RABBIT

The rabbit is not very popular for coxiella diagnosis in laboratories. In rare instances it is used for the direct demonstration of coxiellae from necrotic tissues. Three rabbit does, one third into pregnancy, infected intranasally with coxiella isolated from sheep, delivered dead fetuses at term. Swabs taken immediately at the time of birth yielded coxiella after four and five egg passages. *C. burnetii* was isolated from a stillborn's liver and heart after three egg passages.[136]

E. MONKEYS

The cynomolgus monkey (*Maccaca fascicularis*) was reported[139] to be a suitable model for studying the pathogenesis of Q fever. Aerosol exposure to 10^5 50% mouse infectious doses of *C. burnetii* caused clinical disease 4 to 7 d after infection. Anorexia, depression, cough, dyspnea, fever, and tachypnea were observed. Coxiellemia was demonstrated by day 7 and lasted through day 13, with a mean duration of 7.7 d. Radiographs showed granular infiltration of lungs as early as 5 d after infection and resolution of pneumonia beginning by day 16. Hematologic values indicated neutrophilia and leukopenia. Pathological changes included interstitial pneumonia and subacute hepatitis. *C. burnetii* could be seen in sections of lung, spleen, liver, kidney, heart, and testes. The indirect FA test revealed phase II antibodies from day 7 (1:300) and to phase I by day 14 (1:50).

F. CHICKEN EMBRYO

The extensive growth of *C. burnetii* in the yolk sac of chicken embryos has been known since the time of discovery of the organism.[137,140] It is also known that the growth mechanism of *C. burnetii* is different in chicken embryos from that in animals: in the latter the infection is generalized, and multiplication takes place in mononuclear phagocytes; the agent remains in phase I. *In ovo* the agent grows almost exclusively in the yolk sac endoderm cells, and phase variation takes place during serial passages.[142] Little is known about biological details of this peculiar host-parasite relationship in the yolk sac which could explain this behavior of *C. burnetii*. In chick endodermal cells cultivated *in vitro* *C. burnetii* grows inside large vacuoles without visible damage to the cells.[144]

The growth curve *in ovo* is somewhat unusual in that a rapid rise to a maximum on day 7 is followed by a rapid fall; within 24 h only half the coxiella population remains.[142] The strict localization of coxiella growth in the yolk sac during the first 6 d of infection was verified, and after day 7 spillover of the coxiella to chorioallantoic membrane and gut of the embryo was noted first, then to amniotic membrane and amniotic fluid, liver, spleen, muscles, and heart.[145]

Electron-microscopic studies on the growth of coxiellae *in ovo* were reported by Anacker et al.[146] and Khavkin et al.[147]

IV. WILDLIFE COXIELLOSIS

A. COXIELLOSIS IN TICKS

The taxonomic placement of *C. burnetii* with the rickettsias, which are commonly arthropod transmitted, led to the search for the vector hosts among the ectoparasites of naturally infected warm-blooded animals. Several natural reservoirs were reported. In Australia such a natural reservoir was found in the marsupial bandicoot (*Isoodon torosus*) and the tick *Haemaphysalis humerosa*.[160] Like other animals, the bandicoot suffers no apparent ill effect from the infection by *C. burnetii* but coxiella can be found in the blood for some time after experimental exposure so that ticks have ample opportunity to become infected during feeding. Spread of coxiellosis by contact within the bandicoot population is not very likely, since few coxiellae are shed with urine and feces, and at birth the marsupial placenta is very small and is immediately eaten, so the contamination of the environment is minimal.

A similar parasitic cycle exists in Australian kangaroos (*Macropus major* and *Macropus rufus*) and their associated ticks (*Amblyomma triguttatum*).[161] A number of different tick species can serve as host-vectors for *C. burnetii*. Smith[162,163,164] succeeded in the experimental transmission of *C. burnetii* from guinea pig to guinea pig with *Ixodes holocyclus*, *Haemaphysalis bispinosa*, and *Rhipicephalus sanguineus*. This is important, since the various stages of ticks feed preferentially on certain warmblooded host species at the exclusion of others. *H. humerosa* is a bandicoot tick that does not feed on cattle. *I. holocyclus*, on the other hand, parasitizes the bandicoot as well as cattle and humans and is a plausible natural vector for the infection of livestock from which other tick species feeding on cattle may acquire their coxiella infection.

In 1958, there was a remarkable group of cases on a pineapple farm near Brisbane. Between October and November, all six persons living on a farm — father, mother, and four sons — developed Q fever. There were no cows or sheep on the farm, and no evidence of infection was found in the dairy herd from which they purchased milk. A suspected source of infection was discovered in the marsupial bandicoots (*Isoodon macrourus*) which were plentiful on the farm. *C. burneti* was isolated from 3 of 20 which were examined by inoculating their organs into mice and guinea pigs. Ticks came under suspicion, but there was no history of tick bite, and *C. burnetii* was not isolated from 296 ticks (mostly *Ixodes holocyclus* and *Haemaphysalis humerosa*) collected from the bandicoots. At the time of illness, the weather had been very dry, and this had led to much activity at and around the water hole. Bandicoots had been very numerous near the water hole, and the suggested mode of infection was inhalation of particles of dust or water contaminated in some way by them.[9]

The first isolation of *C. burnetii* in North America was from a guinea pig on which 50 *Dermacentor andersoni* from western Montana had fed.[166] Natural infections have subsequently been reported from *D. occidentalis* in Oregon and California, from *Amblyomma americanum* in eastern Texas, from *Hemaphysalis leporis-palustris* and *Ixodes dentatus* from Virginia, and from *Otobius megnini* collected in southern California.[167-169] Experimental transmission of Q fever by *D. andersoni* was reported by Parker and Davis,[170] who also demonstrated transovarial transference of coxiella.

These reports inspired a worldwide search of Q fever wildlife reservoirs and compilations of lists of naturally and experimentally susceptible species of ticks and other ectoparasites.[1,5] Detailed investigations of coxiella infection of ticks led to the following observations:

1.   *C. burnetii* multiplies principally in the cells of the middle gut or stomach and is excreted with the feces which the ticks expel during the feeding. Thus, the skin of the host becomes heavily contaminated with coxiellae and can infect other animals and humans. This is one mechanism for explaining Q fever epidemics in wool and hide-processing plants.[171,172] Dried feces from infected ticks is extremely rich in virulent coxiellae and can maintain its infectiousness for about 2 years.[173] Moreover, an infected tick may remain infectious for several years,[174] even for the rest of its life.[174] Possible direct coxiella transmission through the bite of some ticks is shown by the infectiousness of their saliva.

2.   Coxiellae from ticks are in phase I and are usually of marked virulence.[175,176]

3.   Infection of the ovaries of certain ticks (*Dermacentor andersoni, Hyalomma savignyi, Rhipicephalus sanguineus, Amblyomma cayennense, Ornithodorus gurneyi, Ornithodorus moubata*) may lead to germinative transmission[179] to the offspring, thus establishing a closed self-perpetuating reservoir within the tick population.[177,178] This rather somber scenario drawn by parasitologists has been used by some animal health officials as an argument that in areas where such a wildlife reservoir exists, control of livestock coxiellosis is hopeless and impractical,[2,7,16,180] a position no longer acceptable because of the possible immune protection of herds through vaccination.

The African tick *Ornithodorus moubata* can easily be maintained in the laboratory and was therefore a species of choice for experiments. It was used to isolate coxiellae from persons who had been treated with antibiotics.[181,182]

Although it is very likely that infected ticks were the original transmitters of *C. burnetii* to domestic animals, it was very early realized that ticks are not a necessary element in the epidemiology of livestock coxiellosis, as Lennette et al.[36] remarked in the early 1950s: "all the field studies from this laboratory have pointed away from arthropods as being responsible for the dissemination of *Coxiella burnetii* among domestic livestock", a view shared also by Babudieri[3] and many others.

The tick *Dermacentor marginatus*, known in southern Germany as the sheep tick, was strongly believed by some to be an important factor in the occurrence of Q fever in humans and livestock in the southern part of the country, in an area congruent with the biotope of the tick.[16,180] An epidemiological study[183] was undertaken to better localize the situation of wildlife reservoirs in the area, and 2967 adult *Dermacentor* ticks were collected from vegetation and from local sheep. From 1210 ticks it was possible to prepare three smears of hemolymph, which were stained either by the Gimenez method, or by immunofluorescence for *C. burnetii* or for *Rickettsia slovaca*, a member of the spotted fever group recently discovered in local ticks. Rickettsias were found by histological staining in 36 ticks, of which 34 were identified as *Rickettsia slovaca* and only 2 as *C. burnetii* by immunofluorescent staining; the latter two ticks were taken from sheep. This is certainly a meager yield in a reputedly endemic and enzootic area of Q fever.

The transmission of Q fever to humans by tick bite has been rarely reported, once in woolsorters[184] and once in an outdoor man.[185]

B. COXIELLOSIS IN OTHER ARTHROPODS

The rickettsial nature of *C. burnetii* and its presence in ticks stimulated the search for other intermediate hosts and vectors. Epidemiological investigations of Q fever outbreaks paid special attention to the arthropod fauna as possible sources of infection; these efforts were, however, not very encouraging and have become marginal activities in present day Q fever research.

During the discovery stage of *C. burnetii* in Australia, researchers tried in vain to isolate coxiella from various species of blood-sucking ectoparasites such as lice, fleas, mites (the chicken mite *Liponyssus bursa*), and two parasitic fly species, of which one was the sheep ked *Melophagus ovinus*[184] Similarly fruitless were the attempts of Huebner et al.[186] to find coxiellae in flies (*Musca domestica, Siphona irritans*), mosquitoes (*Culex quinquefasciatus*), and several species of mites (*Gohiera fusca, Histiosoma* ssp.) captured on coxiella-infected premises. Epidemiological studies of a Q fever epidemic at the School of Agriculture in Davis, California[187] included attempts to isolate the agent from arthropods collected from cows, sheep, and rodents. Forty-five test series were made by inoculating guinea pigs with suspensions of triturated arthropods. The latter included flies (*Haematobia serrata, Stomoxys calcitrans, Musca domestica, Oestrus ovis*), mosquitoes (*Culex tarsalis, C. stigmatosoma*), fleas (*Pulex irritans, Echidnophaga gallinacea, Diamanus montanus*), lice and mallophages (*Enderleinellus osborni, Trichodectes spp.*), mites (*Atridolaelaps* ssp., *Bdellonyssus bagoti*) and ticks (*Ornithodorus parkeri*). The various species pools contained from 2 to 100 animals. The results were all negative. During another Q fever outbreak in northern California,[188] a similar series of 17 isolation attempts was unsuccessful. The arthropods involved included *Melophagus ovinus, Culex* spp., *Aedes* spp., *Anopheles freeborni*, and *Chrysops* spp.

A very extensive search was made by Irons et al.[189] in Zavala County, Texas, an area where Q fever was endemic. Ectoparasites collected from 2261 animals of 16 genera and 28 species, mostly rodents caught in the wild, were examined for coxiella infection; 9320 ectoparasites belonging to 34 genera, mostly mites, ticks, and fleas were collected. *C. burnetii* was not demonstrated in any parasite pool examined.

An exceptional observation was reported by Giroud and Jadin[155] and Giroud et al.,[156] who examined body lice from natives in Belgian colony Ruanda Burundi 3 months after a Q fever epidemic. The lice were triturated and injected into guinea pigs. A few guinea pigs in the third passage developed CF and agglutinating antibodies, from which the researchers concluded that coxiellae had been present in the louse sample injected.

Negative were a number of transmission experiments under laboratory conditions. Smith[157] found that *C. burnetii* neither persisted nor multiplied in the cat flea *Ctenocephalides felis.* Philip[158] and Weyer[159] reported that *Aedes aegypti* were incapable of transmitting *C. burnetii*; apparently the organisms either died in the gut or were rapidly expelled.

Purely mechanical transmission of *C. burnetii* was demonstrated by Philip[158] with *Musca domestica* and *Triatoma infestans*; the first caught in a cage which housed an infected guinea pig, the latter were fed on infected mice. In both species the infection was not transmissible beyond 3 d after the infecting meal.

C. COXIELLOSIS IN VERTEBRATE FAUNA

The second component of the wildlife cycle is the vertebrate fauna on which the ticks feed, or which becomes infected by feeding on contaminated animal carcasses. Rodents and lagomorphs[190-195] and larger game have been principally named in this context. Skinning of rabbits snared in Nova Scotia, Canada, caused acute Q fever in two brothers.[190] In the Vaud canton of Switzerland, where coxiellosis is enzootic in livestock, a survey on sera of 87 roe deer (*Capreolus capreolus*) produced one positive (1.4%), two questionable reactors, and 16 anticomplementary sera could not be interpreted. In the same survey, 1/44 (2.3%) ravens and 1/42 (3.3%) rodents (*Apodemus, Microtus, Chlethrionomya, Arvicola, Pitymys*) seroreacted.[53] Giroud and Capponi[154] examined serologically 361 rodents from 11 species in France, but found only reactors among the species *Apodemus sylvaticus, Eliomys quercinus,* and *Clethrionomys glareolus.*

Both feral and domestic birds are included in the natural host spectrum of *C. burnetii.*[115-119] *C. burnetii* has been isolated from spleen and liver of wild birds, and ectoparasites (*Ornithomyia biloba*) of swallows.[117] Serological evidence of coxiellosis is highest in birds living in close proximity with infected livestock and lowest in birds living away from human dwellings and domestic animals: eusynanthropic species had a 15.8% antibody prevalence, synanthropic species 4.3%, and exanthropic species 1.8%.[117] Transmission of *C. burnetii* among birds is assumed to be mediated by ticks and other ectoparasites.

According to a study in California,[119] feeding habits, residence, and seasonal migration patterns play a role in the exposure of birds to *C. burnetii.* Carrion-feeders (vultures, ravens, crows), with the highest antibody prevalence, are heavily exposed through contact with infected dead animals or placentas on grazing ranges. Carnivorous birds (sparrow hawks, red-tailed hawks) may acquire the infection from infected prey (mainly birds and rodents), while granivorous and insectivorous species that feed and roost in or near cattle barns and stockyards may be exposed to infectious dust (pigeons, cowbirds, and blackbirds); 38% of birds collected on a dairy feedlot reacted in the coxiella capillary tube agglutination test and 13% of birds caught on a sheep range. The differences were attributed to animal concentration and the degree of contamination of the environment in the two settings.

Migrating species acquire their infection mostly during the lambing and calving seasons, that is, birds visiting in spring and summer have a greater risk of infection than those in winter. The extent to which birds transmit Q fever to man and coxiellosis to animals is uncertain. Babudieri and Moscovici[115] regards the epidemiological importance of the bird fauna as moderate. Rather than carriers of disease, birds are indicators of infection in the localities they frequent. Otherwise, it is not possible to explain the absence of coxiellosis in northern countries that are breeding grounds of many bird species which spend winter in places of known Q fever endemicity.

In many instances it is not certain whether livestock or wildlife is the primary or the secondary coxiella reservoir in an area. A study on *C. burnetii* in poikilotherms in Uttar Pradesh, India, found agglutinating antibodies in water snakes (*Natrix natrix*), rat snakes (*Ptyas korros*), pythons (*Python molurus*), and tortoises (*Kachuga sp.*). However, an open fish pond near the university dairy farm was the main source of water snakes and tortoises; coxiellosis was present in the animals of the dairy farm. In rural India, ponds are often shared by cattle, buffalo, dogs, sheep, goats, camels, and people for bathing, swimming, irrigation, and drinking, which makes it difficult to identify the polluting sources.[201] Giroud and Capponi[154] report coxiellosis on a trout farm, where the fish were fed coxiella-contaminated feed. They isolated also coxiella from intestinal parasites of fish (*Tilapia nilotica, Barbus altialanis*) caught in Lake Kivu, Zaire; however, near the lake was also an abattoir where coxiella-infected cattle were slaughtered.[1] In Romania, rich in forests, game, and ticks, the primary reservoir is nonetheless attributed to domestic animals.[19] In northern California also, livestock, rather than wildlife, is regarded as the coxiella reservoir.[196-198]

V. CONCLUSIONS

In retrospect, the history of 50 years of the zoonosis Q fever brings us, in a circuitous way, back to the beginning, the abattoir fever caused by coxiella-infected cattle. It went through the early Australian-American period dominated by the search of the arthropod vector and the wildlife reservoirs, then the Mediterranean period which placed heavy emphasis on sheep and goats as sources of the human disease, and finally the view that dairy cows, and to a lesser degree milk goats and camels, are the perpetual reservoirs from which accessory hosts, including humans, dogs, cats, sheep, and other vertebrates may become infected. California observations[214] suggested that if the level of bovine coxiellosis reaches 5 to 10% in a given area, endemicity may be expected. This hypothesis seems supported not only in California, but also in Atlantic Canada, the province of Ontario, the Netherlands, West Germany, and Switzerland; in each case, the dairy herd infection rates and the human Q fever statistics agree with it. Discrepancies point to weaknesses in medical or veterinary diagnostics. The 100 to 200 cases of Q fever per year in Great Britain indicate endemicity in humans, while veterinary statistics do not identify the livestock reservoir responsible for it. Yet, about 80,000 gallons of goats' milk are marketed untreated every year through farm gate sales and health food shops without any consideration of a possible risk of contamination by *C. burnetii*.[215] On the other hand, if the nationwide presence of *C. burnetii* in dairy herds of the U. S. as described by Luoto in 1960,[69] is reality, then the low rates of human Q fever infection reported by many states are very questionable. The habitat of *C. burnetii* has undeniably moving boundaries, and corrections of epidemiological or enzootiological statistics should be made with data obtained from recent studies using modern serological methods, such as ELISA. A remaining question is to which extent beef cattle alone contribute to the persistence of *C. burnetii* in specified geographic locations, such as the extensive cattle ranges of the western U. S. and western Canada or the open air ranchos in South America.

The contagiousness to humans of sheep acutely infected with coxiella has been observed and reported many times. The new information concerns strictly the role of sheep as coxiella reservoir, which leads to several subsidiary questions:

1.   Do sheep become chronic carriers of coxiella?
2.   Do sheep breeds which are exploited for milk and cheese production behave like dairy cows and goats?
3.   If time between successive lambing seasons were the only determining factor for the nonestablishment of a coxiella reservoir in sheep, would the artificial promotion of ewe fertility toward two pregnancies per year, as described by Rauch et al.,[27] bring about the establishment of a permanent coxiella reservoir in sheep?
4.   Can it be assumed that in the absence of bovine and caprine coxiellosis in a given area or in the absence of contact between sheep flocks and bovines or caprines, coxiellosis of ewes is unlikely and need not to be monitored?

Finally, what can be expected in the future? In this regard, much depends on the human factor. In many countries Q fever and animal coxiellosis are forgotten diseases or, at least, are treated with benign neglect by public and animal health authorities.[2,7,70,122] However, people preoccupied with health and safety on the working place are echoing the sentiment that "while well-recognized hazards such as brucellosis are receding, others such as ornithosis, Q fever, and rabies are of increasing importance".[216] Improvements made in recent years in diagnostic methodology, such as indirect FA tests or ELISA,[15,109,203-207] facilitate serological surveillance of livestock and have improved sensitivity and given greater confidence in the results. Studies on preventive vaccination of livestock have given increased hope that coxiellosis can be stopped at the farm level.[92,93,209-212] How and when this will be done depends on the determination and ability of those in charge of public and animal health.

REFERENCES

1. **Thiel, N.,** *Das Q-Fieber und seine geographische Bedeutung,* Dunker & Humblot, Berlin, 1974.
2. **Stoenner, H. G.,** Q fever, in *CRC Handbook Series in Zoonoses Section A. Bacterial, Rickettsial and Mycotic Diseases,* Vol. 2, CRC Press, Boca Baton, FL, 1980, 337.
3. **Babudieri, B.,** Q fever: a zoonosis, *Adv. Vet. Sci.,* 5, 81, 1959.
4. **Ormsbee, R. A.,** Q fever rickettsia, in *Viral and Rickettsial Infections of Man,* 4th ed., Horsfall, F. L. and Tamm, I., Eds., Lippincott, Philadelphia, 1965, 1144.
5. **Reusse, U.,** Die Bedeutung des Q-Fiebers als Zoonose, *Z. Tropenmed. Parasitol.,* 11, 223, 1960.
6. **Ludwig, Ch.,** Das Q-Fieber, in *Handbuch der Virusinfektionen bei Tieren,* Vol. 3/2, Röhrer, H., Ed., Gustav Fischer, Jena, East Germany, 1968.
7. **Sawyer, L. A., Fishbein, D. B., and McDade, J. E.,** Q fever: current concepts, *Rev. Infect. Dis.,* 9, 935, 1987.
8. **Aitken, I. D., Bögel, K., Cracera, E., Edlinger, E., Houwers, D., Krauss, H., Rády, M., Rehacek, J., Schiefer, H. G., Schmeer, N., Tarasevich, I. V., and Tringali, G.,** Q fever in Europe: current aspects of aetiology, epidemiology, human infection, diagnosis and therapy, *Infection,* 15, 323, 1987.
9. **Derrick, E. H.,** The changing pattern of Q fever in Queensland, *Pathol. Microbiol.,* 24 (Suppl.), 73, 1961.
10. **Derrick, E. H., Pope J. H., and Smith, D. J. W.,** Outbreaks of Q fever in Queensland associated with sheep, *Med. J. Aust.,* 1, 585, 1959.
11. **McKelvie, P.,** Q fever in a Queensland meatworks, *Med. J. Aust.,* 1, 590, 1980.
12. **Marmion, B. P., Kyrkou, M., Worswick, D., Esterman, A., Ormsbee, R. A., Wright, J., Cameron, S., Feery, B., and Collins, W.,** Vaccine prophylaxis of abattoir-associated Q fever, *Lancet,* 2, 1411, 1984.
13. **Somma-Moreira, R. R., Caffarena, R. M., Somma, S., Pérez, G., and Monteiro, M.,** Analysis of Q fever in Uruguay, *Rev. Infect. Dis.,* 9, 386, 1987.
14. **Lang, G. H.,** Serosurvey on the occurrence of *Coxiella burnetii* in Ontario cattle, *Can. J. Public Health,* 79, 56, 1988.
15. **Lang, G. H.,** Serosurvey of *Coxiella burnetii* infection in dairy goat herds in Ontario. A comparison of two methods of enzyme-linked immunosorbent assay, *Can. J. Vet. Res.,* 52, 37, 1988.
16. **Rojahn, A.,** Anmerkungen zu zwei Veröffentlichungen über Q-Fieber, *Tierarztl. Umsch.,* 34, 845, 1979.
17. **Edlinger, E. A.,** Q fever in France, *Zentralbl. Bakteriol. Parasitenkd. Infektionskr. Hyg. Abt. 1 Orig. Reihe A,* 267, 26, 1987.
18. **Tarasevich, I. V.,** Epidemiology (1980—1985) and non-specific prophylaxis of Q fever in the USSR (survey), *Zentralbl. Bakteriol. Parasitenkd. Infektionskr. Hyg. Abt. 1 Orig. Reihe A,* 267, 1, 1987.
19. **Weise, H. J.,** Zur Epidemiologie des Q-Fiebers in der BRD, *Bundesgesundheitsblatt,* 6/7, 71, 1971.
20. **Tringali, G. and Mansueto, S.,** Epidemiology of Q fever in Italy and in other mediterranean countries, *Zentralbl. Bakteriol. Parasitenkd. Infektionskr. Hyg. Abt. 1 Orig. Reihe A,* 267, 20, 1987.
21. **Curet, L. B. and Paust, J. C.,** Transmission of Q fever from experimental sheep to laboratory personnel, *Am. J. Obstet. Gynecol.,* 112, 566, 1972.
22. **Schachter, J., Sung, M., and Meyer, K.,** Potential danger of Q fever in a university hospital environment, *J. Infect. Dis.,* 123, 301, 1971.
23. **Meiklejohn, G., Reimer, L. G., Graves, P. S., and Helmick, C.,** Cryptic epidemic of Q fever in a medical school, *J. Infect. Dis.,* 144, 107, 1981.
24. **Hall, C. J., Richmond, S. J., Caul, E. O., Pearce, N. H., and Silver, I. A.,** Laboratory outbreak of Q fever acquired from sheep, *Lancet,* 1, 1004, 1982.
25. **Simor, A. E., Brunton, J. L., Salit, I. E., Vellend, H., Ford-Jones, L., and Spence, L. P.,** Q fever hazard from sheep used in research, *Can. Med. Assoc. J.,* 130, 1013, 1984.
26. **Ruppanner, R., Brooks, D., Franti, C. E., Behymer, D. E., Morrish, D., and Spinelli, J.,** Q fever hazards from sheep and goats used in research, *Arch. Environ. Health,* 37, 103, 1982.
27. **Rauch, A. M., Tanner, M., Pacer, R. E., Barrett, M. J., Brokopp, C. D., and Schonberger, L. B.,** Sheep-associated outbreak of Q fever-Idaho, *Arch. Intern. Med.,* 147, 341, 1987.
28. **Enright, J. B., Franti, C. R., and Longhurst, W. M.,** *Coxiella burnetii* in a wildlife-livestock environment: antibody response of ewes amd lambs in an endemic Q fever area, *Am. J. Epidemiol.,* 94, 62, 1971.
29. **Brooks, D. L., Ermel, R. W., Franti, C. E., Ruppanner, R., Behymer, D. E., Williams, J. C., and Stephenson, J. C.,** Q fever vaccination of sheep: challenge of immunity in ewes, *Am. J. Vet. Res.,* 47, 1235, 1986.
30. **Schaal, E. and Goetz, W.,** Q fever infections and their cause in the population of the area Simmerath/Eifel from the standpoint of veterinary medicine, *Dtsch. Tiererztl. Wochenschr.,* 81, 477, 1974.
31. **Schaal, E. H.,** Zur Euterbesiedlung mit *Coxiella burnetii* beim Q Fieber des Rindes, *Dtsch. Tiererztl. Wochenschr.,* 89, 393, 1982.
32. **Grist, N. R.,** The persistence of Q-fever infection in a dairy herd, *Vet. Rec.,* 71, 839, 1959.
33. **Stoenner, H. G., Lackman, D. B., Benson, W. W., Mather, J., Casey, M., and Harvey, K. A.,** The role of dairy cattle in the epidemiology of Q fever in Idaho, *J. Infect. Dis.,* 109, 90, 1961.

34. **von Kruedener, R. and Schreyer, K.**, Verlauf einer Q-Fieberinfektion bei Schafherden, *Fortschr. Veterinaermed.,* 17, 70, 1972.

35. **Joubert, L., Fontaine, M., Bartoli, M., and Garrigue, G.**, La fièvre Q ovine. Zoonose d'actualité de type professionnel, rural et militaire, *Rev. Med. Vet.,* 127, 361, 1976.

36. **Lennette, E. H., Holmes, M. A., and Abinanti, F. R.**, Q fever studies: observations on the pathogenesis of the experimental infection induced in sheep by the intravenous route, *Am. J. Hyg.,* 55, 254, 1952.

37. **Grant, C. G., Ascher, M. S., Bernard, K. W., Ruppanner, R., and Vellend, H.**, Q fever and experimental sheep, *Infect. Control,* 6, 122, 1985.

38. **Metzler, A. E., Nicolet, J., Bertschinger, H. U., Bruppacher, R., and Gelzer, J.,** Die Verbreitung von *Coxiella burnetii.* Eine epidemiologische Untersuchung bei Haustieren und Tierärzten, *Schweiz. Arch. Tierheilk.,* 125, 507, 1983.

39. **Caminopetros, J.,** La bronchopneumonie épidémique hivernoprintaniaire humaine et animale (chèvre, mouton), fièvre Q ou grippe des Balkans, à Rickettsia burneti var. caprina. Les charactères particuliers de l'infection animale, *Ann. Inst. Pasteur,* 77, 750, 1949.

40. **Caminopetros, J.,** La Q-fever en Grèce: le lait source de l'infection pour l'homme et les animaux, *Ann. Parasitol. Paris,* 23, 107, 1948.

41. **Clark, W. H. and Romer, M. S.,** Q fever in California. IX. An outbreak aboard a ship transporting goats, *Am. J. Hyg.,* 54, 35, 1951.

42. **Russo, P. and Malo, N.,** Q fever in the Vienne department of France: antibody kinetics and abortion in goats, *Rec. Med. Vet.,* 157, 585, 1981.

43. **Ruppanner, R., Riemann, H. P., Farver, T. B., West, G., Behymer, D. E., and Wijayasinghe, C.,** Prevalence of *Coxiella burnetii* and *Toxoplasma gondii* among dairy goats in California, *Am. J. Vet. Res.,* 39, 867, 1978.

44. **Jellison, W. L., Welsh, H. H., Elson, B. E., and Huebner, R. J.,** Q fever studies in southern California. XI. Recovery of *Coxiella burnetii* from milk of sheep, *Public Health Rep.,* 65, 395, 1950.

45. **Sienko, D. G., Bartlett, P. C., McGee, H. B., Wentworth, B. B., Herndon, J. L., and William, N. H.,** Q fever: a call to heighten our index of suspicion, *Arch. Intern. Med.,* 148, 609, 1988.

46. **Losos, G. J.,** Q fever, in *Infectious Tropical Diseases of Domestic Animals,* Longman, Harlow, England, 1986, chap. 22.

47. **Spicer, A. J., Crowther, R. W., Vella, E. E., Bengtsson, E., Miles, R., and Pitzolis, G.,** Q fever and animal abortion in Cyprus, *Trans. R. Soc. Trop. Med. Hug.,* 71, 16, 1977.

48. **Polydorou, K.,** Q fever in Cyprus: a short review, *Br. Vet. J.,* 137, 470, 1981.

49. **Kilchberger, G. and Wiesmann E.,** Abortus-Epidemie bei Ziegen bedingt durch *Rickettsia burneti, Schweiz. Arch. Tierheilkd.,* 91, 553, 1949.

50. **Waldhalm, D. G., Stoenner, H. G., Simmons, R. E., and Thomas, L. A.,** Abortion associated with *Coxiella burnetii* infection in dairy goats, *J. Am. Vet. Med. Assoc.,* 137, 1580, 1978.

51. **Palmer, N. C., Kierstaed, M., and Key, D. W.,** Placentitis and abortion in goats and sheep in Ontario caused by *Coxiella burnetii, Can. Vet. J.,* 24, 60, 1983.

52. **Lang, G. H.,** unpublished data.

53. **de Meuron, P.-A.,** A propos d'une recrudescence de la fièvre Q dans le bétail du canton de Vaud, *Schweiz. Arch. Tierheilkd.,* 127, 735, 1985.

54. **Schaal, E. H. and Schäfer, J.,** Zur Verbreitung des Q-Fiebers in einheimischen Rinderbeständen, *Dtsch. Tiererztl. Wochenschr.,* 91, 52, 1984.

55. **Krauss, H., Schmeer, N., and Schiefer, H. G.,** Epidemiology and significance of Q fever in the Federal Republic of Germany, *Zentralbl. Bakteriol. Parasitenkd. Infektionskr. Hyg. Abt. 1 Orig. Reihe A,* 267, 42, 1987.

56. **Richardus, J. H., Donkers, A., Dumas, A. M., Schaap, G. J. P., Akkermans, J. P. W. M., Huisman, J., and Valkenburg, H. A.,** Q fever in the Netherlands: a sero-epidemiological survey among human population groups from 1968 to 1983, *Epidemiol. Infect.,* 98, 211, 1987.

57. **Schaap, G. J. P. and Donkers, A.,** Q-koorts bij Nederlands melkvee, *Ned. Tijdschr. Geneeskd.,* 125, 243, 1981.

58. **Houwers, D. J. and Richardus, J. H.,** Infection with *Coxiella burnetii* in man and animals in the Netherlands, *Zentralbl. Bakteriol. Parasitenkd. Infektionskr. Hyg. Abt. 1 Orig. Reihe A,* 267, 30, 1987.

59. **Durand, M. and Strohl, A.,** L'infection bovine par l'agent de la fièvre Q en 1977, *Rev. Med. Vet.,* 129, 491, 1978.

60. **Coche, B.,** La fièvre Q bovine en France. Aspects pratiques et importance de la sérologie, *Point Vet.,* 12, 95, 1981.

61. **Limouzin, C., Baccot, P., and Durand, M. P.,** Epidémiologie de la fièvre Q bovine en France, *Bull. Acad. Vet. Fr.,* 58, 283, 1985.

62. **Miege, R.,** L'éco-pathologie de la fièvre Q des bovins en Haute-Savoie. Note 1—Revue bibliographique, *Rev. Med. Vet.,* 134, 235, 183.

63. **Miege, R.,** L'écopathologie de la fièvre Q des bovins en Haute-Savoie. Note 2, *Rev. Med. Vet.* 134, 623, 1983.

64. **Tainturier, D.,** Métrites en série chez la vache provoquées par la fièvre Q, *Rec. Med. Vet.,* 163, 195, 1987.

65. **Connolly, J. H.,** Q fever in Northern Ireland, *Br. Med. J.,* 1, 547, 1968.

66. **Hillary, I. B., Dooley, S., and Meenan, P. N.,** Q fever in the Republic of Ireland, *Ir. J. Med. Sci.,* 149, 59, 1969.
67. **Kindmark, C. O.,** Domestic Q fever in Sweden, Proc. 3rd Int. Symp. Rickettsiae and Rickettsial Diseases, Smolenice/Bratislava, September 10—14, 1984.
68. **Winn, J. F. and Elson, B. E.,** Effects of feeding colostrum containing *Coxiella burnetii* antibodies to newborn calves of two categories, *Am. J. Trop. Med. Hyg.,* 1, 821, 1952.
69. **Luoto, L.,** Report on the nationwide occurrence of Q fever infections in cattle, *Public Health Rep.,* 75, 135, 1960.
70. **D'Angelo, L. J., Baker, E. F., and Schlosser, W.,** Q fever in the United States 1948—1977, *J. Infect. Dis.,* 139, 613, 1979.
71. **Biberstein, E. L., Behymer, D. F., Bushnell, R., Crenshaw, G., Riemann, H. P., and Franti, C. E.,** A survey of Q fever (*Coxiella burnetii*) in California dairy cows, *Am. J. Vet. Res.,* 35, 1577, 1974.
72. **Lang, G. H.,** Q fever, an emerging public health problem in Canada, *Can. J. Vet. Res.,* 53, 1, 1989.
73. **Marrie, T. J., Van Buren, J., Fraser, J., Haldane, E. V., Faulkner, R. S., Williams, J. C., and Kwan, C.,** Seroepidemiology of Q fever among domestic animals in Nova Scotia, *Am. J. Public Health,* 75, 763, 1985.
74. **McKiel, J. A.,** Q fever in Canada, *Can. Med. Assoc. J.,* 91, 573, 1965.
75. **Lorbacher de Ruiz, H.,** Q Fever in Columbia S.A.: a serological survey of human and bovine populations, *Zentralbl. Veterinaermed. Reihe B,* 24, 287, 1977.
76. **Behymer, D. E.,** Q fever in *Bovine Medicine and Surgery,* Vol. 1., Amstutz, H. E., Ed., American Veterinary Publications, Santa Barbara, CA, 1980.
77. **Jellison, W. L., Ormsbee, R., Beck, M. D., Huebner, R. J., Parke, R. R., and Bell, E. J.,** Q fever studies in southern California. Natural infection in a dairy cow, *Public Health Rep.,* 63, 1611, 1948.
78. **Derrick, E. H., Smith, D. J. W., and Brown, H. E.,** Studies in the epidemiology of Q fever. The role of the cow in the transmission of human infection, *Aust. J. Exp. Biol. Med. Sci.,* 20, 105, 1942.
79. **Blanc, G., Martin, L. A., and Bruneau, J.,** Q fever expérimentale de quelques animaux domestiques, *Ann. Inst. Paster,* 77, 99, 1949.
80. **Parker, R. R., Bell, E. J., and Lackman, D. B.,** Experimental Q fever in cattle. Observations on four heifers and two milk cows, *Am. J. Hyg.,* 48, 191, 1948.
81. **Bell, E. J., Parker, R. R., and Stoenner, H. G.,** Experimental Q fever in cattle, *Am. J. Public Health,* 39, 478, 1949.
82. **Stoenner, H. G.,** Experimental Q fever in Cattle. Epizootiologic aspects, *J. Am. Vet. Med. Assoc.,* 118, 170, 1951.
83. **Badiali, L.,** Infezione sperimentale da *Coxiella burnetii* nei bovini, *Zooprofilassi,* 7, 155, 1952.
84. **Gibert, P.,** Recherches à propos de la fièvre Q dans l'espèce bovine, *Rev. Med. Vet.,* 106, 299, 1955.
85. **Zotov, A. P., Chumakov, M. P., Markov, A. A., Stepanova, N. I., and Petrov, A. N.,** Experimental reproduction of Q fever and serological investigations, *Veterinariya Moskwa,* 33, 44, 1956.
86. **Plommet, M., Capponi, M., Gestin, J., and Renoux, G.,** Fièvre Q expérimentale des bovins, *Ann. Rech. Vet.,* 4, 325, 1973.
87. **Becht, H. and Hess, E.,** Zur Epizootiologie, Diagnostic und Bekämpfung des Q-Fiebers beim Rind, *Schweiz. Arch. Tierheilkd.,* 106, 389, 1964.
88. **Becht, H.,** Der differentialdiagnostische Nachweis von Brucellen und C. burneti in Köster positiven Nachgeburtsausstrichen mit fluoreszierenden Antikörpern, *Schweiz. Arch. Tierheilkd.,* 107, 392, 1965.
89. **Baumgartner, H.,** Die Verbreitung des Q-Fiebers in den Milchvieh-beständen des Emmentals und Oberaargaues, *Schweiz. Arch. Tierheilkd.,* 108, 401, 1966.
90. **Siegrist, J. J. and Hess, E.,** Rickettsioses chez les animaux domestiques, *Bull. Off. Int. Epizoot.,* 70, 315, 1968.
91. **Henner, S., Lugmayr, D., Schmittdiel, E., and Trixl, H.,** Enzootische Aborte infolge von Salmonellen und Rickettsien (Q-Fieber) beim Rind, *Tieraerztl. Umsch.,* 12, 675, 1977.
92. **Woernle, H. and Müller, K.,** Q fever: prevalence, control using vaccine and/or antibiotic treatment, *Tieraerztl. Umsch.,* 41, 201, 1986.
93. **Schmeer, N., Wieda, J., Frost, J. W., Herbst, W., Weiss, R., and Krauss, H.,** Diagnosis, differential diagnosis and control of bovine Q fever in a dairy herd with infertility, *Tieraerztl. Umsch.,* 42, 287, 1987.
94. **Rady, M., Glavits, R., and Nagy, G.,** Demonstration in Hungary of Q fever associated with abortions in cattle and sheep, *Acta Vet. Hung.,* 33, 169, 1985.
95. **Thomson, G. W.,** Coxiella placentitis and abortion in cattle, *Can. Vet. J.,* 27, A4, 1986.
96. **Giroud, P. and Jadin, J.,** Comportement sérologique des animaux domestiques vis-à-vis des antigénes rickettsiens en milieu contaminé de fièvre Q, *C. R. Accd. Sci. Paris,* 230, 86, 1950.
97. **Somma-Moreira, R. R., Caffarena, R. M., Somma, S., Pérez, G., and Monteiro, M.,** Analysis of Q fever in Uruguay, *Rev. Inf. Dis.,* 9, 386, 1987.
98. **Somma-Moreira, R. R., Russi-Cahill, J. C., Hortal de Peluffo, M., and Caffarena, R. M.,** Serological investigations of *Coxiella burnetii* infection in swine in Uruguay (Abstr. 131), *Vet. Bull.,* 57, 29, 1987.
99. **Marmion, B. P. and Stoker, M. G.,** Epidemiology of Q fever in Great Britain, *Br. Med. J.,* 2, 809, 1958.
100. **Schonell, M. E., Brotherston, J. G., and Burnett, A. C.,** Occupational infections in an Edinbourgh abattoir, *Br. Med. J.,* 2, 148, 1966.

101. **Willeberg, P., Ruppanner, R., Behymer, D. E., Haghighi, S., Kaneko, J. J.,Franti, C. E.,** Environmental exposure to *Coxiella burnetii*: a seroepidemiologic survey among domestic animals, *Am. J. Epidemiol.*, 11, 37, 1980.

102. **George, J. and Marrie, T. J.,** Serological evidence of *Coxiella burnetii* infection in horses in Atlantic Canada, *Can. J. Vet. Res.*, 28, 425, 1987.

103. **Sabban, M. S., Hussein, N., Sadek B., and ElDahab, H.,** Q fever in the United Arab Republic, *Bull. Off. Int. Epizoot.*, 69, 745, 1968.

104. **Blinov, P. N.,** Experimental Q fever in horses. *Veterinariya Moskwa* 34, 34, 1957. Quoted in **Ludwig, Ch.,** in *Handbuch der Virusinfektionen bei Tieren,* Vol. 3/2, Röhrer, H., Ed., Gustav Fischer, Jena, East Germany, 1968.

105. **Krauss, H.,** Die Bedeutung von Rickettsien und Chlamydien bei kleinen Haustieren als Erreger von Zoonosen, *Berl. Muench. Tieraerztl. Wochenschr.*, 95, 480, 1982.

106. **Mantovani, A. and Benazzi, P.,** The isolation of *Coxiella burnetii* from *Rhipicephalus sanguineus* on naturally infected dogs, *J. Am. Vet. Med. Assoc.*, 122, 117, 1953.

107. **Mirri, A.,** La fièvre Q chez les animaux en Italie, *Bull. Off. Int. Epizoot.*, 36, 97, 1951.

108. **Badali, C. and Venturi, R.,** Ricerche sulla febbra Q in Emilia Romagna: anticorpi fissanti il complemento per la Rickettsia burneti nel siereo di cani, *Riv. Ital. Ig.*, 13, 271, 1953. as cited by **Babudieri, B.,** *Adv. Vet. Sci.*, 5, 81, 1959.

109. **Werth, D., Schmeer, N., Müller, H. P., Karo, M., and Krauss, H.,** Nachweis von Antikörpern gegen *Chlamydia psittaci* and *Coxiella burnetii* bei Hunden und Katzen: Vergleich zwischen Enzymimmuntest, Komplementbindungsreaktion, Agargelpräzipitationstest und indirekter Immunperoxidase Technik, *J. Vet. Med.*, B34, 165, 1987.

110. **Gramsch H.,** Über einen Fall von Q-Fieber in Berlin, *Dtsch. Med. Z.*, 3, 148, 1952; as cited by **Babudieri, B.,** *Adv. Vet. Sci.*, 5, 81, 1959.

111. **Rauch, A. M., Tanner, M., Pacer, R. E., Barrett, M. J., Brokopp, C. D., and Schonberger, L. B.,** Sheep-associated outbreak of Q fever — Idaho, *Arch. Intern. Med.*, 147, 341, 1987.

112. **Randhawa, A. S., Jolley, W. B., Dieterich, W. H., and Hunter, C. C.,** Coxiellosis in pound cats, *Feline Pract.*, 4, 37, 1974.

113. **Kosatsky, T., Rideout, V., and Lavigne, P.,** Household outbreak of atypical pneumonia attributed to Q fever — Nova Scotia, *Can. Dis. Wkly. Rep.*, 8, 169, 1982.

114. **Gillespie, J. H. and Baker, J. A.,** Experimental Q fever in cats, *Am. J. Vet. Res.*, 13, 91, 1952.

115. **Babudieri, B. and Moscovici, L.,** Experimental and natural infection of birds by *Coxiella burneti, Nature (London)*, 169, 195, 1952.

116. **Zhmaeva, Z. M. and Pchelkina, A. A.,** Domestic birds as carriers of the rickettsia of Q fever in the Turkmen SSR, *J. Microbiol. Epidemiol. Immunobiol.*, 28, 347, 1957. Quoted by **Giroud, P. and Capponi, M.** in Ref. 154.

117. **Syrucek, L. and Raska, K.,** Q fever in domestic and wild birds, *Bull. W. H. O.*, 15, 329, 1956.

118. **Raska, K. and Syrucek, L.,** Epidemiology of Q fever in Czechoslovakia, *Zentralbl. Bakteriol. Parasitenkd. Infektionskr. Hyg. Abt. 1*, 167, 267, 1956.

119. **Enright, J. B., Longhurst, W. M., Wright, M. E., Dutson, V. J., Franti, C. E., and Behymer, D. E.,** Q fever antibodies in birds, *J. Wildl. Dis.* 7, 14, 1971.

120. **Riemann, H. P., Behymer, D. E., Franti, C. E., Crabb, C., and Schwab, R. G.,** Survey of Q fever agglutinins in birds and small rodents in Northern California 1975—76, *J. Wildl. Dis.*, 15, 515, 1979.

121. **Rarota, J. A, Yadav, M. P., and Sethi M. S.,** Seroepidemiology of Q fever in poultry, *Avian Dis.*, 22, 167, 1978.

122. **LeBer, C.,** Q fever in Ontario, *Can. Dis. Wkly. Rep.*, 14/17, 70, 1988.

123. **Elisberg, B. L. and Bozeman, F. M.,** The rickettsiae, in *Diagnostic Procedures for Viral and Rickettsial Diseases of Man,* Lennette, E. H. and Schmidt, N. Eds., American Society of Microbiologists, Washington, D.C., 1979.

124. **Parker, R. R. and Steinhaus, E. A.,** American and Australian Q fevers: persistence of the infectious agents in guinea pig tissues after defervescence, *Public Health Rep.*, 8, 3, 1943.

125. **Tigertt, W. D., Benenson, A. S., and Gochenour, W. S.,** Airborne Q fever, *Bacteriol. Rev.*, 25, 285, 1961.

126. **Cox, H. R.,** Rickettsia diaporica and American Q fever, *Am. J. Trop. Med.*, 20, 463, 1940.

127. **Lillie, R. D.,** Pathologic histology in guinea pigs following intraperitoneal inoculation with the virus of Q fever, *Public Health Rep.*, 57, 296, 1942.

128. **Herzberg, K.,** Epidemische Bronchopneumonie des Menschen. Kultur und Darstellung des Erregers, *Zentralbl. Bakteriol. Parasitenkd. Infectionskr. Hyg. Abt. I Orig.*, 152, 1, 1947.

129. **Herzberg, K.,** Ergänzende Befunde über den Erreger der Viruspneumonie des Menschen. III. Mitteilung, *Z. Hyg. Infektionskr.*, 128, 361, 1948.

130. **Germer, W. D.,** Das Q-Fieber beim Meerschweinchen. Ein Beitrag zur allgemeinen Infektionslehre bei einem nur intrazellulär gedeihenden Erreger, *Arch. Gesamte Virusforsch.*, 5, 336, 1954.

131. **Giroud, P. and Gaillard, J. A.,** Developement des rickettsies aux depens du cytoplasme des cellules hôtes au cours de l'infection due à Rickettsia burneti, *Bull. Soc. Pathol. Exot.*, 46, 16, 1953.

132. **Rychlo, A. and Pospisil, R.,** Zur Morphologie und Pathogenese des experimentellen Q-Fiebers beim Meerschweinchen, *Pathol. Microbiol.,* 23, 489, 1960.
133. **Scavo, R.,** Lesioni anatomo-pathologiche della Rickettsiosi burneti in cavie, *Bull. Soc. Ital. Biol. Sper.,* 28, 329, 1952.
134. **Bock, M.,** Experimentelle Untersuchungen zur Epidemiologie des Q-Fiebers, *Z. Tropenmed. Parasitol.,* 5, 348, 1954.
135. **Stoenner, H. G., and,Lackman, D. B.,** The biologic properties of *Coxiella burnetii* isolated from rodents collected in Utah. *Am. J. Hyg.,* 71, 45, 1960.
136. **Quignard, H., Geral, M. F., Pellerin, J. L., Milon, A., and Lautié R.,** La fièvre Q chez les petits ruminants. Enquête épidémiologique dans la region midi-pyrenées, *Rev. Med. Vet.,* 133, 413, 1982.
137. **Burnet, F. M., and Freeman, M.,** Experimental studies on the virus of Q fever, *Med. J. Aust.,* 2, 299, 1937.
138. **Perrin, T. K., Bengtson, I. A.,** The histopathology of experimental Q fever in mice, *Public Health Rep.,* 57, 790, 1942.
139. **Gonder, J. C., Kishimoto, R. A., Kastello, M. D., and Pedersen, C. E., Jr.,** Cynomolgus monkey model for experimental Q fever infection, *J. Infect. Dis.,* 139, 191, 1979.
140. **Cox, H. R., and Bell, E. J,** The cultivation of *Rickettsia diaporica* in tissue culture and in the tissues of developing chicken embryos, *Public Health Rep.* 54, 2171, 1939.
141. **Burnet, F. M., and Freeman, M.,** Studies of the x-strain (Dyer) of Rickettsia burneti. I. Chorioallantoic membrane infections, *J. Immunol.,* 40, 405, 1940.
142. **Ormsbee, R. A.,** The growth of *Coxiella burnetii* in embryonated eggs, *J. Bacteriol.,* 63, 73, 1952.
143. **Stoker, M. G. P., and Fiset, P.,** Phase variation of the Nine Mile and other strains of *Rickettsia burnetii, Can. J. Microbiol.,* 2, 310, 1956.
144. **Weiss, E., and Pietryk, H.,** Growth of *Coxiella burnetii* in monolayer cultures of chick embryo entodermal cells, *J. Bacteriol.,* 72, 235, 1957.
145. **Leyk W., and Scheffler R.,** Multiplication of *Coxiella burnetii* in different tissues and organs of embryonated eggs, *Zentralbl. Bakteriol. Parasitenkd. Infektionskr. Hyg. Abt.'1 Orig. Reihe A,* 231, 519, 1975.
146. **Anacker, R. L., Fukushi, K., Pickens, E. G., and Lackman, D. B.,** Electron-microscopic observations of the development of *Coxiella burneti* in the chicken yolk sac, *J. Bacteriol.,* 88, 1130, 1964.
147. **Khavkin, T., Sukhinin, V., and Amosenkova, N.,** Host-parasite interaction and development of infraforms in chicken embryos infected with *Coxiella burnetii* via the yolk sac, *Infect. Immun.,* 32, 1281, 1981.
148. **Tigertt, W. D., Benenson, A. S., and Gochenour, W. S.,** Airborne Q fever, *Bacteriol. Rev.,* 25, 285, 1961.
149. **Burnet, F. M., Freeman, M.,** Experimental studies on the virus of Q fever, *Med. J. Aust.,* 2, 299, 1937.
150. **Raoult, D., Etienne, J., Massip, P., Iacono, S., Prince, M. A., Beaurain, P., Benichou, S., Auvergnat, J. C., Mathieu, P., Bachet, P., and Serradimigni, A.,** Q fever endocarditis in the south of France, *J. Infect Dis.,* 155, 570, 1987.
151. **Lang, G. H.,** Q fever, *Vet. Rec.,* 123, 582, 1988.
152. **Gillespie, J. H., and Timoney, J. F.,** Q fever, in *Hagan and Bruner's Infectious Diseases of Domestic Animals,* 7th ed., Comstock, Ithaca, 1981.
153. **Burgdorfer, W.,** Q fever, in *Diseases Transmitted from Animals to Man,* Hubbert, W. T., McCulloch, W. F., and Schnurrenberger, P.R., Eds., pp. 387—392, Charles C Thomas, Springfield, IL, 1975, 387.
154. **Giroud, P. and Capponi, M.,** *La fièvre Q ou maladie de Derrick and Burnet,* Editions Médicales Flammarion, Paris, 1966.
155. **Giroud, P. and Jadin, J.,** Le pou dans la Fièvre Q au Ruanda-Urundi (Congo Belge). Conservation naturelle de l'antigène de la fièvre Q sur le pou, *Bull. Soc. Pathol. Exot.,* 43, 674, 1950.
156. **Giroud, P., Jadin, J., and Jezierski, A.,** Le pou peut-il jouer un rôle dans la transmission de la fièvre Q?, *C. R. Soc. Biol.,* 145, 569, 1951.
157. **Smith, D. J. W.,** Studies on the epidemiology of Q fever. IV. The failure to transmit Q fever with the cat flea *Ctenocephalides felis, Aust. J. Exp. Biol. Med. Sci.,* 18, 119, 1940.
158. **Philip, C. B.,** Observations on experimental Q fever, *J. Parasitol.,* 34, 457, 1948.
159. **Weyer F.,** Zur Übertragung des Q Fiebers, *Zentralbl. Bakteriol. Parasitenkd. Infektionskr. Hyg. Abt. 1 Orig.,* 154, 165, 1949.
160. **Smith, D. J. W. and Derrick, E. H.,** Studies on the epidemiology of Q fever. 1. The isolation of six strains of *Rickettsia burneti* from the tick *Haemaphysalis humerosa, Aust. J. Exp. Biol. Med. Sci.,* 18, 1, 1940.
161. **Pope, J. H., Scott, W., and Dweyer, R.,** *Coxiella burnetii* in kangaroos and kangaroo-ticks in Western Queensland, *Aust. J. Exp. Biol.,* 38, 17, 1960.
162. **Smith, D. J. W.,** Studies on the epidemiology of Q fever. III. The transmission of Q fever by the tick *Haemaphysalis humerosa, Aust. J. Exp. Biol. Med. Sci.,* 18, 103, 1940.
163. **Smith, D. J. W.,** Studies on the epidemiology of Q fever. VIII. The transmission of Q fever by the tick *Rhipicephalus sanguineus, Aust. J. Exp. Biol. Med. Sci.,* 19, 119, 1941.
164. **Smith, D. J. W.,** Studies on the epidemiology of Q fever. X. The transmission of Q fever by the tick *Ixodes holocyclus, Aust. J. Exp. Biol. Med. Sci.,* 20, 213, 1942.

165. **Smith, D. J. W.,** Studies on the epidemiology of Q fever. XI. Experimental infection of the ticks *Haemaphysalis bispinosa* and *Ornithodorus* sp. with *Rickettsia burneti, Aust. J. Exp. Biol. Med. Sci.,* 20, 295, 1942.

166. **Davis, G. E. and Cox, H. R.** A filter-passing infectious agent isolated from ticks. I. Isolation from Dermacentor andersoni, reaction in animals, and filtration experiments, *Public Health Rep.,* 53, 2259, 1938.

167. **Cox, H. R.,** A filter-passing infectious agent isolated from ticks. III. Description of organism and cultivation experiments, *Public Health Rep.,* 53, 2270, 1938.

168. **Cox, H. R.,** Rickettsia diaporica and American Q fever, *Am. J. Trop. Med. Hyg.,* 20, 463, 1940.

169. **Davis, G. E.,** Rickettsia diaporica: recovery of three strains from Dermacentor andersoni collected in southeastern Wyoming: their identity with Montana strain 1, *Public Health Rep.,* 54, 2219, 1939.

170. **Parker, R. R. and Davis, G. E.,** A filter-passing infectious agent isolated from ticks. II. Transmission by *Dermacentor andersoni, Public Health Rep.,* 53, 2267, 1938.

171 **Abinanti, F. R., Welch, H. H., Winn, J. F., and Lennette, E. H.,** Presence and epidemiologic significance of *Coxiella burnetii* in sheep wool, *Am. J. Hyg.,* 61, 362, 1955.

172. **Windel, T.,** Queensland-Fieber in einem ledererzeugenden Betrieb, *Leder,* 3, 71, 1952.

173. **Philip, C. B.,** Observations on experimental Q fever, *J. Parasitol.,* 34, 457, 1948.

174. **Weyer, F.,** Aetiologie und Epidemiologie der Rickettsiosen des Menschen, *Ergeb. Mikrobiol.,* 32, 74, 1959.

175. **Brezina, R. and Rehacek, J.,** A study of the phase variation phenomenon by experimental infection of the tick Dermacentor marginatus Sulzer with Coxiella burneti, *Acta Virol.,* 5, 250, 1961.

176. **Pautov, V. M. and Morozov, Y. I.,** Reactivation of non-pathogenic *Coxiella burnetii* in the tick *Alveonasus (Ornithodorus) canestrini,* (Abstr. 76), *Vet. Bull.,* 45, 32, 1975.

177. **Weyer, F.,** Die Beziehungen des Q-Fiebererregers (Rickettsia burnetii) zu Arthropoden, *Z. Tropenmed.,* 4, 344, 1953.

178. **Weyer, F.,** Über die Lebensdauer von Rickettsien im Kot der Laus, *Arch. Inst. Pasteur Tunis,* 36, 411, 1959.

179. **Burgdorfer, W. and Varma, M. G. R.,** Transstadial and transovarial development of disease agents in arthropods, *Ann. Rev. Entomol.,* 12, 347, 1967.

180. **Liebisch, A.,** Die Rolle einheimischer Zecken (Ixodidae) in der Epidemiologie des Q-Fiebers in Deutschland, *Dtsch. Tieraerztl. Wochenschr.,* 63, 274, 1976.

181. **Weyer, F.,** Beobachtungen über das Verhalten des Q-Fieber-Erregers (*Coxiella burnetii*) in der Lederzecke *Ornithodorus moubata, Z. Tropenmed. Parasitol.,* 26, 219, 1975.

182. **Burgdorfer, W.,** *Ornithodorus moubata* als Testobjekt bei Q-Fieberfällen in der Schweiz, *Acta Trop.,* 8, 44, 1951.

183. **Liebisch, A., Burgdorfer, W., and Rahman, M. S.,** Epidemiologische Untersuchungen an Schafzecken (*Dermacentor marginatus*) auf Infektionen mit Rickettsien, *Dtsch. Tieraerztl. Wochenschr.,* 85, 121, 1978.

184. **Anon.,** Annual Report on the Health and Medical Services of the State of Queensland for the year 1937—1938. "Q-Fever Investigations" pp. 45—54; cited by **Parker, R. R., Bell E. J., and Stoenner, H. G.,** Q-Fever: a brief survey of the problem, *J. Am. Vet. Med. Assoc.,* 114, 55, 1949.

185. **Eklund, C. M., Parker, R. R., and Lackman, D. B.,** A case of Q fever probably contracted by exposure to ticks in nature, *Publ. Health Reports* 62, 1413, 1947.

186. **Huebner, R. J., Jellison, W. L., Beck, R. R., Parker, R. R., and Shepard, C. C.,** Q fever studies in Southern California. I. Recovery of Rickettsia burneti from raw milk, *Public Health Rep.,* 63, 214, 1948.

187. **Clark, W. H., Bogucki, A. S., Lennette, E. H., Dean, B. H., and Walker, J. R.,** Q fever in California. VI. Description of an epidemic occurring in Davis, California, in 1948, *Am. J. Hyg.,* 54, 15, 1951.

188. **Clark, W. H., Romer, M. S., Holmes, M. A., Welsh, H. H., Lennette, E. H., and Abinanti, F. R.,** Q fever in California. VIII. An epidemic of Q fever in a small rural community in Northern California, *Am. J. Hyg.,* 54, 25, 1951.

189. **Irons, J. V., Eads, R. B., Johnson, C. W., Walker, O. L., and Norris, M. A.,** Southwest Texas Q fever studies, *J. Parasitol.,* 38, 1, 1952.

190. **Marrie, T. J., Schlech, W. F., III, Williams, J. C., and Yates, L.,** Q fever pneumonia associated with exposure to wild rabbits, *Lancet,* 1, 427, 1986.

191. **Perez-Gallardo, F., Clavero, G., and Hernandez, S.,** Investigaciones sobre la epidemiologia de la fiebre Q en España: los conejos de monte y los lirones come reservoirs de la *Coxiella burneti, Rev. Sanid. Hig. Publica,* 26, 81, 1952.

192. **Blanc, G., Martin, L., and Maurice, A.,** Le mérion (Meriones shawi) de la region de Goulimine est un reservoir du virus de la Q fever marocaine, *C. R. Acad. Sci. Paris,* 224, 1673, 1947.

193. **Stoenner, H. G., Holdenreid, R., Lackman, D., and Osborn, J. S.,** The occurrence of *Coxiella burnetii.* Brucella and other pathogens among the fauna of the Great Salt Lake in Utah, *Am. J. Trop. Med. Hyg,* 8, 590, 1959.

194. **Sidwell, R. W., Lundgren, D. L., Bushman, J. B., and Thorpe, B. D.,** The occurrence of a possible epizootic of Q fever in fauna of the Great Salt Lake Desert of Utah, *Am. J. Trop. Med. Hyg.,* 13, 754, 1964.

195. **Burgdorfer, W., Pickens, E. G., Newhouse, V., and Lackman, D. B.,** Isolation of *Coxiella burnetii* from rodents in western Montana, *J. Infect. Dis.,* 112, 181, 1963.

196. **Enright, J. R., Behymer, D. E., Franti, C. E., Dutson, V. J., Longhurst, W. M., Wright, M. E., and Goggin, J. E.,** The behaviour of Q fever rickettsiae isolated from wild animals in Northern California, *J. Wildl. Dis.,* 7, 83, 1971.

197. **Enright, J. B., Longhurst, T. W., Franti, C. E., Wright, M. E., Dutson, V. J., and Behymer, D. E.,** Some observations on domestic sheep and wildlife relationships in Q   fever, *Bull. Wildl. Dis. Assoc.,* 5, 276, 1969.

198. **Enright, J. B., Frantsi, C. E., and Behymer, D. E.,** Coxiella burnetii in a wildlife-livestock environment. Distribution of Q   fever in mammals, *Am. J. Epidemiol.,* 94, 79, 1971.

199. **Corboz, L. and Zurgilgen, H.,** Isolation of *Coxiella burnetii* from milk and colostrum of naturally infected cows after parenteral and intramammary treatment with oxytetracycline, *Experientia,* 38, 1371, 1982.

200. **Addo, P. B.,** A serological survey for evidence of Q fever in camels in Nigeria, *Br. Vet. J.,* 136, 519, 1980.

201. **Yadav, M. P. and Sethi, M. S.,** Poikilotherms as reservoirs of Q fever (*Coxiella burnetii*) in Uttar Pradesh, *J. Wildl. Dis.,* 15, 15, 1979.

202. **Polydorou, K.,** Q fever control in Cyprus — recent progress, *Br. Vet. J.,* 141, 427, 1985.

203. **Schmeer, N.,** Enzyme-linked Immunosorbent Assay (ELISA) for detection of IgG_1, IgG_2 and IgM antibodies in Q fever infected cattle, *Zentralbl. Bakteriol. Parasitenkd. Infektionskr. Hyg. Abt. 1 Orig. Reihe A,* 259, 20, 1985.

204. **Krauss, H., Semler, B., Schmeer, N., and Sommer, M.,** Immunoglobulin classes and subclasses of antibodies to Chlamydia psittaci and Coxiella burnetii in sheep after vaccination and infection, in *Chlamydial Diseases of Ruminants* Aitken, I.D., Ed., Rep. EUR 10056, Commission of European Community, Luxembourg, 1986.

205. **Schmeer, N., Adami, M., Doepfer, B., Herbst, W., and Schmuck, W.,** Humoral immune response of goats, rabbits and guinea pigs after vaccination with a Q   fever vaccine, *Berl. Muench. Tieraerztl. Wochenschr.,* 98, 20, 1985.

206. **Cracea, E., Constantinescu, S., Dimitrescu, A., Stefanescu, M., and Szegli, G.,** Q fever serum diagnostic by immunoenzymatic (ELISA) test, *Arch. Roum. Pathol. Exp. Microbiol.,* 42, 283, 1983.

207. **Behymer, D. E., Ruppanner, R., Brooks, D., Williams, J. C., and Franti, C. E.,** Enzyme immunoassay for surveillance of Q fever, *Am. J. Vet. Res.,* 46, 2413, 1985.

208. **Schmittdiel, F., Bauer, K., Steinbrecher, H., and Jüstl, W.,** Vaccination of Q fever infected cattle: the effect on excretion of *Coxiella burnetii, Tieraerztl. Umsch.,* 36, 159, 1981.

209. **Schmeer, N., Müller, P., Langel, J., Krauss, H., Frost, J. W., and Wieda, J. W.,** Q   fever vaccines for animals, *Zentralbl. Bakteriol. Parasitenkd. Infektionskr. Hyg. Abt. 1 Orig. Reihe A,* 267, 79, 1987.

210. **Schaal, E. H.,** Erfahrungen und neuere Erkenntnisse by der *Coxiella burnetii*-Infection (Q fever). II, *Tieraerztl. Praxis,* 11, 141, 1983.

211. **Woernle, H., Limouzin, C., Müller, K., and Durand, M. P.,** La fièvre Q bovine. Effets de la vaccination et de l'antibiothérapie sur l'évolution clinique et l'excrétion de Coxiella dans le lait et les sécrétions utérines, *Bull. Acad. Vet. Fr.,* 58, 91, 1985.

212. **Roth, C. D. and Bauer, K.,** The prevalence of Q fever in cattle in Northern Bavaria and possible effects of vaccination, *Tieraerztl. Umsch.,* 41, 197, 1986.

213. **Blanc, G.,** Epidémiologie de la fièvre Q (coxiellose), *Maroc Med.* 33, 354, 1954.

214. **Lennette, E. H., Dean, B. H., Abinanti, F. R., Clark, W. H., Winn, J. F., and Holmes, M. A.,** Q fever in California. V. Serologic survey of sheep, goats and cattle in three epidemiologic categories from several geographic areas, *Am. J. Hyg.,* 54, 1, 1951.

215. **Roberts, D.,** Microbiological aspects of goat's milk. A Public Health Laboratory Service survey, *J. Hyg. Camb.,* 94, 31, 1985.

216. **Constable, P. J. and Harrington, J. M.,** Risks of zoonoses in a veterinary service, *Br. Med. J.,* 284, 246, 1982.

217. *Animal Health Yearbook,* FAO/WHO, Food and Agriculture Organization, Rome, 1987.

218. **Rosati, T.,** Seminario sull' epidemiologia della febbre Q in Italia, *Ann. Sanita Pubblica,* 13, 88, 1952. Cited by **Babudieri, B.,** Q fever: a zoonosis, *Adv. Vet. Sci.,* 5, 81, 1959.

219. **Caporale, G., Mirri, A., and Rosati, T.,** La febbre Q quale zoonosi, *Atti Soc. Ital. Sci. Vet.,* 7, 13, 1953. Cited by **Babudieri, B.,** Q fever: a zoonosis, *Adv. Vet. Sci.,* 5, 81, 1959.

Chapter 3

EPIDEMIOLOGY OF Q FEVER

Thomas J. Marrie

TABLE OF CONTENTS

I. OVERVIEW

The initial description of Q fever as an illness in a Brisbane meat works[1] was a portent of the future. Q fever has continued to be a zoonosis. The epidemiology of this infection in man is linked to the epidemiology of *Coxiella burnetii*, the etiologic agent, in animals. The epidemiology of Q fever in animals is discussed in detail by Lang in Chapter 2. Of all the animals infected by *C. burnetii*,[2] man is the only host to develop illness regularly as a result of infection.[3] Worldwide the most common animal reservoirs for spread of Q fever to man are cattle, sheep, and goats.[2] These and other domestic ungulates, when infected, shed the desiccation-resistant organism in urine, feces, and, especially, in birth products.[4] The heavily infected placenta contaminates the environment at the time of parturition. Air samples are positive for up to 2 weeks following parturition, and viable organisms are present in the soil for periods of up to 150 d.[4] There are several factors which make *C. burnetii* ideally suited to this method of transmission.

1.   The organism has the ability to withstand harsh environmental conditions, probably as the result of formation of a spore stage.[5] It survives for 7 to 10 months on wool at 15 to 20°C, for more than 1 month on fresh meat in cold storage, and for more than 40 months in skimmed milk at room temperature.[6]
2.   It is extraordinarily virulent for man. It has been estimated that a single organism can cause disease.[6] In contrast, massive doses of *C. burnetii* can be introduced into the trachea of sheep without the development of pulmonary lesions or overt disease.[7] Transient rickettsiaemia occurs, and then the organism becomes dormant only to undergo extensive multiplication in the placenta when the animal becomes pregnant.[4]

In man, the epidemiology of Q fever varies from country to country and may reflect differences in strains of Q fever[8] and in the biology of the host. For example, a self-limited febrile illness is the most common manifestation of Q fever in Australia,[9] whereas in Nova Scotia pneumonia is the major manifestation of this illness,[10] while in Ontario it is granulomatous hepatitis.[11] Likewise the frequency with which chronic Q fever occurs varies from country to country. In England and Wales there were 92 (11%) cases of endocarditis among 839 Q fever infections.[12] In Spain, 6% of 249 cases of Q fever were chronic.[13] In Maritime Canada there were 8 cases (4.4%) of endocarditis among 178 cases of Q fever.[14] In contrast, in the U.S. Q fever endocarditis is very rare.[15-17] There were apparently no cases of Q fever endocarditis among the 228 cases of Q fever reported from the U.S. from 1978 through 1986.[18]

There were only three cases of endocarditis among the 1164 (0.25%) cases of Q fever reported to the Centers for Disease Control between 1948 and 1977.[17]

II. MODES OF TRANSMISSION

A. AEROSOLS

The most important route whereby *C. burnetii* infects man is through the inhalation of contaminated aerosols. In this setting a dose response effect is evident. Those who receive the largest dose have the shortest incubation period. This has been shown experimentally. Thus, guinea pigs exposed to 10^5 infectious dose units had an incubation period of 7 d, whereas those exposed to 10^1 infectious dose units had an incubation period of 13 d.[20] Similarly, monkeys exposed to a low dose of *C. burnetii* had no clinical evidence of infection, whereas the incubation period for those exposed to the higher dose was 7 d.[21] Finally, studies in human volunteers have shown that inhalation from an aerosol of whole egg slurry results in Q fever.[20] We have made similar observations in humans who were exposed to an infected parturient cat. Those who cleaned up the products of conception had the shortest incubation period and the most severe illness.[22] Thus, the incubation period ranged from 7 to 30 d[22] according to the intensity of the exposure.

The aerosol mode of spread and the common reservoirs of *C. burnetii* result in the following being at highest risk for infection: abattoir workers, farmers, and veterinarians. For abattoir workers the highest risk of infection exists on the cattle slaughter floor.[23,24]

Indirect exposure to contaminated aerosols is also important in the epidemiology of Q fever. British residents who lived along a road over which farm vehicles traveled developed Q fever as a result of exposure to contaminated straw, manure, or dust from farm vehicles.[25] Of residents of a Swiss valley who lived along a road over which sheep traveled to and from mountain pastures, 415 developed Q fever.[26] *C. burnetii* aerosols can travel considerable distances. Early and subsequent studies implicated residence in a dairy area[27] as a risk factor for Q fever. In one rural town of 3000, 41 cases of Q fever occurred. The only sources implicated were sheep grazing on either side of the town: 10.5% of the sheep sampled were seropositive for Q fever.[28] In another study, an outbreak of an influenza-like illness due to Q fever occurred in a small rural community of northern California; nearly 1.5% of the population of 3000 was ill.[29] Of the 41 cases, 40 were male, and cases clustered by place of occupation. The authors hypothesized that contact with infected environments led to the infection.

In our case control study,[22] the following were risk factors by univariate analysis: rural residence, working on a farm, slaughtering or dressing animals, and contact with cats, cattle, sheep, tick-infested animals, and newborn animals. By multivariate analysis exposure to newborn animals (chiefly kittens) and stillborn kittens, rural residence, and slaughtering or dressing animals were significant risk factors. The odds ratio for exposure to stillborn kittens was 37. Approximately one third of the 51 persons with Q fever in our study had no identifiable risk factors. Such was the case for a small outbreak of Q fever among naval personnel, one of whom developed bronchopneumonia following a training course in Wiltshire and the Black Mountains of South Wales.[30] Similarly, the source of an outbreak of Q fever involving 25 people in Oxfordshire, 18 of whom were postal workers, remained unknown.[31]

We investigated an outbreak of Q fever in a truck repair plant[32] in which 16 of the 32 employees were affected. One of the employees had a cat which gave birth to kittens 2 weeks prior to the outbreak. The cat refused to let the kittens suckle, and the employee, after donning his work clothes, fed the kittens from a bottle and then went to work. The differential attack rate for the upstairs employees were the cat owner worked was higher 13/19 (67%) than that for those who worked downstairs 3/12 (25%) (p <0.01). This observation combined with the absence of any other source led us to hypothesize that the contaminated clothing of the cat owner served as the vehicle whereby *C. burnetii* was introduced into the truck repair plant. An incident described by Marmion and Stoker[33] is remarkably similar to the truck repair plant outbreak. Of 30 people who performed a play in a village church at Easter in 1949, 10 became ill with Q fever. The only source was indirect contact with sheep in that a shepherd who had a role in the play came to rehearsals in his working clothes, accompanied by his sheepdog. Twenty-four samples of dust from the clothing of shepherds and others who came in contact with the two known infected flocks of sheep were obtained with a suction device. *C. burnetii* was isolated from one specimen of dust collected from the clothing of a shepherd. Contaminated clothing from the

Rocky Mountain Laboratory in Montana, a Q fever research facility, led to cases of Q fever among laundry workers.[34]

Four employees of an exotic bird and reptile importing company in New York developed Q fever 3 weeks after they unpacked a shipment of 500 ball pythons from Ghana.[35] Three types of ticks were removed from the snakes; however, *C. burnetii* was not isolated from the ticks or from the pythons.

The reservoir(s) of *C. burnetii* in an area very much influences the local epidemiology. In Baddeck, Nova Scotia, an infected parturient cat who bled per vaginum for 3 weeks prior to delivery led to an outbreak of Q fever affecting 2.8% of the population of the town.[36]

Two other outbreaks illustrate the amazing capability of *C. burnetii* as an airborne pathogen. Of 52 (29%) crew members of a ship that transported goats, 15 became ill with Q fever. Many goats gave birth during the trip. Thirty-two percent of the goats had antibodies to *C. burnetii*. The authors implicated drinking raw goats milk as a potential mode of spread, although it is likely that aerosols accounted for all the cases.[37] Sixteen people who attended a birthday party became ill with Q fever. At 3 p.m. on the day of the party the hostess' cat gave birth to kittens in the bedroom closet. The hostess closed the closet and bedroom doors and prepared for the party which began at 6 p.m. None of the guests entered the bedroom; however, they all spent some time in the kitchen which adjoined the bedroom.[22] *C. burnetii* was isolated from the cat's uterus. We have attributed an urban outbreak of Q fever to an infected cat. A group of Halifax, Nova Scotia, poker players developed Q fever when they were exposed to an infected parturient cat during the course of their poker game.[38] This outbreak illustrates how Q fever can affect urban as well as rural dwellers.

The full extent of aerosol transmission of *C. burnetii* is shown in the several outbreaks of Q fever that have occurred in institutions.

B. INSTITUTION-ACQUIRED Q FEVER

Johnson and Kadull[39] reported 50 laboratory-acquired cases of Q fever that occurred between 1950 and 1965. In addition, they reviewed all the laboratory-associated cases reported up to that time. It is noteworthy that the first laboratory-acquired cases were reported from the laboratories of the scientists who made the initial isolations of *C. burnetii*.[40-43] During the years 1938 to 1955, 266 laboratory workers developed Q fever. Of the 50 cases of Johnson and Kadull,[39] 37 worked directly in the laboratory, while the 13 others included clerical workers, engineering personnel, animal handlers, steam fitters, carpenters, repairmen, and janitors.

The late 1960s saw a change in the epidemiology of institution-acquired Q fever. At that time infected sheep used in research were recognized as a source of Q fever.[44] Since then, there have been several large outbreaks of Q fever as a result of the use of infected pregnant sheep in research.[45-47] These outbreaks involved large numbers of people most (63 to 70%) of whom did not have direct contact with the sheep.[45,47] Instead, they worked along the route over which the sheep were transported to the laboratory.[45, 47]

C. INGESTION

Epidemiological studies have suggested that ingestion of raw (presumably contaminated milk) is a risk factor for acquisition of Q fever.[27,48,49] *C. burnetii* was not killed by pasteurization techniques used in the 1940s and 1950s.[50] However, techniques currently in use do kill *C. burnetii*. Transmission of *C. burnetii* to humans via the gastrointestinal tract could not be demonstrated in two studies. Of 11 Portuguese volunteers who ingested food contaminated with *C. burnetii*, only two developed complement-fixing antibodies.[51] In a study from Milwaukee, 34 human volunteers consumed unpasteurized raw milk naturally infected with *C. burnetii*. None became ill, and none developed antibodies that could be detected by the complement-fixation, the capillary agglutination, or the radioisotope precipitation test.[52] A study carried out at the Idaho State Penitentiary showed that seroconversion could occur from ingesting raw milk, but clinical disease could not.[53]

Kumar and co-workers[54] found that 5 of 97 samples of breast milk from Indian women were positive for *C. burnetii* (3) or antibody to *C. burnetii* (3) — one sample contained both antigen and antibody.

It is likely that ingestion is important in the maintenance of Q fever in the animal population. Cats experimentally infected via the oral route did not become ill whereas cats infected via the subcutaneous route became febrile and lethargic.5

D. PERCUTANEOUS TRANSMISSION

All 29 Portuguese volunteers who were infected intradermally developed signs of disease.[51]

A 24-year-old male crushed ticks between his fingers while hiking in the mountains of Montana. Sixteen days later he became ill and was shown to have Q fever. The authors concluded that the only source for the Q fever was the ticks that he crushed. This could be an example of percutaneous infection, but one cannot rule out exposure to an aerosol of *C. burnetii*.[56]

There is one report of transmission of Q fever via a blood transfusion.[57]

E. VERTICAL TRANSMISSION

C. burnetii has been isolated from human placentas.[58] The suggestion has been made that mothers previously infected with *C. burnetii* may reactivate their infection at the time of pregnancy in the same fashion as cows and sheep.[59] It is likely that vertical transmission is unimportant in the overall epidemiology of Q fever in man.

F. PERSON-TO-PERSON TRANSMISSION

This mode of transmission is distinctly unusual. Harman[60] reported cases of pneumonia that occurred among two pathologists, a nurse, and an autopsy attendant, all of whom were present at an autopsy on a patient who died with *C. burnetii* pneumonia.

Gerth et al.[61] described 12 cases of Q fever at the Institute of Pathology of Tübingen University, West Germany. All those who fell ill had taken part in postmortem examinations of a case demonstration at the institute. In none of the 12 patients, who underwent autopsy at this time, was there clinical evidence of Q fever, and it was not possible to determine the source of infection retrospectively.

Mann et al.[62] described what they believe is the first example of person-to-person transmission of *C. burnetii* in a family. The index case, a 79-year-old male, was admitted with *C. burnetii* pneumonia in June 1985. His wife had had Q fever 1 month previously. Their married daughter lived 6 miles away; she and her daughter had Q fever in March 1985. The son-in-law was a shepherd at the local veterinary sheep station and had been lambing during the period when he and his wife became ill. He probably had been recently infected as well since his CF phase II titer was 1:64. He worked in coveralls that were washed at home, and he changed before visiting his wife's parents' house. The assumption is that one of the three (son-in-law, daughter, granddaughter) transmitted the infection to the 79-year-old man and his wife. There were apparently no other cases of Q fever in the area, and there was not a high density of domestic animals in the area.

There is one instance reported wherein hospital staff were infected by a patient.[63]

We have not observed person-to-person transmission at our hospital even though we have provided care to 30 seriously ill patients with *C. burnetii* pneumonia. In addition, none of the members of our cardiac surgery team who have replaced the heart valves of seven patients with *C. burnetii* endocarditis have developed Q fever.

Because of the rarity of person-to-person transmission of Q fever there is no need to isolate patients hospitalized with this illness;[64] however, precautions should be taken in the handling or the postmortem examination of infected cadavers.[65]

Marmion and Stoker[66] best summarized the epidemiology of Q fever in 1958.

Perhaps the most lasting impression left by a survey of the literature on Q fever is of the remarkable diversity

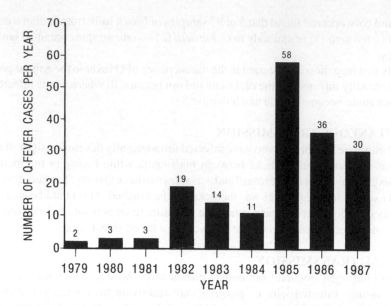

FIGURE 1. Number of cases of Q fever diagnosed in Nova Scotia from 1979 to 1987.

of the ecology of *R. burnetii*. Its pattern of maintenance in animal or insect hosts differs strikingly from one part of the world to another, and the particular local variant which has evolved may have much significance for the infection of man. Thus, for example, the infection of sheep in the central valley of California or in southern Europe is accompanied by appreciable morbidity among the human beings in contact, whereas the presence of the organism in rabbits and ticks of the Nefitik Forest, Casablanca, is an inapparent infection of seemingly little consequence to visitors to this popular holiday resort.

Early in the study of Q fever it was recognized that in some areas there always seemed to be human cases of Q fever, while in other areas with the same rate of infection among animals spread to man does not occur.[6]

The evolution of Q fever in Nova Scotia illustrates many of the features of the epidemiology of this illness. Q fever was first recognized there in 1979. During the next 2 years sporadic cases were observed (Figure 1). In 1982, Kosatsky[67] recognized the first outbreak of Q fever in the province: 10 members of a Jordan Bay family and two friends developed Q fever between April 26 and May 6, 1982. All 12 were having dinner on April 17, 1982 when one of the children announced "our cat is giving birth to kittens." They all trooped out into the yard to watch, and 9 to 19 d later they became ill with fever, headache, malaise, and a nonproductive cough. Q fever was diagnosed serologically. Since then, an additional 17 incidents of cat-associated Q fever have been observed. Thirteen of these are detailed elsewhere.[22] In general, these tend to be small outbreaks involving anywhere from 1 to 16 persons. They usually occur in rural areas and include exposure to the products of feline conception. The stillbirth rate is high: 70% among cats epidemiologically associated with outbreaks of Q fever,[22] while the stillbirth rate for other cats is 10%.[68]

C. burnetii has been isolated from the products of conception of these cats.[22] The next question to be answered is how do the cats become infected? *C. burnetii* has been isolated from a wide variety of insects and animals, including 22 species of ticks, 1 insect, 8 different mammals, several species of birds, lice, and bedbugs.[2,3,6] There are several species of ticks in Nova Scotia: *Dermacentor variabilis* (Say), *D. albipictus*, *Ixodes angustus*, *I. uriae*, *I. cookei*, *I. leporisplaustris*, and *Rhicephalus sanguineus*.[69] *C. burnetii* has been recovered from *D. variabilis* collected in Nova Scotia.[70] It is our hypothesis that cats became infected by eating the infected mice.

Another unusual epidemiological feature of Q fever in Nova Scotia has been the association of five cases of Q fever with exposure to infected rabbits.[71]

Recently (April 1988), an outbreak of Q fever following exposure to an infected cow has been observed. This cow gave birth to a stillborn calf, had retention of the placenta for 3 d, and a vaginal discharge for another 7 d. Four members of the family were hospitalized with Q fever, two of whom had pneumonia, and three others developed fever, headache, and malaise due to *C. burnetii* infection. This suggests that Q fever in Nova Scotia is now occurring following exposure to one of the traditional reservoirs: cattle.

There are still many unanswered questions about the epidemiology of Q fever in Nova Scotia. How did it spread to this province? Why is the rate of pneumonia so high? Why are cats the major reservoir for infection? There are suggestions that the strain of *C. burnetii* in Nova Scotia is different from other strains.[72] The epidemiology of Q fever is summarized pictorially in Figure 2.

In countries where cattle, sheep, and goats serve as the reservoirs for Q fever, there is a seasonal variation in the cases among humans. Most cases occur during the lambing season.[6] We have found that in Nova Scotia, Q fever occurs throughout the year, with cases occurring during the spring, summer, and fall (Figure 3).

The mean age of 174 Nova Scotians with acute Q fever was 40.17 years; the range was 17 to 80 years. The percent of patients with Q fever according to each decade of age is given in Figure 4. Most (68.9%) of the cases occurred between 20 to 49 years of age.

Of 1684 Nova Scotians tested for antibodies to *C. burnetii* phase II antigen, 230 (14%) were positive. A bimodal distribution of seropositivity was seen, with the lowest rate of seropositivity occurring among the 30- to 39-year age group. The highest rate of 44% occurred among the 70 to 79 year olds; however, only 39 subjects in this age group were tested.

We also tested 100 children, from 2 months to 16 years of age (Table 1). Only 5 were seropositive, the youngest of whom was 9 years of age.

The rate of positivity among males and females in our serosurveys was equal. However, males outnumbered females, by 2:1, among the acute Q fever cases. There were 117 males (69%) and 53 females (31%).

III. GEOGRAPHIC DISTRIBUTION OF *C. BURNETII*

Kaplan and Bertagna[73] reviewed the literature to 1955 and published the results of a WHO-assisted survey of the distribution of Q fever in 32 countries. They found that at that time Q fever existed in 51 countries on five continents (Table 2). Ireland, the Netherlands, New Zealand, and Poland did not have Q fever. In these studies the complement-fixation test was the main serological test used.

In the brief examples that follow, the highlights of the epidemiology of Q fever in individual countries are given.

A. AFRICA
1. Nigeria
Cases of Q fever have been reported in humans.[74] A unique feature of Q fever in this country is the high rate of seropositivity among pet dogs.[75] With use of the capillary agglutination test, 30.3% of male dogs tested and 28.8% of the female dogs tested were positive.

2. Tanzania
Serum samples from humans, cattle, sheep, goats, and a variety of game animals were tested using the complement-fixation test.[76] Twenty-eight (3.9%) of the 724 humans tested had a positive antibody titer; 13% of the cattle, 17% of the sheep, and 13.6% of the goats tested were positive. Some of the following game animals were seropositive: Coke's hartebeest, topi, baboon, Thomson's gazelle, impala, and wildebeest. The warthog, common waterbuck, zebra,

EPIDEMIOLOGY OF Q FEVER

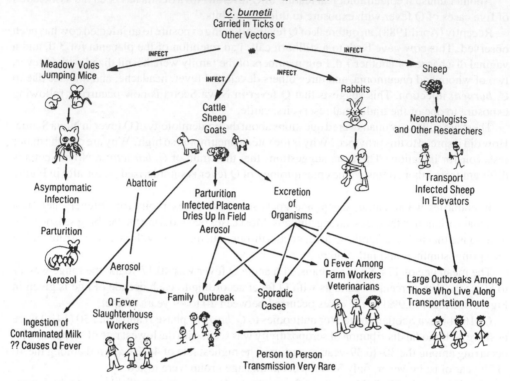

FIGURE 2.   Pictorial summary of the epidemiology of Q fever.

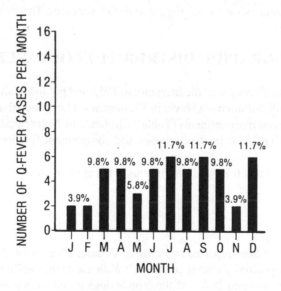

FIGURE 3.   Number of cases of Q fever per month in Nova Scotia.[22]

elephant, buffalo, giraffe, hare, leopard, jackal, Roch hyrax, and Lichtenstein's hartebeest were seronegative. Clinical cases of Q fever had not been recognized in Tanzania at the time of Hummels report in 1974.[76]

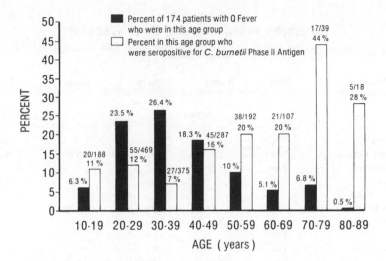

FIGURE 4. A comparison of the age distribution of 174 patients with acute Q fever with 230 Nova Scotians who had antibodies detected to *Coxiella burnetii* phase II antigen as part of various serosurveys. The age distribution of the patients with acute Q fever is expressed as the percentage of the total in each decade of age, whereas the age distribution of seropositive persons is given as the percentage of subjects in this age group who were positive. The number positive/the number tested is also shown.

TABLE 1
Results of Testing Serum Samples from 100 Nova Scotia Children for Antibodies to *Coxiella burnetii* Phase I and II Antigens Using a Microimmunofluorescence Test

Age (years)	No. tested	No. positive[a]
<1	5	0
1	6	0
2	9	0
3	8	0
4	9	0
5	6	0
6	5	0
7	4	0
8	7	0
9	4	1
10	13	1
11	5	1
12	4	0
13	4	0
14	7	0
15	2	2
16	2	0

[a] A titer of >1:8 was considered positive.

B. AMERICA
1. Canada

In 1956, Marc-Aurele et al.[77] described the first case of Q fever in Canada. Later that year an outbreak of Q fever involving slaughterhouse workers from Princeville, Quebec occurred.[78] Only nine cases of Q fever were reported from then until 1976.[14,79-82]

TABLE 2
Geographic Distribution of Q Fever as of 1955[a]

Infected Species

Country	Man	Cattle	Sheep	Goats	Ticks
Africa					
Algeria	Yes				Yes
Belgian Congo	Yes	Yes	Yes	Yes	Yes
Cameroons	Yes	Yes			
Egypt	Yes		Yes	Yes	Yes
Upper Volta, Ubangi Shari	Yes	Yes	Yes	Yes	
Kenya	Yes	Yes			
Libya	Yes				
Madagascar	Yes				
Morocco	Yes	Yes	Yes	Yes	Yes
Portuguese Guinea and Jao Tome		Yes			Yes
Southern Rhodesia	Yes				
Sudan					Yes
South Africa	Yes	Yes			
America					
Argentina		Yes			
Brazil	Yes				
Canada	Yes				
Martinique	Yes				
Mexico	Yes	Yes	Yes	Yes	
Panama	Yes				
United States	Yes	Yes	Yes		Yes
Venezuela	Yes				
Asia					
Ceylon	Yes	Yes	Yes	Yes	
China	Yes				
India	Yes	Yes	Yes	Yes	
Indonesia		Yes			
Iran	Yes	Yes	Yes	Yes	
Israel	Yes	Yes			
Japan	Yes	Yes	Yes	Yes	
Jordan		Yes	Yes	Yes	
Lebanon		Yes			
Malaya	Yes	Yes		Yes	
Pakistan		(Camel)			
Turkey	Yes	Yes	Yes	Yes	Yes
U.S.S.R.	Yes	Yes			
Europe					
Austria	Yes		Yes	Yes	
Bulgaria	Yes		Yes	Yes	
Cyprus	Yes		Yes	Yes	
Denmark	No	No			
Finland	No	No			
France	Yes	Yes	Yes		
West Germany	Yes	Yes	Yes	Yes	

TABLE 2 (continued)
Geographic Distribution of Q Fever as of 1955[a]

Infected Species

Country	Man	Cattle	Sheep	Goats	Ticks
			Europe (continued)		
Greece	Yes		Yes	Yes	
Hungary	Yes				
Iceland	No		No		
Italy	Yes	Yes	Yes	Yes	
Netherlands	No	No			
Norway		No			
Poland	?No				
Portugal	Yes	Yes	Yes	Yes	Yes
Roumania	Yes				
San Marino	Yes	Yes	Yes		
Spain	Yes	Yes			Yes
Sweden	No	No			
Switzerland	Yes	?Yes	?Yes	Yes	
U.S.S.R.					
(Moscowoblast)	Yes	Yes			
United Kingdom					
(inc. No. Ireland)	Yes	Yes	Yes	Yes	Yes
Yugoslavia	Yes	Yes	Yes		
			Oceania		
Australia	Yes	Yes			Yes
New Zealand	No				

[a] Modified from Reference 73.

In 1979 we began to study atypical pneumonia in Nova Scotia and unexpectedly began to diagnose cases of *C. burnetii* pneumonia.[83,84] At about this time Vellend et al.[11] noted cases of Q fever in Ontario. Many of the patients had pyrexia of unknown origin due to *C. burnetii* granulomatous hepatitis. An outbreak of Q fever occurred at the Hospital for Sick Children in Toronto when infected pregnant ewes were transported through the hospital to a laboratory.[47] Cases have also been reported from Alberta and Quebec.[85,86] An imported case of Q fever endocarditis was reported from Ottawa by Saginur et al.[87] A total of 328 cases were reported from Canada during the period 1980 to 1987 (Table 3) with most being reported from Ontario and Nova Scotia. The number of cases reported from Nova Scotia and Prince Edward Island are incomplete. Marrie has recorded 170 cases of acute Q fever in Nova Scotia and 19 cases in Prince Edward Island over this time period.[14]

Few recent seroepidemiological studies are available for Canada. In a study carried out in Nova Scotia in 1984,[88] 11.8% of 997 blood donors had antibodies to phase II *C. burnetii* antigen using the microimmunofluoresence test. About the same percentage of Prince Edward Island blood donors possessed such antibodies.[88] The prevalence of antibodies among slaughterhouse workers and veterinarians in Nova Scotia was significantly higher than that of the general population.[89] Thus, 35% of the 96 slaughterhouse workers and 49% of the 65 veterinarians tested had antibodies to phase II *C. burnetii* antigen. Twenty percent of cats tested in Nova Scotia had antibodies to *C. burnetii* phase II antigen; none of the dogs tested had such antibodies.[90] In Manitoba, a prairie province where no cases of Q fever have been recognized clinically, 15.9% of a sample of 503 blood donors were seropositive. In New Brunswick, where cases were

TABLE 3
Q Fever Canada 1980—1987

Province	No. of cases	% of total
British Columbia	1	0.3
Alberta	9	2.74
Saskatchewan	0	—
Manitoba	0	—
Ontario	198	60.37
Quebec	29	8.84
New Brunswick	2	0.61
Nova Scotia	86[a]	26.22
Prince Edward Island	3[b]	0.92
Newfoundland	0	—
TOTAL	328	

[a] 170 cases identified during that period only 86 reported.

[b] 19 identified — only 3 reported.

diagnosed for the first time in 1986, 4.2% of 966 blood donors had antibodies to *C. burnetii* phase II antigen.[91]

2. United States

The first outbreak of Q fever in the U.S. occurred in March 1946 among stock handlers and slaughterhouse workers in Amarillo, TX.[24] The attack rate among 146 employees was 40%, and two patients died.[92] In a series of studies over the next several years Q fever was shown to be endemic in California.[27] During the period 1948 to 1977, 1164 cases of Q fever were reported from the U.S.[19] Sixty-seven percent of these cases were reported from California. Sawyer et al.[18] summarized the data from the 24 states that require reporting of Q fever for the period 1978 through 1986. There were 228 cases, a mean of 28.5 cases per year. California accounted for 120, or 52.6% of the total; Colorado with 36 cases, Idaho with 27, and Oregon with 14 accounted for most of the remainder. Sawyer and co-workers tested serum samples for antibodies to *C. burnetii* from 959 individuals who had hepatitis and 673 persons whose serum had been submitted for testing against a panel of respiratory pathogens. Six (0.6%) of the 959 in the hepatitis group and 4 (0.6%) in the respiratory group were positive for Q fever. These workers concluded that at present Q fever occurs infrequently in the U.S. It is likely, however, that cases of Q fever go undiagnosed, as exemplified by a recent report from Michigan.[93] The first two reported cases of Q fever in Michigan were in 1984. Both patients lived in adjacent rural counties and had multiple exposures to goats. A subsequent investigation revealed that 28 of 86 (32%) goat-owning family members had antibodies to phase II *C. burnetii* antigen by the immune adherence hemagglutination test. In contrast, 7 of 47 (14.8%) community members who did not own goats were seropositive. Factors associated with seropositivity among goat owners were number of goats on the farm, number of goats positive for Q fever antibodies on the farm, and number of goat births occurring on the farm in the 12 months previous to the study. A multiple regressions analysis revealed that the number of goats on the farm best predicted the seropositivity of a household. When the number of goats was held constant, the addition of one seropositive goat predicted a 10% increase in household seropositivity.[93]

3. Central America

Peacock et al.[94] used the complement fixation and microagglutination tests to examine serum samples from humans from six Central American countries. The results obtained with the complement fixation test were as follows: Costa Rica 7/391 (1.8%) positive; El Salvador 7/348 (2%); Guatemala 6/550 (1.1%); Honduras 9/348 (2.6%); Nicaragua 2/312 (0.6%); Panama 6/336 (1.8%). Cases of Q fever have been reported from Panama both in expatriates and in Panamanians.[95,96] A serosurvey of Panamanians carried out in 1968 to 1969 using a complement fixation technique revealed that 9.4% of 1059 subjects had antibodies to *C. burnetii*.[97] Positive reactions were most common among those 15 to 19 years old.

C. SOUTH AMERICA

1. Colombia

There has been a marked increase in the number of cattle in this country that are infected with *C. burnetii*.[98] Fifty-eight percent of 357 cattle were seropositive using the complement fixation test. Fifty-four percent of 153 slaughterhouse workers possessed such antibodies. Despite this high prevalence of seropositivity among humans no clinical cases had been reported up to 1977.[98] de Ruiz[98] speculates that this may be due to misdiagnosis, properties of the strain of *C. burnetii* present in Colombia, or immunization of the majority of the population through the ingestion of raw milk.

2. Uruguay

The first case of Q fever in this country was described in 1956. There have been 14 outbreaks since 1975 involving 1358 persons, 814 of whom had serologically confirmed Q fever.[49] All cases occurred among workers in the meat industry. Serological surveys have shown that 4.2 to 5% of those studied possessed antibodies to *C. burnetii*. The seropositivity rate among cattle, horses, and swine varied according to the year of the study but ranged from 0.9 to 30%.[99]

D. ASIA

1. People's Republic of China

Q fever was first reported from Beijing in 1951.[100] At least ten provinces have reported cases of this illness. Studies utilizing the complement fixation test have shown a seroprevalence rate ranging from 1.6 to 28.7%. Cattle, sheep, and goats are infected and are epidemiologically the most important. An outbreak at a meatpacking plant has been reported from Inner Mongolia. *C. burnetii* has been isolated from three species of ticks in China.[100]

2. India

Human Q fever was first recognized in this country in 1941, at which time an outbreak of a febrile illness occurred among 400 Gurkha troops stationed at Dehra Dun. This diagnosis was made on clinical grounds only. There are few case reports of human Q fever in the Indian literature although there are a large number of serological reports. The positive seroprevalence rate ranges from 0 to 41%. The age group 2 to 15 has the highest rate. Generally, the rate is higher in farming communities and for those in close contact with domestic livestock.[101] *C. burnetii* has been isolated from cows and from *Rhipicephalus* ticks in India. The seropositivity rates for various animals in various parts of India are as follows: goats, 0 to 60.0%; sheep, 0 to 49%; cattle, 0 to 35.7%; buffalo, 0 to 32.4%; dogs, 0 to 34.6%; fowl, 0 to 13.2%; pigs, 0 to 9.5%; mules, 23%; camels, 11.9 to 24.9%. These studies were carried out using a variety of techniques including complement fixation test, capillary agglutination test, and the microagglutination test.[101]

3. Israel

Q fever is endemic in Israel. Infections are often subclinical. Of recent admissions to an agricultural school, 4.3% had a complement fixation titer of ≥1:8 to *C. burnetii* antigen, while 7.9% of more senior students had such a titer.[102]

4. Iran

The first clinical cases of Q fever were reported between 1970 and 1973. Caughey and Harootunian[103] skin tested 318 Iranians using *C. burnetii* antigen supplied by the Rocky Mountain Laboratory, MT. Forty-three (13.5%) had a positive result.

5. Saudi Arabia

Clinical illness due to Q fever was found to be common among Americans in Saudi Arabia but uncommon among the Saudis.[104] Seventy percent of adult Saudis had a positive antibody titer to *C. burnetii* using the radioisotope precipitation test. Gelpi[104] concluded that Q fever was holoendemic in eastern Saudi Arabia and that subclinical infection develops in childhood.

E. EUROPE

1. Cyprus

Q fever was first reported here in 1951. In 1974 and 1975, 78 British soldiers stationed in Cyprus became infected.[105,106] This outbreak in humans coincided with an outbreak of abortions among sheep and goats. Soldiers probably became infected by inhaling dust from brush contaminated with *C. burnetii* infected parturition products from sheep and goats. These areas were frequented by soldiers performing their military duties. The only case among military dependents was in a woman who frequented the golf course near a grazing area.[107] No cases occurred among the Cypriots who lived there prior to the 1974 invasion or among the refugees.[108] Eight percent of adult shepherds had complement fixing antibodies to *C. burnetii*, indicating a mostly immune local population.[108]

Control measures consisting of destruction of aborted material, isolation of infected dams, and disinfection of the premises has reduced the incidence of Q fever among sheep and goats.[109]

2. Czechoslovakia

Nineteen of 429 (4.4%) human sera tested in 1972 had complement fixation antibodies to *C. burnetii*.[110] Various species of ticks were also positive. *C. burnetii* has been isolated from placentas of four women in Prague.[58]

In a recent review,[111] Reháček states that most parts of Czechoslovakia have experienced Q fever. Some aspects of the epidemiology of Q fever in this country are different from those found elsewhere. For example, an outbreak involving 100 people occurred in a cotton-processing plant in the summer months of 1980. The investigators felt that contamination of this imported cotton at its place of growth and collections (countries of the Near East and some Central Asiatic republics of the U.S.S.R.) was responsible for the outbreak. However, Q fever is endemic in several parts of the country such as the southern part of Central Slovakia. Here the tick *Dermacentor marginatus* is the main vector as well as the reservoir for *C. burnetii*. The incidence of Q fever among humans in Czechoslovakia has decreased recently probably due to the vaccination of cattle.[111]

3. Finland

Fourteen cases of Q fever were reported from Finland in 1981.[112] However, all of these patients were felt to have acquired the illness outside Finland. The three who became ill 69, 75, and 88 d after arriving in Finland were believed to have acquired their infection from their clothing or from souvenirs. Kerttula et al.[113] during a study of community-acquired pneumonia requiring hospitalization in Helsinki, Finland, noted that 1 of 162 persons studied seroconverted to *C. burnetii*. This individual also seroconverted to *Chlamydia* species. No other details were given regarding this case.

4. Sweden

Kindmark et al.[114] in 1985 reported the first case of Q fever in Sweden. This 56-year-old man with pneumonia had cleaned out the basement of a house 10 d before the onset of his illness. The basement included a room for dust bins which was shared by a pet shop in the same house.

5. Italy

During World War II, epidemics of Q fever involved allied troops and Italian communities. During a period of reorganization in agriculture during 1949 to 1955, epidemics of Q fever occurred continuously almost everywhere in Italy.[115] From 1960 to 1980 Q fever became endemic because of the change from an agricultural to an industrial society and occurred mainly in the spring and summer. At present the major manifestation of Q fever in humans in Italy is atypical pneumonia. Seventeen cases of Q fever were reported in Italy in 1984.

A serosurvey carried out in Sicily, using an immunofluoresence test, revealed that 21 of 685 (3.1%) persons had an antibody titer of $\geq$1:40 to *C. burnetii*.

6. France

While Q fever is a notifiable disease in France, it is rare for clinicians to do so; thus, information is scanty. A serosurvey carried out in 1980 in Dijon using the immunofluoresence test revealed that 4.4% of randomly selected persons living in the country were seropositive, while 26% of 86 animals herders possessed antibodies to *C. burnetii*.[116] The rate of positive results (an IFA titer $\geq$1:20) among 2614 serum samples submitted to the National Reference Center for Rickettsioses was 23.4%. Acute Q fever could be confirmed in 132 persons. Other reports from France suggest that acute and chronic Q fever are common in this country.[117,118]

7. Germany

Q fever epidemics appeared among German troops in Greece and Italy during World War II.[119] The first case apparently acquired in Germany was reported in 1948, and from then until the present at least 5300 people have developed Q fever in 31 epidemics and multiple small outbreaks.[119]

Heinrich et al.[120] studied the inhabitants of a village in southern Germany where Q fever is endemic, 136/715 (19%) studied were seropositive. Males accounted for 63% of the seropositive cases. The risk factors for seropositivity were males aged 14 to 19 years, women aged 40 to 49 years, those residents less than 200 m from the next farmhouse, agricultural workers, and those in contact with animal products and farm animals.

In contrast to most countries, in Germany *coxiellosis* is responsible for infertility in cattle.[119]

The role of pets in the epidemiology of Q fever in Germany is raised by the finding that 13% of 1127 dogs and 26% of 108 cats tested using an ELISA technique had antibodies to *C. burnetii*.[121]

8. Hungary

A total of 87 cases have been reported since the first case in 1977.[122] The major manifestation of *C. burnetii* infection in Hungary is granulomatous hepatitis.

9. Netherlands

Cases have been identified from 1979 onward.[123] Using an IFA test Richardus et al.[123,124] found that 31% of 208 male blood donors and 13.7% of 151 female blood donors had an IgG antibody titer of $\geq$1:16 to *C. burnetii* phase II antigen. Antibody was present at all ages between 1 and 64. This group of workers[124] reported 18 cases of Q fever in infancy. In this country cattle and sheep are infected, goats rarely. All 26 cats tested using an ELISA technique were negative, but no positive control was available.[125]

10. Northern Ireland

In 1966 a 21-year-old nurse developed pneumonia 2 to 3 weeks after washing her sheepdog, which was soiled with bits of sheep placenta. In the investigations that followed it was found that 23 patients had had Q fever (retrospective and prospective serological tests). Abattoir workers, veterinary surgeons, and farm workers were seropositive. As a result of previous studies it was concluded that Q fever was introduced into Northern Ireland sometime between April 1957 and February 1962, probably by the importation of infected ewes from Great Britain. Connolly speculated[126] that birds, especially the collared dove (*Stretopelia decaocto*), may have had a role in the spread of Q fever in Northern Ireland. This dove spread from Europe to County Down in 1959 and appeared in the Belfast area in 1963. Birds were often seen feeding on sheep placentas in the fields of Northern Ireland. *C. burnetii* has been isolated from pigeons and swallows.[126]

Hilelary and Meenan,[127] from 1971 to 1974, tested 1587 paired serum samples from patients with febrile illnesses. Eighty-six had complement fixation antibodies to *C. burnetii* phase II antigen.

11. Portugal

Q fever is common in Portugal. For example, 176 cases were recently reported from one hospital in this country.[145] It is noteworthy that only 4% had pulmonary involvement whereas all 176 had hepatomegaly and/or abnormal liver function tests. Ninety-seven patients had a liver biopsy performed, 79 of whom had the distinctive granulomas associated with Q fever. In the remaining 18, the changes were nonspecific.

12. Rumania

Q fever was reported in Rumania in 1947 following the importation of sheep from Australia.[128] Twenty-two cases were reported in 1984 and 10 in 1985. During the 1981 to 1985 period most of the cases were from urban areas (72.1%) and were not related to a known Q fever epidemic focus.[128]

13. Spain

During the period 1981 to 1985, 249 cases of Q fever were documented.[13] A predominance of cases occurred in the cattle raising regions of the North-Basque and Nawana provinces.[13,129] Tellez et al.[13] found that outbreaks are more likely to occur in areas of high incidence of asymptomatic infection. Males between 15 and 44 years of age predominated. Both the outbreaks they noted occurred in young people, the first affecting school students <17 years of age. Fifteen cases of chronic Q fever were noted, 6% of the total number of cases. Three of these cases have been reported in detail elsewhere.[130] All three had endocarditis.

14. Switzerland

Each year 30 to 90 cases are reported.[131] During the autumn of 1983, an outbreak involving 415 people occurred in the Val de Bagnes 3 weeks after 12 flocks of sheep descended from the Alpine pastures to the valley.[26] A study of 2000 blood donors in 1982 showed that 3.5% had an antibody titer of $\geq 1{:}10$. Among forestry workers the prevalence was 7.6% and for veterinarians it was 25.7%. Contact with sheep and consumption of unboiled and not commercially processed milk proved to be mutually independent risk factors.[132]

15. United Kingdom

From 1967 to 1974 an average of 59 cases of Q fever were reported each year, range 48 to 78. However, there was a major increase in the numbers of cases reported in 1975 (104), 1976 (117) and 1977 (98).[133] Cases have been reported from all regions of England and Wales. *C. burnetii* endocarditis accounts for 11% of all cases of Q fever in this country.[12,134]

A probable source of the infection was identified for 202 of the 1351 (14.9%) patients with Q fever.[135] Farm livestock was the source for 115 (57%) of the patients; 53 (26%) acquired the disease overseas, mainly in Mediterranean and Middle Eastern countries; 19 (9%) worked in abattoirs, and in 15 (7.4%) the apparent source was raw milk. In general, Q fever is more common here in spring and early summer, suggesting a link with lambing.[136]

16. U.S.S.R.

While Q fever was first diagnosed in this country in 1952, since 1957 mandatory reporting has occurred.[137] Recurrent outbreaks of Q fever are an important cause of morbidity in some parts of the country. In the Voroneszh region $3.4 \pm 2.7\%$ of workers in cattle breeding complexes possessed antibodies to *C. burnetii*, while for dairy workers it was $12.2 \pm 2.3\%$. One of the measures taken to control Q fever in the U.S.S.R. has been the massive destruction of *Ixodes* ticks.[137]

F. OCEANIA
1. Australia

Q fever continues in this the land where it was first recognized. In mid-1934, three Brisbane practitioners were present at the Johnsonian Club when one mentioned that he had recently seen two puzzling cases of fever. Both patients worked at the Brisbane Abattoir.[138] The other two doctors recalled similar cases, and they decided that the next patients with "abattoir fever" should be admitted to hospital for investigation. Twenty patients were investigated over the next year but no cause was found.[138] In 1935, Derrick, the Medical Officer of Health for the area, was asked to investigate these cases of fever. He named it "Q" for query fever.[1] His studies in conjunction with those of Burnet and Freeman[139] established the entity of Q fever, defined the etiologic agent, and elucidated aspects of the clinical features and epidemiology.[140-142]

Johnson[138] determined that the annual incidence of Q fever in Queensland in the late 1950s and early 1960s was 13.3/100,000 population. McKelvie in a retrospective study of a Brisbane meatworks for the years 1968 to 1977 found that the average annual incidence was 1%.[23]

Studies by Marmion in a large abattoir in Adelaide from 1978 to 1987 have revealed cases of Q fever every year. A mean rate of 27 to 28 cases per 1000 employees/year was observed[143] — thus Q fever in Australia continues to be as it began, primarily an illness of abattoir workers. The number of cases reported[144] each year from 1960 to 1986 is shown in Table 4. There is no consistent trend. Crude data (i.e., unadjusted for population) do show an apparent peak in 1980. The total number of cases over this 26-year period (no data are available for 1961) was 6479.

TABLE 4
Number of Cases of Q Fever Diagnosed
in Australia Per Year from 1960[a]

	1960	255
	1962	97
January—June	1963	150
July 63—June	1964	168
	1964—1965	128
	1966—1967	257
	1967—1968	121
	1968—1969	106
	1969	148
	1970	140
	1971	168
	1972	140
	1973	124
	1974	117
	1975	218
	1976	275
	1977	568
	1978	296
	1979	566
	1980	621
	1981	433
	1982	344
	1983	208
	1984	262
	1985	202
	1986	367
	TOTAL	6479

[a] Data for 1961 were not published.

REFERENCES

1. **Derrick, E. H.,** "Q" fever, new fever entity: clinical features, diagnosis and laboratory investigation, *Med. J. Aust.*, 2, 281, 1937.
2. **Babudieri, B.,** Q fever: a zoonosis, *Adv. Vet. Sci.*, 5, 81, 1959.
3. **Stoker, M. G. P. and Marmion, B. P.,** The spread of Q fever from animals to man. The natural history of a rickettsial disease, *Bull. W.H.O.*, 13, 781, 1955.
4. **Welsh, H. H., Lennette, E. H., Abinanti, F. R., and Winn, J. F.,** Air-borne transmission of Q fever: the role of parturition in the generation of infective aerosols, *Ann. N.Y. Acad. Sci.*, 70, 528, 1957.
5. **McCaul, T. F. and Williams, J. C.,** Developmental cycle of *Coxiella burnetii*: structure and morphogenesis of vegetative and sporogenic differentiations, *J. Bacteriol.*, 147, 1063, 1981.
6. **Christie, A. B.,** Q fever, in *Infectious Diseases: Epidemiology and Clinical Practice*, Christie, A. B., Ed., Churchill Livingstone, New York, 1980, 800.
7. **Abinanti, F. R., Welsh, H. H., Lennette, E. H., and Brunetti, O.,** Q fever studies. XVI. Some aspects of the experimental infection induced in sheep by the intratracheal route of inoculation, *Am. J. Hyg.*, 57, 170, 1953
8. **Samuel, J. E., Frazier, M. E., and Mallavia, L. P.,** Correlation of plasmid type and disease caused by *Coxiella burnetii, Infect. Immun.*, 49, 775, 1985.
9. **Derrick, E. H.,** The course of infection with *Coxiella burnetii, Med. J. Aust.*, 1, 1051, 1973.
10. **Marrie, T. J.,** Q fever pneumonia, *Med. Grand Rds.*, 3, 364, 1984.
11. **Vellend, H., Salit, I. E., Spence, L., McLaughlin, B., Carlson, J., Palmer, N., Van Dreumel, A. A., and Hodgkinson, J. R.,** Q fever — Ontario, *Can. Dis. Wkly. Rep.*, 8, 171, 1982.

12. **Palmer, S. R. and Young, S. E. J.,** Q fever endocarditis in England and Wales 1975-81, *Lancet*, 2, 1448, 1982.
13. **Tellez, A., Sainz, C., Echevarria, C., de Carlos, S., Fernandez, M. V., Leon, P., and Brezina, R.,** Q fever in Spain: acute and chronic cases, 1981 — 1985, *Rev. Infect. Dis.*, 10, 198, 1988.
14. **Marrie, T. J.,** Q fever, 1979—1987 — Nova Scotia, *CDWR*, 14, 69, 1988.
15. **Kimbrough, R. C., III, Ormsbee, R. A., Peacock, M., Rogers, W. R., Bennetts, R. W., Raaf, J., Krause, A., and Gardner, C.,** Q fever endocarditis in the United States, *Ann. Intern. Med.*, 91, 400, 1979.
16. **Pierce, M. A., Saag, M. S., Dismukes, W. E., and Cobbs, C. G.,** Q fever endocarditis, *Am. J. Med. Sci.*, 292, 104, 1986.
17. **Gallagher, J. G. and Remington, J. S.,** Q fever endocarditis, *West. J. Med.*, 140, 943, 1984.
18. **Sawyer, L. A., Fishbein, D. B., and McDade, J. E.,** Q fever in hepatitis and pneumonia patients in the United States: results of laboratory-based surveillance, *J. Infect. Dis.*, 158, 497, 1988.
19. **D'Angelo, L. J., Baker, E. F., and Schlosser, W.,** Q fever in the United States, 1948 — 1977, *J. Infect. Dis.*, 139, 613, 1979
20. **Tigertt, W. D. and Benenson, A. S.,** Studies on Q fever in man, *Trans. Assoc. Am. Phys.*, 69, 98, 1956.
21. **Gonder, J. C., Kishimoto, R. A., Kastello, M. D., Pedersen, C. E., Jr., and Larson, E. W.,** Cynomolgus monkey model for experimental Q fever infection, *J. Infect. Dis.*, 139, 191, 1979.
22. **Marrie, T. J., Durant, H., Williams, J. C., Mintz, E., and Waag, D.,** Exposure to parturient cats is a risk factor for acquisition of Q fever in Maritime Canada, *J. Infect. Dis.*, 93, 98, 1988.
23. **McKelvie, P.,** Q fever in a Queensland meatworks, *Med. J. Aust.*, 1, 590, 1980.
24. **Topping, N. H., Shepard, C. C., and Irons, J. V.,** Q fever in the United States: epidemiologic studies of outbreak among stock handlers and slaughterhouse workers, *J. Am. Med. Assoc.*, 133, 813, 1947.
25. **Salmon, M. M., Howells, B., Glencross, E. J. G., Evans, A. D., and Palmer, S. R.,** Q fever in an urban area, *Lancet*, 1, 1002, 1982.
26. **Dupuis, G., Petite, J., Peter, O., and Vouilloz, M.,** An important outbreak of human Q fever in a Swiss Alpine Valley, *Int. J. Epidemiol.*, 16, 282, 1987.
27. **Huebner, R. J. and Bell, J. A.,** Q fever studies in Southern California. Summary of current results and a discussion of possible control measures, *JAMA*, 145, 301, 1951.
28. **Lennette, E. H. and Clark, W. H.,** Observations on the epidemiology of Q fever in Northern California, *JAMA*, 145, 306, 1951.
29. **Clark, W. H., Romer, M. S., Homes, M. A., Welsh, H. H., Lennette, E. H., and Abinanti, F. R.,** Q fever in California. VIII. An epidemic of Q fever in a small rural community in Northern California, *Am. J. Hyg.*, 54, 25, 1951.
30. **Fraser, P. K., Hatch, L. A., Carmichael, R. J., Evans, A. D., Forster, J. M. R., and Goodings, J. E.,** Q fever in naval personnel, *Lancet*, 2, 971, 1960.
31. **Winner, S. J., Eglin, R. P., Moore, V.I.M., and Mayon-White, R. T.,** An outbreak of Q fever affecting postal workers in Oxfordshire, *J. Infect.*, 14, 255, 1987.
32. **Marrie, T. J., Langille, D., Papukna, V., and Yates, L.,** An outbreak of Q fever in a truck repair plant, *Epidemiol. Infect.*, 102, 119, 1989.
33. **Marmion, B. P. and Stoker, M. G. P.,** The varying epidemiology of Q fever in the South East region of Great Britain. II. In two rural areas, *J. Hyg.*, 54, 547, 1956.
34. **Oliphant, J. W., Gordon, D. A., Meis, A., and Parker, R. R.,** Q fever in laundry workers, presumably transmitted from contaminated clothing, *Am. J. Hyg.*, 49, 76, 1949.
35. Center for Disease Control, Q fever: New York, *Morbid. Mortal. Wkly. Rep.*, 27, 321, 1978.
36. **Marrie, T. J., MacDonald, A., Durant, H., Yates, L., and McCormick, L.,** An outbreak of Q fever probably due to contact with a parturient cat, *Chest*, 93, 98, 1988.
37. **Clark, W. H., Lennette, E. H., and Romer, M. S.,** Q fever in California. IX. An outbreak aboard a ship transporting goats, *Am. J. Hyg.*, 54, 35, 1951.
38. **Langley, J. M., Marrie, T. J., Covert, A. A., Waag, D. M., and Williams, J. C.,** Poker players pneumonia — an urban outbreak of Q fever following exposure to a parturient cat, *N. Engl. J. Med.*, 93, 98, 1988.
39. **Johnson, J. E., III and Kadull, P. J.,** Laboratory acquired Q fever. A report of fifty cases, *Am. J. Med.* 41, 391, 1966.
40. **Dyer, R. E.,** Filter-passing infectious agent isolated from ticks; human infection, *Public Health Rep.*, 53, 2277, 1938.
41. **Burnet, F. M. and Freeman, M.,** Note on a series of laboratory infections with the rickettsia of "Q" fever, *Med. J. Aust.*, 1, 11, 1939.
42. **Smith, D. J. W., Brown, H. E., and Derrick, E. H.,** Further series of laboratory infections with the Rickettsia of "Q" fever, *Med. J. Aust.*, 1, 13, 1939.
43. **Cox, H. R.,** *Rickettsia diaporica* and American Q fever, *Am. J. Trop. Med.*, 20, 463, 1940.
44. **Schachter, J., Sung, M., and Meyer, K. F.,** Potential danger of Q fever in a university hospital environment, *J. Infect. Dis.*, 123, 301, 1971.

45. **Meiklejohn, G., Reimer, L. G., Graves, P. S., and Helmick, C.**, Cryptic epidemic of Q fever in a medical school, *J. Infect. Dis.*, 144, 107, 1981.
46. **Hall, C. J., Richmond, S. J., Caul, E. O., Pearce, N. H., and Silver, I. A.**, Laboratory outbreak of Q fever acquired from sheep, *Lancet*, 1, 1004, 1982.
47. **Curet, L. B. and Paust, J. C.**, Transmission of Q fever from experimental sheep to laboratory personnel, *Am. J. Obstet. Gynecol.*, 114, 566, 1972.
48. **Marmion, B. P., Stoker, M. G. P., Walker, C. B. V., and Carpenter, R. G.**, Q fever in Great Britain — epidemiological information from a serological survey of healthy adults in Kent and East Anglia, *J. Hyg.*, 54, 118, 1956.
49. **Luoto, L. and Pickens, E. G.**, A resume of recent research seeking to define the Q fever problem, *Am. J. Hyg.*, 74, 43, 1961.
50. **Wentworth, B. B.**, Historical review of the literature on Q fever, *Bacteriol. Rev.*, 19, 129, 1955.
51. Editorial, Experimental Q fever in man, *Br. Med. J.*, 1, 1000, 1950
52. **Krumbiegel, E. R. and Wisniewski, H. J.**, Q fever in Milwaukee. II. Consumption of infected raw milk by human volunteers, *Arch. Environ. Health*, 21, 63, 1970.
53. **Benson, W. W., Brock, D. W., and Mather, J.**, Serologic analysis of a penitentiary group using raw milk from a Q fever infected herd, *Public Health Rep.*, 78, 707, 1963.
54. **Kumar, A., Yadav, M. P., and Kakkar, S.**, Human milk as a source of Q fever infection in breast-fed babies, *Indian J. Med. Res.*, 73, 510, 1981.
55. **Gillespie, J. H. and Baker, J. A.**, Experimental Q fever in cats, *Am. J. Vet. Res.*, 13, 91, 1952.
56. **Eklund, C. M., Parker, R. R., and Lackman, D. B.**, Case of Q fever probably contracted by exposure to ticks in nature, *Public Health Rep.*, 62, 1413, 1947.
57. Editorial Comment on Q fever transmitted by blood transfusion — United States, *Can. Dis. Wkly. Rep.*, 3, 210, 1977.
58. **Syrůcek, L., Sobéslavský, O., and Gutvirth, I.**, Isolation of *Coxiella burnetii* from human placentas, *J. Hyg. Epidemiol. Mikrobiol. Immunol.*, 2, 29, 1958.
59. **Fiset, P., Wisseman, C. L., Jr., and El-Bataine, Y.**, Immunologic evidence of human fetal infection with *Coxiella burnetii*, *Am. J. Epidemiol.*, 101, 65, 1975.
60. **Harman, J. B.**, Q fever in Great Britain; clinical account of eight cases, *Lance*, 2, 1028, 1949.
61. **Gerth, H. J., Leidig, U., and Reimenschneider, T.**, Q-Fieber-Epidemie in einem Institut fur Humanpathologie, *Dtsch. Med. Wochenschr.*, 107, 1391, 1982.
62. **Mann, J. S., Douglas, J. G., Inglis, J. M., and Leitch, A. G.**, Q fever: person to person transmission within a family, *Thorax*, 41, 974, 1986.
63. **Deutch, D. L. and Peterson, E. T.**, Q fever: transmission from one human being to others, *JAMA*, 143, 348, 1950.
64. **Grant, C. G., Ascher, M. S., Bernard, K. W., Ruppaner, R., and Vellend, H.**, Q fever and experimental sheep, *Infect. Control*, 6, 122, 1985.
65. **Ormsbee, R. A.**, Q fever rickettsia, in *Viral and Rickettsial Infections of Man*, Horsfall, F. L., Jr. and Tamm, I., Eds. 4th ed., J. B. Lippincott, Philadelphia, 1972, 1144.
66. **Marmion, B. P. and Stoker, M. G. P.**, The epidemiology of Q fever in Great Britain. An analysis of the findings and some conclusions, *Br. Med. J.*, 2, 809, 1958.
67. **Kosatsky, T.**, Household outbreak of Q fever pneumonia related to a parturient cat, *Lancet*, 2, 1447, 1984.
68. **Pratt, P. W.**, *Feline Medicine*, Am. Vet. Publications, Santa Barbara, 1983, 521.
69. **Martell, A. M., Yescott, R. E., and Dodds, D. G.**, Some records for Ixodidae of Nova Scotia, *Can. J. Zool.*, 47, 183, 1969.
70. **Williams, J. C.**, personal communication, April 1988.
71. **Marrie, T. J., Schlech, W. F., Williams, J. C., and Yates, L.**, Q fever pneumonia associated with exposure to wild rabbits, *Lancet*, 1, 427, 1986.
72. **Williams, J. C.**, personal communication, 1988.
73. **Kaplan, M. M. and Bertagna, P.**, The geographical distribution of Q fever, *Bull. W.H.O.*, 13, 829, 1955.
74. **Addo, P. B., Greenwood, B. M., and Schnurrenberger, P. R.**, A serological investigation of Q fever in clinical patients, *J. Trop. Med. Hyg.*, 80, 197, 1977.
75. **Okoh, A. E.**, Canine diseases of public health significance in Nigeria, *Int. J. Zoon.*, 10, 33, 1983.
76. **Hummel, P. H.**, Incidence in Tanzania of CF antibody to *Coxiella burnetii* in sera from man, cattle, sheep, goats and game, *Vet. Rec.*, 98, 501, 1976.
77. **Marc-Aurele, J., Gregoire, F., and Comeau, M.**, Clinical report on Q fever. First case in Canada, *Can. Med. Assoc. J.*, 75, 931, 1956.
78. **Pavilanis, V., Duval, L., Foley, A. R., and L'Heureux, M.**, An epidemic of Q fever at Princeville, Quebec, *Can. J. Public Health*, 49, 520, 1958.
79. **MacLean, D. M., Rance, C. P., and Walker, S. J.**, Q fever infection in an Ontario family, *Can. Med. Assoc. J.*, 83, 1110, 1960.

80. **Herbert, F. A., Morgante, O., Burchak, E. C., and Kadis, V. W.**, Q-fever in Alberta — infection in humans and animals, *Can. Med. Assoc. J.*, 93, 1207, 1965.
81. **McKiel, J. A.**, Q fever in Canada, *Can. Med. Assoc. J.*, 91, 573, 1964.
82. **Somlo, F. and Kovalik, M.**, Acute thyroiditis in a patient with Q fever, Can. Med. Assoc. J., 95, 1091, 1966.
83. **Marrie, T. J., Haldane, E. V., Noble, M. A., Faulkner, R. S., Martin, R. S., and Lee, S. H. S.**, Causes of atypical pneumonia: results of a 1-year prospective study, *Can. Med. Assoc. J.*, 125, 1118, 1981.
84. **Marrie, T. J., Haldane, E. V., Noble, M. A., Faulkner, R. S., Lee, S. H. S., Gough, D., Meyers, S., and Stewart, J.**, Q fever in Maritime Canada, *Can. Med. Assoc. J.*, 126, 1295, 1982.
85. **Russell, M. L. and Raven, J. B.**, Q fever: an outbreak in Alberta, 1985, personal communication.
86. **Dobija-Domaradzki, M., Hausser, J.-L., and Gosselin, F.**, Coexistence chez un même patient de la maladie des légionnaires et de la fièvre Q, *Can. Med. Assoc. J.*, 130, 1022, 1984.
87. **Saginur, R., Silver, S. S., Bonin, R., Carlier, M., and Orizaga, M.**, Q fever endocarditis, *Can. Med. Assoc. J.*, 133, 1228, 1985.
88. **Marrie, T. J., Van Buren, J., Faulkner, R. S., Haldane, E. V., Williams, J. C., and Kwan, C.** Seroepidemiology of Q fever in Nova Scotia and Prince Edward Island, *Can. J. Microbiol.*, 30, 129, 1984.
89. **Marrie, T. J. and Fraser, J.**, Prevalence of antibodies to *Coxiella burnetii* among veterinarians and slaughterhouse workers in Nova Scotia, *Can. Vet. J.*, 26, 181, 1985.
90. **Marrie, T. J., Van Buren, J., Fraser, J., Haldane, E. V., Faulkner, R. S., Williams, J C., and Kwan, C.**, Seroepidemiology of Q fever among domestic animals in Nova Scotia, *Am. J. Public Health*, 75, 763, 1985.
91. **Marrie, T. J.**, Seroepidemiology of Q fever in New Brunswick and Manitoba, *Can. J. Microbiol.*, 34, 1043, 1988.
92. **Irons, J. V. and Hooper, J. M.**, Q fever in United States, Clinical data on an outbreak among stock handlers and slaughterhouse workers, *J. Am. Med. Assoc.*, 133, 815, 1947.
93. **Sienko, D. G., Bartlett, P. C., McGee, H. B., Wentworth, B. B., Herndon, J. L., and Hall, W. N.**, Q fever. A call to heighten our index of suspicion, *Arch. Intern. Med.*, 148, 609, 1988.
94. **Peacock, M. G., Ormsbee, R. A., and Johnston, K. M.**, Rickettsioses of Central America, *Am. J. Trop. Med.*, 20, 941, 1971.
95. **de Rodaniche, E. C. and Rodaniche, A.**, Q fever, Report of a case and study of the etiologic agent, *Arch. Hosp. Santo. Tomas.*, 2, 327, 1947.
96. **de Rodaniche, E. C. and Rodaniche, A.**, Studies of Q fever in Panama, *Am. J. Hyg.*, 49, 67, 1949.
97. **Kourany, M. and Johnson, K. M.**, A survey of Q fever antibodies in a high risk population in Panama, *Am. J. Trop. Med. Hyg.*, 29, 1007, 1980.
98. **de Ruiz, H. L.**, Q fever in Columbia, S.A., A serological survey of human and bovine populations, *Zentralbl. Veterinaermed. Reihe B*, 24, 287, 1977.
99. **Somma-Moreira, R. E., Caffarena, R. M., Somma, S., Pérez, G., and Monteiro, M.**, Analysis of Q fever in Uruguay, *Rev. Infect. Dis.*, 9, 386, 1987.
100. **Fan, M.-Y., Walker, D. H., Yu, S.-R., Liu, Q.-H.**, Epidemiology and ecology of rickettsial diseases in the People's Republic of China, *Rev. Infect. Dis.*, 9, 823, 1987.
101. **Stephen, S. and Achyutha, Rao, K. N.**, Q fever in India: a review, *J. Indian Med. Assoc.*, 74, 200, 1980.
102. **Alkan, W. J., Alkalay, L., Klingberg, W., Goldwasser, R. A., Stolar, R., and Klingberg, M. A.**, A study of Q fever in Central Israel, *Scand. J. Infect. Dis.*, 5, 17, 1973.
103. **Caughey, J. E. and Harootunian, S. H.**, Q fever in Iran, *Lancet*, 2, i, 638, 1976.
104. **Gelpi, A. P.**, Q fever in Saudi Arabia, *Am. J. Trop. Med. Hyg.*, 15, 785, 1966.
105. **Crowther, R. W. and Spicer, A. J.**, Abortion in sheep and goats in Cyprus caused by *Coxiella burnetii*, *Vet. Rec.*, 99, 29, 1976.
106. **Spicer, A. J., Crowther, R. W., Vella, E. E., Bengtsson, E., Miles, R., and Pitzolis, G.**, Q fever and animal abortion in Cyprus, *Trans. R. Soc. Trop. Med. Hyg.*, 71, 16, 1977.
107. **Leedom, J. M.**, Q fever: an update, in *Current Clinical Topics in Infectious Diseases*, Remington, J. S. and Swartz, M. W., Eds., McGraw-Hill, New York, 1980, 304.
108. **Polydorou, K.**, Q fever in Cyrpus: a short review, *Br. Vet. J.*, 137, 470, 1981.
109. **Polydorou, K.**, Q fever control in Cyprus — recent progress, *Br. Vet. J.*, 141, 427, 1985.
110. **Reháček, J., Palanova, A., Zupancicova, M., Urvologyi, J., Kovacova, E., Jarabek, L., and Brezina, R.**, Study of rickettsioses in Slovakia. I. *Coziella burnetii* and rickettsiae of the spotted fever (SF) group in ticks and serological surveys in animals and humans in certain selected localities in the Lucenec and V. Krtis districts, *J. Hyg. Epidemiol. Microbiol. Immunol.*, 19, 105, 1975.
111. **Reháček, J.**, Epidemiology and significance of Q fever in Czechoslovakia, *Zentralbl. Bakteriol. Parasitenkd. Infektionskr. Hyg. Abt. I Orig. Reihe A.*, 267, 16, 1987.
112. **Lumio, J., Penttinen, K., and Pettersson, T.**, Q fever in Finland: clinical, immunological and epidemiological findings, *Scand. J. Infect. Dis.*, 13, 17, 1981.
113. **Kerttula, Y., Leinonen, M., Koskela, M., and Makela, P. H.**, The aetiology of pneumonia. Application of bacterial serology and basic laboratory methods, *J. Infect.*, 14, 21, 1987.

114. **Kindmark, C.-O., Nyström-Rosander, C., Friman, G., Peacock, M., and Vene, S.**, The first human case of domestic Q fever in Sweden, *Acta Med. Scand.*, 218, 429, 1985.

115. **Tringali, G. and Mansueto, S.**, Epidemiology of Q fever in Italy and in other Mediterranean countries, *Zentralbl. Bakteriol. Parasitenkd. Infektionskr. Hyg. Abt. I Orig. Reihe A*, 267, 20, 1987.

116. **Edlinger, E. A.**, Q fever in France, *Zentralbl. Bakteriol. Parasitenkd. Infektionskr. Hyg. Abt. I Orig. Reihe A*, 267, 26, 1987.

117. **Raoult, D., Piquet, P. H., Gallais, H., Micco, C., and Drancourt, M.**, *Coxiella burnetii* infection of a vascular prosthesis, *N. Engl. J. Med.*, 315, 1358, 1986.

118. **Raoult, D., Etienne, J., Massip, P.**, et al., Q fever endocarditis in the south of France, *J. Infect. Dis.*, 155, 570, 1987.

119. **Krauss, H., Schmeer, N., and Schiefer, H. G.**, Epidemiology and significance of Q fever in the Federal Republic of Germany, *Zentralbl. Bakteriol. Mikrobiol. Hyg. [A]*, 267, 42, 1987.

120. **Heinrich, R., Naujoks-Heinrich, S., Saebisch, R., Seuffer, R., Grauer, W., Jacob, R., and Schomerus, H.**, Seroprävalenz des Q-Fiebers in einem Endemiegebiet Süddeutschlands, *Dtsch. Med. Wochenschr.*, 108, 1318, 1983.

121. **Werth, D., Schmeer, N., Müller, H. P., Karo, M., and Krauss, H.**, Nachweis von Antikörpern gegen *Chlamydia psittaci* and *Coxiella burnetii* bei Hunden und Katzen: Vergleich zwischen Enzymimmuntest, Immunoperoxidase-Technik, Komplementbindungsreaktion, und Agargelpräzipittationstest, *J. Vet. Med. B.*, 34, 165, 1987.

122. **Rády, M., Glávits, R., and Nagy, G.**, Epidemiology and significance of Q fever in Hungary, *Zentralbl. Bakteriol. Parasitenkd. Infectionskr. Hyg. Abt. I Orig. Reihe A.*, 267, 10, 1987.

123. **Richardus, J. H., Donkers, A., Dumas, A. M., Schaap, G. J. P., Akkermans, J. P. W. M., Huisman, J., and Valkenburg, H. A.**, Q fever in the Netherlands: a sero-epidemiological survey among human population groups from 1968 to 1983, *Epidemiol. Inf.*, 98, 211, 1987.

124. **Richardus, J. H., Dumas, A. M., Huisman, J., and Schaap, G. J.**, Q fever in infancy: a review of 18 cases, *Pediatr. Infect. Dis.*, 4, 369, 1985.

125. **Houwers, D. J. and Richardus, J. H.**, Infections with *Coxiella burnetii* in man and animals in The Netherlands, *Zentralbl. Bakteriol. Parasitenkd. Infektionskr. Hyg. Abt. I Orig. Reihe A*, 267, 30, 1987.

126. **Connolly, J. H.**, Q fever in Northern Ireland, *Br. Med. J.*, 1, 547, 1968.

127. **Hillary, I. B and Meenan, P. N.**, Q fever in the Republic of Ireland, *Ir. J. Med. Sci.*, 145, 10, 1976.

128. **Cracea, E.**, Q fever epidemiology in Roumania, *Zentralbl. Bakteriol. Parasitenkd. Infektionskr. Hyg. Abt. I Orig. Reihe A*, 267, 7, 1987.

129. **Errasti, C. A., Baranda, M. M., Hernandez Almaraz, J. L., De La Hoz Torres, C., Martinez-Gutierrez, E., Villate Navarro, J. L., and Sobradillo Peña, V.**, An outbreak of Q fever in the Basque country, *Can. Med. Assoc. J.*, 131, 48, 1984.

130. **Fernández-Guerrero, M. L., Muelas, J. M., Aguado, J. M., Renedo, G., Fraile, J., Soriano, F., and De Villalobos, E.**, Q fever endocarditis on porcine bioprosthetic valves — clinicopathologic features and microbiologic findings in three patients treated with doxycycline, cotrimoxazole and valve replacement, *Ann. Intern. Med.*, 108, 209, 1988.

131. *Bulletin l'Office Federal de la Sante Publique*, 52, 697, 1984.

132. **Gelzer, J., Abelin, T., Bertschinger, H. U., Bruppacher, R., Metzler, A. E., Nicolet, J.**, Wie verbreitet ist Q-Fieber in der Schweiz?, *Schweiz. Med. Wochenschr.*, 113, 892, 1983.

133. Q fever — United Kingdom, *Morbid. Mortal. Wkly. Rep.*, 28, 230, 1979.

134. **Young, S. E.**, Aetiology and epidemiology of infective endocarditis in England and Wales, *J. Antimicrob. Chemother.*, 230 (Suppl. A), 7, 1987.

135. **Aitken, I. D.**, Q fever in the United Kingdom and Ireland, *Zentralbl. Bakteriol. Parasitenkd. Infektionskr. Hyg. Abt. I Orig. Reihe A*, 267, 37, 1987.

136. **Kirby, F. D.**, Zooneses in Britain, *J. R. Soc. Health*, 105, 77, 1985.

137. **Tarasevich, I. V.**, Epidemiology (1980-1985) and nonspecific prophylaxis of Q fever in the USSR (survey), *Zentralbl. Bakteriol. Parasitenkd. Infektionskr. Hyg. Abt. I Orig. Reihe A*, 267, 1, 1987.

138. **Johnson, D. W.**, Epidemiology of Q fever in Queensland: a seven-year survey, *Med. J. Aust.*, 1, 121, 1966.

139. **Burnet, F. M. and Freeman, M.**, Experimental studies on virus of "Q" fever, *Med. J. Aust.*, 2, 299, 1937.

140. **Derrick, E. H. and Smith, D. J. W.**, studies on the epidemiology of Q fever. II. The isolation of three strains of *Rickettsia burneti* from bandicoot *Isoodon torosus*, *Aust. J. Exp. Biol. Med. Sci.* 18, 99, 1940.

141. **Derrick, E. H., Pope, J. H., and Smith, D. J.**, Outbreaks of Q fever in Queensland associated with sheep, *Med. J. Aust.*, 1, 585, 1959.

142. **Derrick, E. H.**, Epidemiology of Q fever, *J. Hyg. (London)*, 433, 1942.

143. **Marmion, B. P.**, personal communication, 1988.

144. Commonwealth of Australia notifiable disease statistics yearly reports 1960; 1961-1986.

145. **Mendes, M. R., Carmona, H., Malva, A., and Dias Sousa, R.**, Review of 176 cases of Q fever in Portugal. Preliminary report, Abstr. No. 314, *International Congress for Infectious Diseases*, Rio de Janeiro, April 17-21, 1988.

Chapter 4

EXPERIMENTAL STUDIES OF THE INFECTIOUS PROCESS IN Q FEVER

Theodor Khavkin

TABLE OF CONTENTS

I. INTRODUCTION AND SCOPE

This chapter describes the direct interactions of *Coxiella burnetii,* the agent of Q fever, with cells, tissues, and the defense mechanisms of the host, in other words, the Q-fever infectious process. The concept of the infectious process is not identical to that of the infectious disease. It comprises the complex of pathophysiological, morphological, and immunological events that underlie the pathogenesis of both the infectious disease, with overt morbidity and lethality, and inapparent and latent forms of infection. At the same time, it does not include the details of the clinical manifestations of the disease.[1] Neither the pathogenic features of microorganisms nor the pathogenesis of infectious diseases can be fully comprehended without an understanding of the kinetics of host-pathogen interactions, i.e. the morphogenesis of the infectious process.

In Q-fever studies, morphogenesis attracted less attention than other aspects of the infectious process.[2,3] This is due, in part, to the scarcity of autopsy and biopsy observations and, in part, to the lack of a morphogenetic approach in experimental studies. The goal of this chapter is to summarize available data and to outline the areas that have been insufficiently studied, hoping that this will stimulate further investigations. The focus is on the characteristics of experimental models of the disease and on the basic events in the infectious process that are important for the understanding of the pathogenic features of *C. burnetii,* and, hence, the pathogenesis of Q fever. Nosologic models of clinical forms of Q fever, such as pneumonia, endocarditis, and hepatitis, are considered in other chapters of this book. Biochemical and physiological aspects of the host-pathogen interactions are mentioned only briefly, as they are discussed in Volume II of this book.

II. HOST-PARASITE SYSTEMS IN *C. BURNETII* INFECTION AND BASIC EXPERIMENTAL MODELS

A. CONCEPTS OF PATHOGENICITY AND VIRULENCE AS APPLIED TO *C. BURNETII*

C. burnetii, exerts diverse degrees of pathogenicity and virulence toward different hosts. Pathogenicity and virulence are sometimes used as synonyms or "near synonyms"[4] (Smith[4] p.5) to describe the bacterium's capacity to cause disease. In this chapter, these terms are used in a different sense. Pathogenicity is used as a general capacity of the microorganism to produce infection. Virulence refers to the deleterious effects elicited by the various strains of the pathogen on a given animal species.[5] This distinction, of course, cannot always be made. However, it is worth applying this distinction for descriptive and comparative purposes, especially in relation to such an ubiquitous organism as *C. burnetii.* From this point of view, the infection caused by *C. burnetii* can be divided into three natural and two laboratory host-parasite systems: (1) endosymbiosis, (2) inapparent infection in animals, (3) animal infection with overt morbidity and lethality, (4) infected cell culture, and (5) infection of the chicken embryo. The value and limitations of each system, as experimental model, should be weighed when experimental studies are planned.

B. ENDOSYMBIOSIS (ARTHROPOD MODEL)

Rickettsias are considered to be nonpathogenic for arthropods, specifically for ticks,[6] which constitute a major reservoir for vertebrate infection with *C. burnetii.* Animal inoculations[7,8] and immunohistological studies of the tick *Hyalomma asiaticum*[9,10] have shown that in ixodid and argasid ticks, *C. burnetii,* indeed, harmlessly resides in all cell types and in the hemolymph. At the same time, the cyclic nature of the infection and certain limitations in the extent of multiplication of *C. burnetii* suggest that the tick is capable of controlling the infectious process.[7] Thus, the infection acquires a generalized nature only when the tick feeds on its animal host. During starvation, the infection turns into the inapparent state, with an indefinite persistence of dormant rickettsias in the body. Mechanisms that control the infectious process in the tick are unknown.

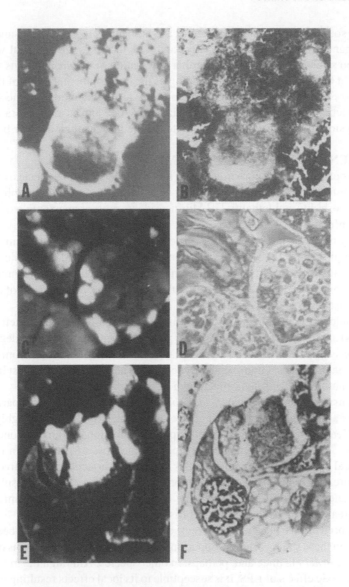

FIGURE 1.   Cells of the digestive epithelium (A—D), and a salivary gland (E, F) of the tick *Hyalomma asiaticum,* 11 weeks post challenge, 8 d of feeding. Paraffin sections were incubated with anti-*Coxiella* fluorescent antibody (A, C, and E) and then either Giemsa stained (B and F) or examined with the phase contrast optics (D). Specific fluorescence of *C. burnetii* colonies is seen in A, C, and E. Phase contrast optics reveals blood remnants in the *Coxiella*-containing digestive cell (D). (From Balashov, Yu. A., Daiter, A. B., and Khavkin, T. N., *Parazitologiya,* 6, 22, 1972. With author's permission.)

During feeding, *C. burnetii* multiplies extensively in cells of the digestive epithelium of the midgut (Figures 1 A to D). These cells are specialized for the uptake, intraphagolysosomal digestion of host's blood, subsequent transport of nutrients to the tick's internal environments, and for extrusion of undigestible residual material into the gut.[11] *C. burnetii,* ingested with the blood, proliferates within digestive epithelium cells and then is spread in two directions: internally, throughout the body, and externally into the gut lumen with the feces.

Among the internal organs affected by *C. burnetii* are the salivary glands (Figures 1 E and F) which present one possible source of host infection via inoculation. Organisms shed with the feces present a source for contaminant, airborne, and alimentary infection.[7] It is unclear whether *C. burnetii* organisms released with the saliva and with the feces are identical with regard to metabolism and structural organization. Further elucidation of *C. burnetii* in its two locations in the tick might be helpful for an understanding of the developmental cycle of this pathogen. The infected ixodid tick *Hyalomma asiaticum* presents a suitable model for the study of host-pathogen interactions in ticks. Because of the size of the tick and ease of laboratory handling, features of physiology and morphology are better studied in this than in other ticks.[7,11]

C. INAPPARENT INFECTION (MURINE MODEL)

C. burnetii is pathogenic but nonvirulent or of low virulence for many species. In other words, it induces an infectious process with specific inflammatory and immune phenomena, but without overt morbidity and lethality.[12,13] Murine rodents present an example of animals with an inapparent infection.

In laboratory mice, the infectious process is characterized by long-lasting multiplication of the pathogen in infectious foci, without apparent systemic toxic and metabolic changes. The extent of multiplication is in part controlled by mechanisms of cell-mediated immunity. In part, it is determined by yet unclear features of the pathogen's strains. Mechanisms of cell-mediated immunity are discussed in Chapter 5.

Unphysiologically high concentrations of living or killed coxiellae or their lipopolysaccharides (LPSs) are required to kill mice.[13] Mice possess a comparatively high degree of natural tolerance to the systemic toxic effects of bacterial LPS.[14] Furthermore, LPS from *C. burnetii* is less toxic than that from most Gram-negative bacteria.[15] These peculiarities most likely underlie the inapparent nature of *C. burnetii* infection in mice.

The mechanism of mouse resistance to *C. burnetii* LPS may, at least in part, involve the activities of the adrenal glands. Thus, adrenalectomy, which renders mice highly sensitive to many toxins,[16] leads to the transformation of inapparent infection into an overt one, with almost 100% lethality.[17-19] This is accompanied by lipidosis and loss of glycogen in hepatocytes.[18] These are typical morphological manifestations of the overt infection in sensitive animals, but absent in nonadrenalectomized mice. Furthermore, injection of *Coxiella* LPS results in the lethal shock in adrenalectomized mice. At the same time, both shock and lethal infection can be prevented by treatment of the animal with cortisone.[18]

Recently, inbred murine lines with different sensitivities to *C. burnetii* have been obtained.[13] Further studies in these strains offer the opportunity of highlighting mechanisms of nonspecific resistance to *C. burnetii* in mice and, perhaps, in other species. Although the mouse is tolerant to the systemic toxic effects of LPS, it is susceptible to its local effects resulting in cell damage and the inflammatory response.[20,21] The expression and possible mechanisms of the local damage will be discussed later on. Tolerance to the toxic effects of LPS makes murine models suitable for the study of phenomena that are mostly local rather than systemic in the *C. burnetii* infectious process, except for the immune response. Adrenalectomy or treatment with drugs, such as actinomycin D or galactosamine which render mice sensitive to LPS,[14,16] results in a multitude of side effects that might modulate the infectious process.

A peculiar sequela to either infection or vaccination of mice with *C. burnetii* is splenomegaly. The splenomegaly is a dose-dependent phenomenon.[13,22,23] Therefore, it is utilized for evaluation of the immune response, along with using the cell suspension obtained from the spleen in laboratory tests.

The pathogenic basis for the splenomegaly is little studied.[3] Our observations suggest that, for the most part, extensive proliferation of the immunoglobulin-producing B-cells underlies the pathogenesis of the splenomegaly[20] (Figure 2A). In case of infection, the extensive proliferation of B-cells goes along with the proliferation of the pathogen in the respective cells (Figures 2B and C). The extensive proliferation of B-cells is consistent with the correlation between the splenomegaly and production of anticoxiella antibodies.[22] It also agrees with the literature data concerning the cellular composition in the spleen in mice inoculated with LPS and other

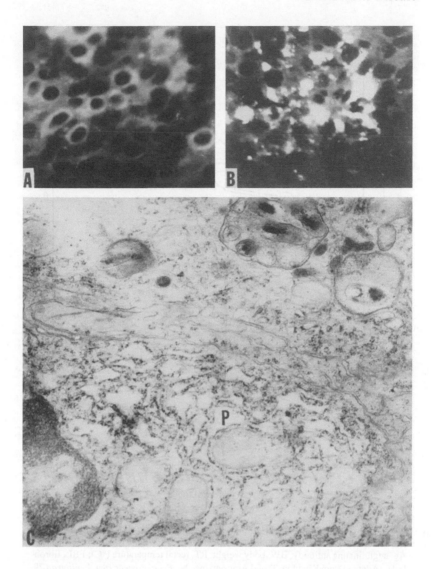

FIGURE 2.   Immunoglobulin-producing and *Coxiella*-bearing cells in an enlarged murine spleen, 7 d post challenge. Specific fluorescence of globulin-producing cells (A) and collections of microorganisms (B) in sections treated with fluorescent antibodies against either mouse IgG (A) or *C. burnetii* (B). (C) Ultrathin section showing a plasma cell (P) and the cell with *Coxiella*-bearing vacuoles.

immunogens.[24,25] The actual kinetics of the change in B-cell-to-T-cell ratio in the spleen induced by *C. burnetii* is not known. Neither are the target antigens for the production of immunoglobulins in the spleen known. It is conceivable that *C. burnetii* stimulates production of immunoglobulins both related and unrelated to its antigens since, LPSs present polyclonal mitogens for B-lymphocytes.[26] Furthermore, some *C. burnetii* antigens possess adjuvant activities.[27]

D. INFECTION WITH OVERT MORBIDITY AND LETHALITY (GUINEA PIG MODEL)

In some animals, including guinea pigs, monkeys, rabbits, and in man *C. burnetii* exerts its virulence by impairing normal functions resulting in disease and death. An array of systemic and metabolic disorders known under the common name the acute phase response[28] underlies the *C. burnetii*-caused disease, Q fever. In Q fever, manifestations of the acute phase response, which

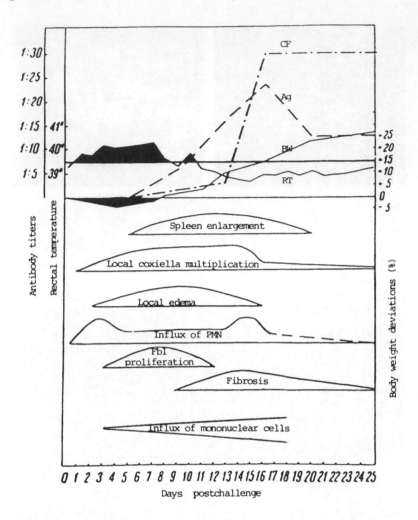

FIGURE 3.   Some local and systemic indicators of the infectious process in the guinea pig subcutaneously challenged with phase I *C. burnetii*. CF, complement-fixing antibody; Ag, agglutinating antibody; BW, body weight; RT, rectal temperature (°C); FBL, fibroblasts. (Adapted from Khavkin, T. and Amosenkova, N., *Proc. Pasteur Inst. Leningrad* 29, 212, 1965. With permission.)

for the most part are associated with systemic activities of LPS,[2,29] constitute the fundamental difference between inapparent and overt infectious processes. The pathophysiological basis for these manifestations, which is similar to those induced by bacterial LPS, is reviewed by Baca and Paretsky.[2]

In the guinea pig, systemic manifestations of the disease, such as fever, weight loss, hypoglycemia, and other metabolic disorders, have a distinctive cyclic nature coinciding with the cyclic development of the infectious foci[20,29] (Figure, 3).

There is little information on the morphological manifestations of the systemic toxic effects of *C. burnetii*. We have observed a transient loss of glycogen and lipidosis of hepatocytes in infected guinea pigs[20] that was similar to the effect of bacterial LPS.[30] The dose-dependent nature of the glycogen deprivation can be utilized in the evaluation of comparative virulence and comparative toxicity of *C. burnetii* strains.

Although the guinea pig succumbs to the systemic toxic effects of *C. burnetii*, its infectious foci are cleared of microorganisms faster and more effectively than in mice. Still, in both mice and guinea pigs infectious foci are cleared incompletely, and acute infection can be revived in the immunocompromised host and during pregnancy.

E. INFECTED CELL CULTURE

The infected cell culture presents an important model for the study of the basic aspects of the rickettsia-cell interactions:[31] attachment, entry, and exit of the parasite, functions of the cell defense mechanisms as well as influence of hormones, lymphokines, and other extrinsic factors on the cell-pathogen interactions. Furthermore, cell cultures provide the opportunity for observations of living cells and for motion picture studies which permit the analysis of the cellular locomotory response to the parasite. Essential information on the intracellular behavior of *C. burnetii* and on its influence on the host cell obtained with the infected cell cultures has been reviewed by Baca and Paretsky.[2]

At the same time, the cell culture model has certain limitations. Some of these limitations have been discussed by Moulder.[32] It is worthwhile to add that the range of cells capable of harboring rickettsias *in vitro* is limited compared to that in *in vitro* systems.[31] This suggests that *in vivo*, only restricted subsets of cells express appropriate receptors to *C. burnetti*. Furthermore, the established cell cultures which are, for the most part, composed of transformed immortalized cells may differ from their ancestors in the body in terms of structural and functional features, such as receptors, endocytic activity, etc. The cell culture model should be cautiously applied for studying processes developing on the organ or organismal level and for mechanisms of latency. Thus, the persistence of vigorously multiplying coxiellae in generations of immortalized cells *in vitro* demonstrates that under favorable culture conditions, *C. burnetii* does not immediately damage the basic activities of the host cell and is not damaged by the cellular defense mechanisms.[33] (Figure 4). This, however, barely mimics the latent infection which implies the persistence of metabolically dormant microorganisms in nonproliferating or slowly proliferating cells.

F. INFECTED CHICKEN EMBRYO

The ability of *C. burnetii* to multiply extensively in the chicken embryo's yolk sac has been recognized since the time of discovery of the organism and has become a general procedure for obtaining large pools of rickettsias.[34] The infectious process induced by this procedure is different from that seen in laboratory animals and in cell cultures. It is limited to the yolk sac[35] and is associated with phenotypic modulation of the pathogen resulting in its conversion into phase II[36] and a large-scale production of infraforms.[37] In the yolk sac, *C. burnetii* inhabits the endodermal epithelium, the layer of cells specialized for taking up the yolk constituents for selective transport of nutrients to the developing embryo.[38] Endodermal cells engulf coxiellae with the yolk. In the cell, *C. burnetii* multiplies within phagolysosomes (Figure 5A), but eventually escapes into the cytoplasm[38] (Figure 5B). The escape results from the natural evolution of the endodermal epithelium rather than from enzymatic activities of the pathogen that is typical of other rickettsias.[39] In phagolysosomes many coxiellae are confined to lipovitelline spheres undergoing digestive degradation. Such spheres are rich in rickettsial infraforms and atypically dividing organisms (Figure 6). It is not clear what environmental conditions in the yolk sac endoderm influence phenotypic modulation of coxiellae and whether infraforms accumulating in the lipovitelline spheres are identical to those described by McCaul and Williams.[37]

III. STEPS OF THE INFECTIOUS PROCESS

A. PORTALS OF ENTRY

C. burnetii infection can be acquired from infected ticks, via aerosol inhalation, and via the gastrointestinal tract.

1. Tick-Borne Infection

Q fever is the only tick-borne rickettsiosis which develops without formation of a primary infectious focus in the bite area. Our histological immunofluorescent study failed to reveal

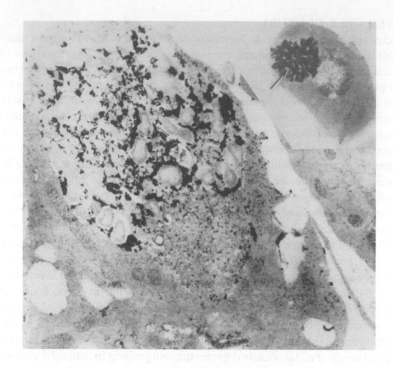

FIGURE 4.   *C. burnetii* -infected cells of two established lines which maintain functional activities (A). An ultrathin section of a murine macrophage-like J774 cell incubated for the demonstration of acid phosphatase. Black deposit of the reaction product is concentrated within the coxiella-bearing vacuole. (B) Giemsa-stained cell of monkey kidney epithelium (MK2) under mitotic division; arrow shows a metaphase plate. (A from Akporiaye, E. T., Rowatt, J. D., Aragon, A. A., and Baca, O., *Infect. Immun.*, 40, 1155, 1983. With permission.)

coxiella-bearing cells in the guinea pig skin in the area of bite by the heavily infected tick *Hyalomma asiaticum*. Lesions produced in the skin by saliva and by mouth parts and cement-like material from either infected or noninfected ticks were identical.[10] These observations suggest that tick-borne Q fever, as louse-borne epidemic typhus, develops at once as a generalized infection. The possibility cannot be ruled out that the dermal port of entry of *C. burnetii* does not necessarily coincide with the bite area, since during feeding, ticks eliminate coxiellae by several ways: with the saliva, feces, and coxal fluids.[7,8] The actual portal of entry of tick-borne coxiellae cannot be detected because of the absence of a primary infectious focus. Balashov[7] considers three ways of *C. burnetii* transmission from the tick to the host: the inoculative one, with the saliva, contaminative with the saliva and feces, and airborne with the feces. He refers to observations by Daiter showing that in the tick's feces, *C. burnetii* remain infectious for 635 d.

2. Airborne Infection

Infection via inhalation is the most common mode of acquiring *C. burnetii* by man and in many cases by animals (see Chapter 3). Inhalation may result in formation of primary pulmonary infectious foci which are not always detectable by clinical and radiological investigations (described in Chapter 6). Animal experiments have shown that the ability of *C. burnetii* to inhabit specific cells of the respiratory parenchyma, pneumocytes, as well as fibroblasts and histiocytes of the pulmonary stroma, most likely underline the pathogenesis of pulmonary lesions in Q fever pneumonia.[21]

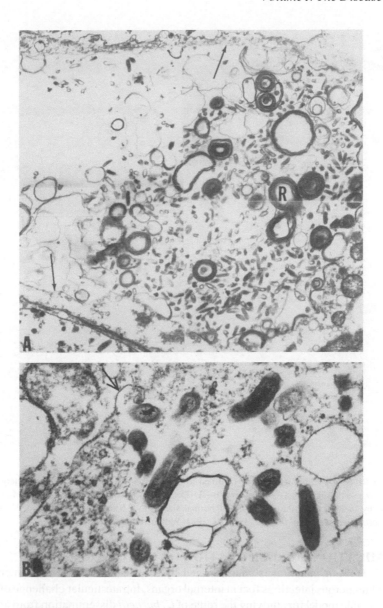

FIGURE 5.   Endodermal cells of the yolk sac of the *C. burnetii*-infected chicken embryo, 6 d post challenge. (A) Cell filled with a large vacuole containing microorganisms and myelin-like yolk residual bodies (R). Arrows point to the membrane of the vacuole. (B) Part of the *Coxiella*-bearing vacuole, with ruptured membrane (arrow). Ultrathin sections. (From Khavkin, T., Sukhinin, V., and Amosenkova, N., *Infect. Immun.*, 32, 1281, 1981. With permission.)

3. Alimentary Infection

The possibility of *C. burnetii* invasion via the gastrointestinal tract has been demonstrated by epidemiological observations of the infection with the milk. Gutzu et al.[40] described cases of a severe gastroenteritis in man, with fever and mesenteric lymphadenitis that most likely had a *C. burnetii* etiology. Coxiellae were found and specifically identified in surgically removed mesenteric lymph nodes. Furthermore, infection of guinea pigs with *C. burnetii* by feeding has also been described.[41] The pathogenic basis for the alimentary invasion of *C. burnetii*, in particular, its interaction with cells and defense apparatus of the intestinal mucosa and submucosa are not known.

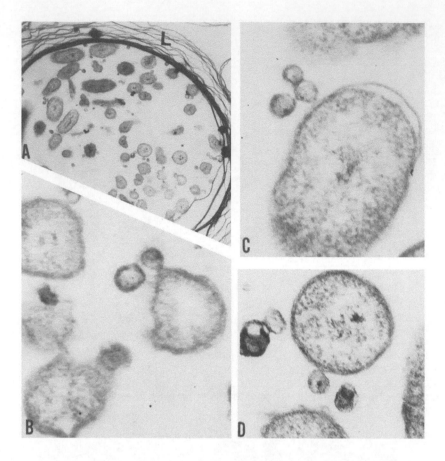

FIGURE 6. A *C. burnetii* colony confined to the lipovitelline sphere (L) which undergoes intralysosomal digestion. (A) Overview; (B—D) Details showing smallest bodies and their formation by budding from the organisms of normal size. (From Khavkin, T., Sukhinin, V., and Amosenkova, N., *Infect. Immun.*, 32, 1281, 1981. With permission.)

B. DISSEMINATION AND ELIMINATION OF THE PATHOGEN

Acute *C. burnetii* infection is characterized by extensive rickettsiemia which results in formation of numerous infectious foci in internal organs. Intratesticular challenge of guinea pig presents a suitable model for studying the route of *C. burnetii* dissemination from an infectious focus.[20] From the testicular focus, coxiellae enter the blood and the lymph via the spermatic vein and epididymal lymph vessels, respectively. Organisms either freely circulate in the plasma, or on the surface of blood cells (Figure 7A), or are confined to the macrophage-like cells released from infectious foci (Figure 7B). The *Coxiella*-bearing cells can also be found within pulmonary veins (Figure 7C). Entry of the infected cells into the pulmonary circulation resembles the clearance mechanism of the reticuloendothelial system of cells containing ingested particles.[42] In *C. burnetii* infection, this may result in formation of the secondary pulmonary infectious foci. Secondary infectious foci can also appear in the liver. They derive from the infected spleen via the portal system. Furthemore, *C. burnetii* enters sinuses of lymph nodes draining infectious foci (Figure 8A) and then invades stromal cells and macrophages of the node (Figures 8 B to D), resulting in formation of the secondary regional lymphadenitis.

From some infectious foci *C. burnetii* may be released into the outer environment. Thus, from infected glomeruli and kidney stroma (Figure 9) and from hepatocytes (Figure 10) *C. burnetii* can be excreted via urine or through the bile, respectively. There is no information on the pathogenic basis for the *Coxiella* excretion from the mammary gland.

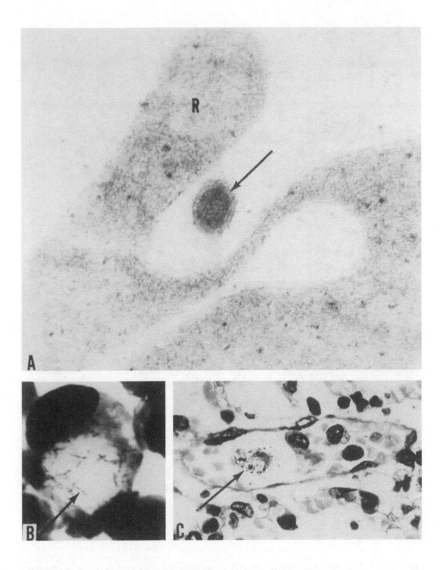

FIGURE 7.   *C. burnetii* (arrows) in peripheral blood: (A) on the surface of erythrocyte (R), within macrophage-like cells within vena cava (B), and within a branch of the pulmonary vein (C). Ultrathin section (A), Giemsa-stained smear (B), and paraffin section (C).

C. INTRACELLULAR PARASITISM
1. Entry, Intravacuolar Confinement, Escape

C. burnetii is an obligate intracellular agent that is adapted to reside in the phagosome-derived vacuoles. Because of its ability to inhabit macrophages and because of its requirement for the acidic environment elicited by the lysosomes,[43] *C. burnetii* has been designated a parasite of macrophage phagolysosomes. Although the parasitism of *C. burnetii* in nonphagocytic cells has been known for some time[44] (Figure 10), its entry into the cell has long been considered a process of passive phagocytosis.[2] Meanwhile, the ability of *C. burnetii* to inhabit epithelial cells *in vivo* and its mode of entry into cultured fibroblasts[45] (Figure 11) suggest that, along with being taken up passively, this organism is capable of initiating its own entry. The role of phase variation of *C. burnetii* and opsonizing antibodies in the uptake of the pathogen by the phagocytic cells has been elucidated by Wisseman et al.[46] The influence of normal serum opsonizing factors and of

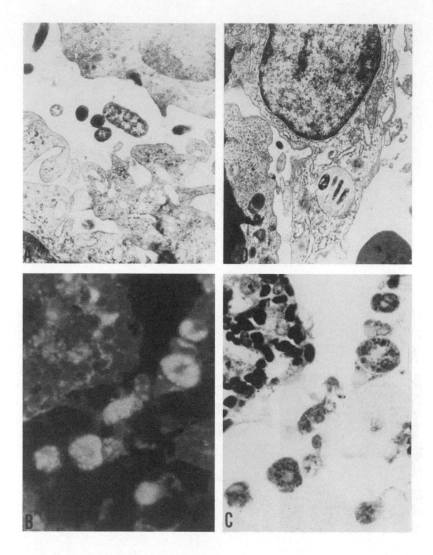

FIGURE 8.   Ultrathin (A, D) and paraffin (B, C) sections of a guinea pig lymph node draining a subcutaneous infectious focus. (A) Free organisms inside a sinus. (B—D) *Coxiella* colonies within sinusoidal stromal cells. Fluorescent and light microscopic images of the same paraffin section treated with anti-coxiella fluorescent antibody (B) and then stained after the Giemsa-Pappenheim technique (C). Ultrastructural appearance of a coxiella-bearing cell (D) is consistent with that of a dendritic cell.

acute phase reactants that might facilitate the entry of *Coxiella* into cells has been discussed by Williams and Vodkin.[47]

In the cell, *C. burnetii* resides in the phagosome-derived vacuole which presents a stable and, at the same time, a dynamic formation, like a cell organella. The vacuole kinetics, however, are little studied. Some cell culture observations indicate that shortly after its formation, the vacuole continues to function as a phagolysosome, as shown by its ability to degrade some ingested microorganisms and to assimilate rickettsial constituents into the host-cell[45] (Figure 12). In infectious foci, this is seen from the presence of degraded leukocytes and proliferating coxiellae inside the same macrophage phagosome. Furthermore, the vacuole wall acquires a two-membrane structure which is different from the one-membrane wall of the original phagosome[48] (Figure 13). A vacuole with a multilayer wall has been shown in the paper by Handley et al.[44]

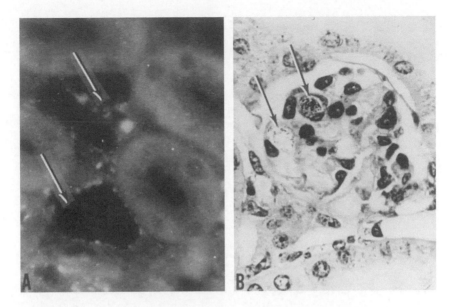

FIGURE 9.   Guinea pig kidney with *Coxiella*-containing cells (arrows) in the medullar stroma (A) and glomerular mesangium (B). Paraffin sections treated with the fluorescent anti-coxiella antibody (A) and stained after the Giemsa-Pappenheim technique (B). There is no nephritis-like glomerular alterations, 13 d post challenge. (B from Khavkin, T. N., *Arkh. Patol. (Mosc.)*, 2, 75, 1977. With permission.)

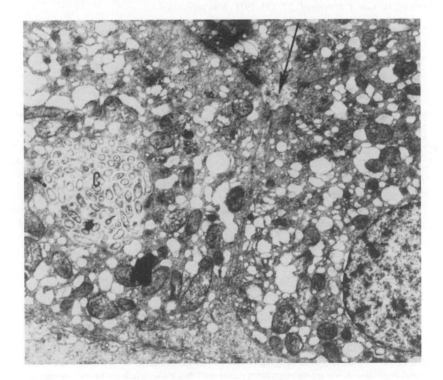

FIGURE 10.   A lobule of the guinea pig liver with a *Coxiella* colony in one of the hepatocytes. Vesicles and widened profiles of endoplasmic reticulum indicate cell damage. Arrow, profile of a biliferous capillary. Ultrathin section. (From Handley, J., Paretsky, D., and Stueckman, J. J., *Bacteriology* 94, 1, 1967. With permission.)

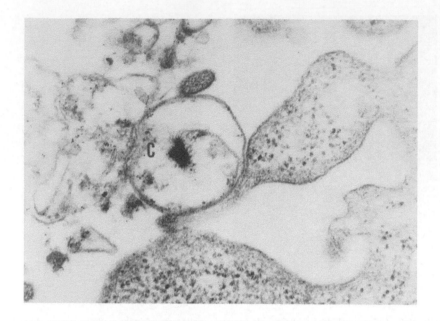

FIGURE 11.   The early step of *C. burnetii* entry into the cell of the murine fibroblast-like L line, 60 min of incubation at 37°. A rickettsial body (C) is attached to the membrane of a cytoplasmic projection extended to the organism. Ultrathin section. (From Burton, P. R., Kordova, N., and Paretsky, D., *Can. J. Microbiol.*, 17, 143, 1971. With permission.)

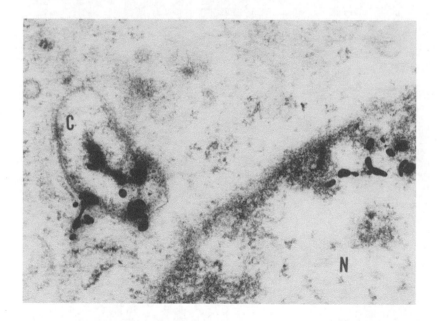

FIGURE 12.   An electronoautoradiograph of a murine L cell infected with *C. burnetii* labeled with tritiated thymidine. Silver grains pinpointing rickettsial DNA are seen in both the microorganism (C) and in the host cell nucleus (N). (From Burton, P. R., Kordova, N., and Paretsky, D., *Can. J. Microbiol.*, 17, 143, 1971. With permission.)

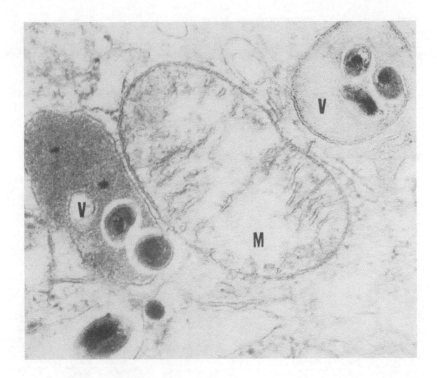

FIGURE 13. *Coxiella*-bearing vacuoles (V) in a murine spleen cell. The vacuole walls are composed of either one or two-membrane (inset) units. M, mitochondrion displaying typical two-membrane wall. Ultrathin section. (From Khavkin, T. and Amosenkova, N., in *Rickettsiae and Rickettsial Diseases*, Burgdorfer W. and Anacker, R. L., Eds., Academic Press, New York, 1981, 335. With permission.)

Observations on living cells as well as histologic and ultrastructural studies have shown a significant variability in the density of coxiellae populations and in viscosity and in volume of the liquid contents within the vacuole[48](Figure 14). Motion picture studies have shown that pinocytic vesicles may fuse with the vacuole, thus providing the vacuole with the liquid material.[48] The presence of Golgi complexes around vacuoles suggests the existence of a continuous exchange of fluid between the cell and the vacuole involving Golgi complexes. Representative sections show that some complexes are oriented by their convex surfaces toward both the vacuole with apparently viable organisms and the phagolysosomes degrading nonviable coxiellae (Figure 15). This Golgi surface is believed to produce vesicles with an acidic content.[49] They may provide the vacuole with a low pH that is favorable for *C. burnetii*.[43] The complete survival mechanisms of *C. burnetii* inside the vacuole are still unknown. As Frehel at al.[50] have pointed out, the means of transfer of nutrients to the vacuoles harboring bacteria and parasites has not been adequately investigated.

Coxiella-bearing vacuoles do not harm the host cell and do not hamper its essential functions and multiplication.[32] The vacuole protects the pathogen from antibodies and some other active substances.[51] At first, infected cells display signs of increased metabolic and functional activities (Figure 16). Eventually, however, the growth of the vacuole results in the reduction of cytoplasm and rupture of the cell. Motion picture study demonstrated that the cell responds to the presence of the vacuole by periodic contractions. This may result in temporary extrusion of the vacuole from the cell[48] (Figure 17). It is conceivable that the influx of fluid and the cell contraction

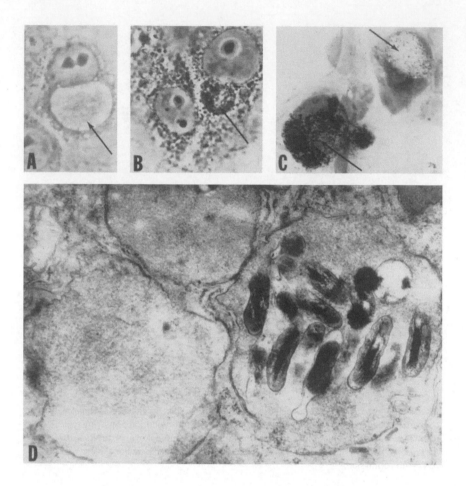

FIGURE 14.   *Coxiella*-bearing vacuoles (arrows) with different amounts of fluid and different densities of organisms, as observed using different techniques: vital examination of cultured cells (A, B), fuchsin staining of omental macrophages (C, Gimenez technique), and transmission electron microscopy of a murine spleen cell (D). In A, organisms are not clearly resolved because of Brownian movement. (A and B from Khavkin, T. and Amosenkova, N., in *Rickettsiae and Rickettsial Diseases*, Burgdorfer, W. and Anacker, R. L., Eds., Academic Press, New York, 1981, 335. With permission.)

contribute to the vacuole rupture resulting in the exit of coxiellae from the cell. Exit, as entry, is required by intracellular parasites for survival as a species.[52] Cell contractions with subsequent extrusion of the parasitophorous vacuole have also been observed in cultures infected with *Toxoplasma gondii*.[53]

2. Interaction with Mononuclear Phagocytes

The ability to resist the defense mechanisms of mononuclear phagocytes is considered the most important pathogenic feature of *C. burnetii*. These defense mechanisms can roughly be divided into oxygen-dependent, associated with activities of oxygen radicals, and oxygen-independent ones.[54] Lysosomal hydrolases represent one of oxygen-independent mechanisms. The resistance of *C. burnetii* to lysosomal hydrolases, was first noticed in animal experiments[55] (Figure 18). It has been studied in detail in cell culture experiments that demonstrated acidophilic features of *C. burnetti*. *C. burnetii* has also been shown to generate superoxide anions and exhibit activities of catalase and superoxide dismutase.[2] These enzymes represent the bacterium's means for disposal of the oxygen intermediates produced by the host cell.[56]

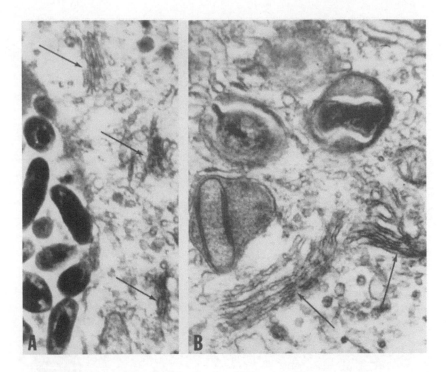

FIGURE 15. Golgi complexes (arrows) facing a vacuole with viable *C. burnetii* (A) and phagol-ysosomes with degraded microorganisms (B). (A from Khavkin, T., and Amosenkova, N., in *Rickettsiae and Rickettsial Diseases,* Burgdorfer, W., and Anacker, R. L., Eds., Academic Press, New York, 1981, 335. With permission.)

In the course of rickettsemia, *C. burnetii* inhabits mostly resident macrophages scattered along the blood and lymph vessels, the adventitial cells. It also inhabits the sinusoidal macrophages in the liver (Kupffer cells), and in the hematopoietic and lymphoid tissues.[20,41,57] The affection of adventitial cells is especially apparent in total spread-preparations of the greater omentum (Figure 19), but is also seen in sections of internal organs (Figure 20). At the same time, *C. burnetii* does not affect vascular endothelium (Figure 19) that is characteristic of some other rickettsias.[58]

The distribution pattern of *C. burnetii* in the body is similar to that of colloidal stains and various suspensions vitally injected into the bloodstream to pinpoint cells of the reticuloendo-thelial system,[59] This pattern is consistent with the concept of passive uptake of *C. burnetii* by mononuclear phagocytes,[2] because the uptake by professional phagocytes does not necessarily require specialized receptors.[60] It also indicates that the resident macrophages scattered along the blood and lymph vessels present the prime target for *C. burnetii*, along with stromal and covering cells in sinuses of the hematopoietic and lymphoid tissues.

In infectious and inflammatory foci, *C. burnetii* also inhabits exudate macrophages. Less is known about the immediate fate of coxiellae in peripheral blood monocytes. These cells possess a strong antimicrobial capacity and can kill pathogens such as *T. gondii, Leishmania donovani, and Chlamydia psittaci*, which also are capable of residing in resident macrophages.[54]

The fate of *C. burnetii* in mononuclear phagocytes *in vivo* in the presence of developing immunity is less well known. Cell culture experiments have shown that coxiellae readily multiply within and eventually destroy unstimulated peritoneal macrophages, but succumb to the antimicrobial mechanisms of immunologically activated cells,[51,61] and that gamma-inter-feron suppresses multiplication of the pathogen in fibroblasts and macrophages.[62,63] *In vivo,*

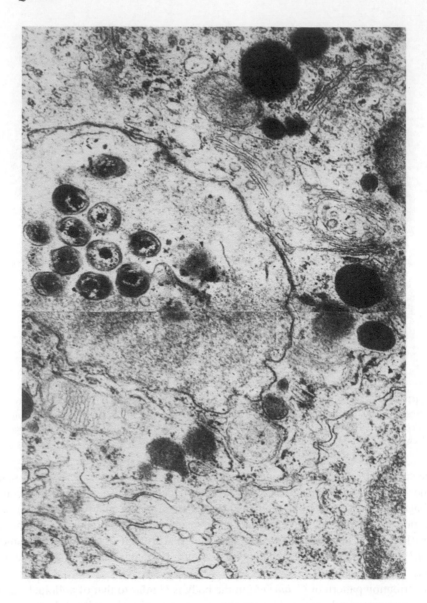

FIGURE 16. Infected murine splenic cell with ultrastructural signs of activation: well-develped mitochondria, polysomes, and Golgi complexes, some facing the *Coxiella*-bearing vacuole. Cytoplasmic projections interwoven with the adjacent cell are consistent with those of interdigitating cells. (From Khavkin, T. and Amosenkova, N., in *Rickettsiae and Rickettsial Diseases,* Burgdorfer, W. and Anacker, R. L., Eds., Academic Press, New York, 1981, 335. With permission.)

however, the *Coxiella*-macrophage interactions appear to be more complicated than in *in vitro* systems. Thus, the continuing persistence of *C. burnetii* in animals can be detected for several months after acute infection or immunization.

The length and extent of the microorganism's proliferation are different in different body areas and upon challenge of animals with different *Coxiella* strains. One can speculate that in the body, the *Coxiella*-macrophage interaction is influenced by at least three groups of factors: (1) macrophage heterogeneity with regard to their enzymatic and bactericidal activities;[64,65] (2) diversity of extracellular factors, such as antibodies, lymphokines, complement fractions which

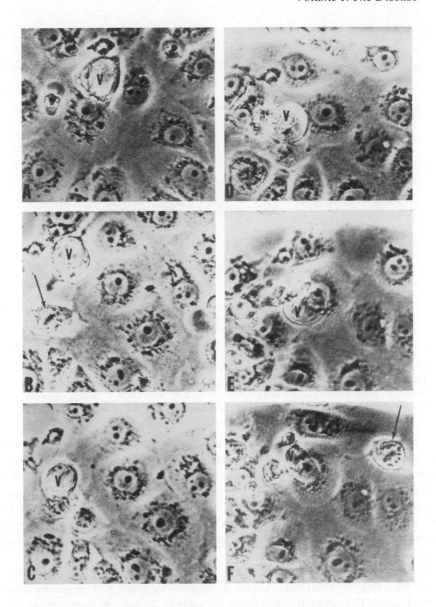

FIGURE 17. Frames from a motion picture showing sequential steps of contraction and relaxation of the same infected cell *in vitro*. Temporary extrusion of the *Coxiella*-bearing vacuole (V) by the host cell is seen. Arrows show mitotic divisions of some noninfected cells. Each division lasted for 60 to 65 min. (From Khavkin, T. and Amosenkova, N., in *Rickettsiae and Rickettsial Diseases*, Burgdorfer, W. and Anacker, R. L., Eds., Academic Press, New York, 1981, 335. With permission.)

may either enhance or suppress degradation of the pathogen; and (3) diverse capability of *C. burnetii* strains to modulate cell-mediated immunity. The immunomodulating capability of *C. burnetii*, and specifically negative modulation, has recently been demonstrated in clinical studies and in animal experiments.[66,67] This feature, which is discussed by Jerrelis[63] and in Chapter 5, can be considered an important determinant of pathogenicity and virulence of *C. burnetii* strains. It might contribute to the survival of the pathogen in the body even when immunity is well established.

The mechanism of coxiellae killing by stimulated macrophages presents an as yet unresolved

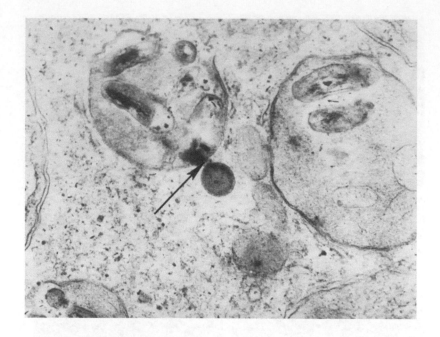

FIGURE 18. Murine splenic cell with the *Coxiella*-bearing vacuole apposed by the lysosome. The extension of an osmiophilic lysosomal contents to the vacuole (arrow) is suggestive of lysosome-vacuole fusion. Ultrathin section. (From Ariel, B. M., Khavkin, T. N., and Amosenkova, N. I., *Pathol. Microbiol.*, 39, 412, 1973. With permission.)

problem. The mechanism most likely, resides outside the activity of lysosomal hydrolases. Thus, Kishimoto et al.[51,61] demonstrated that activated macrophages which can kill coxiellae do not express increased activities of lysosomal enzymes. We have observed some *Coxiella*-macrophage patterns, suggesting the existence of bacteriostatic (i.e., oxygen-independent) mechanisms that may interfere with the coxiellae growth. These patterns are (1) the loss of structural integrity by microorganisms confined to the well-developed *Coxiella*-bearing vacuole in activated macrophages, and (2) the appearance of filamentous forms of the microorganism.

The loss of structural integrity was observed in alveolar macrophages[20] (Figure 21A). This pattern was quite different from that of the intralysosomal degradation of apparently nonviable organisms (Figure 21B), suggesting that inside the vacuole, *C. burnetii* started dying after the period of multiplication. The appearance of filamentous bacterial forms is believed to result from the damage to the replication of the bacterium's DNA.[68] It leads to incomplete cytokinesis of the bacterium. Some rickettsia, such as *R. akari* and *R. prowazekii* not infrequently display filamentous forms under unfavorable conditions. Thus, Kokorin[69] demonstrated the presence of filamentous forms of *R. akari* in cell culture by motion picture photomicroscopy. The filaments were promptly broken into regular rod-shaped rickettsial forms after the culture medium was replaced with fresh medium. *C. burnetii*, which is more resistant than other rickettsiae to the environmental pressure, rarely displays filamentous forms. We observed these forms in both exudate and resident macrophages in guinea pigs, 2 weeks post challenge with phase II *C. burnetii* (Figure 22). Possible involvement of gamma-interferon or other factors in the suppression of the *Coxiella* survival wthin macrophages is to be studied.

Finally, one can speculate that certain types of activated macrophages overhelm the ability of *C. burnetii* to dispose of oxygen radicals produced by the host cell. Thus, killing of *T. gondii* by stimulated macrophages has been shown to involve both oxygen-dependent and independent mechanisms, although this pathogen, like *Coxiella*, possesses antioxidant activities.

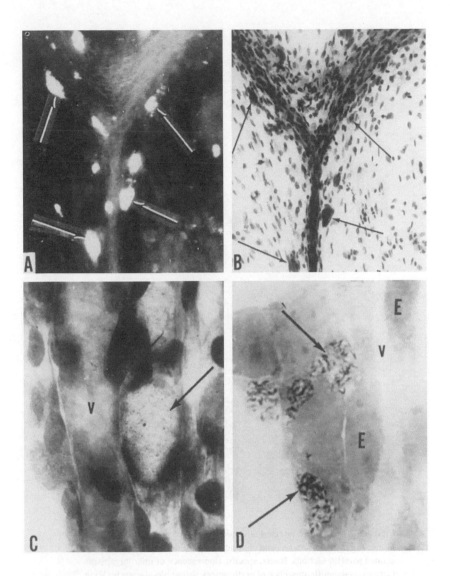

FIGURE 19. *Coxiella*-infected adventitial resident macrophages (arrows) in the guinea pig greater omentum. Overview of the same total spread preparation treated with fluorescent anti-*Coxiella* antibody (A) and then Giemsa stained (B). Specific fluorescence of *Coxiella* colonies along the blood vessel is seen in A. In a venule (C) and capillary (D) only adventitial cells contain microorganisms. Vascular endothelium (E) is not involved. V, vascular lumen. C, Giemsa stain; D, basic fuchsin stain after Gimenez. (From Khavkin, T. N., and Amosenkova, N. I., *Proc. Pasteur Inst. Leningrad*, 29, 222, 1965. With permission.)

3. Interaction with Nonprofessional Phagocytes

The full spectrum of nonphagocytic cells capable of harboring *C. burnetii* is not yet known. In acute infection, typical *Coxiella*-bearing vacuoles were observed in epithelial, mesothelial (Figure 23), and connective tissue cells as well as in yet unspecified stromal cells of the lymphoid and hematopoietic tissues (Figures 8—10). All these cell types collectively referred to as nonprofessional phagocytes express an apparent endocytic activity.[60] It is unclear, however, whether this activity relates to the susceptibility of the cell to *C. burnetii*. Vascular endothelium, which also possesses endocytic activity, and is capable of harboring other rickettsias[58] does not harbor *C. burnetii*. Factors that promote the entry of *C. burnetii* into nonphagocytic cells, such

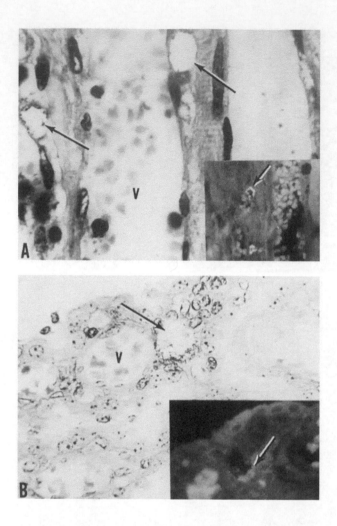

FIGURE 20. *Coxiella*-infected adventitial macrophages (arrows) in the guinea pig myocardium (A) and choroid plexus of the brain (B). Giemsa-stained paraffin sections. Insets, specific fluorescence of microorganisms (arrows) and autofluorescence of erythrocytes within blood vessels (V) in the same paraffin sections treated with fluorescent anti-*Coxiella* antibody.

as specific receptors, and the ultimate fate of the pathogen within these cells require further investigation.

4. Latency and Recrudescence

Latency is observed in all groups of infections, viral, rickettsial, bacterial, and protozoal. At the same time, latency, perhaps, is studied least in infectious pathology, primarily because of technical difficulties. Latency becomes apparent in retrospect, after recrudescence of the acute infection. Recrudescence of *C. burnetii* infection has clinically been observed in immunocompromised patients or provoked in experimental animals by suppression of cell-mediated immunity, which most likely controls the latency. From the point of view of morphogenesis, however, immunosuppression presents only one aspect of the problem. The other aspect implies structural and metabolic adaptation of the pathogen to dormant life. An example of such adaptation is the transformation of *Toxoplasma* tachyzoites into bradizoites during conversion of acute infection into a chronic and eventually into a latent one.[70] Whether a spore-like form

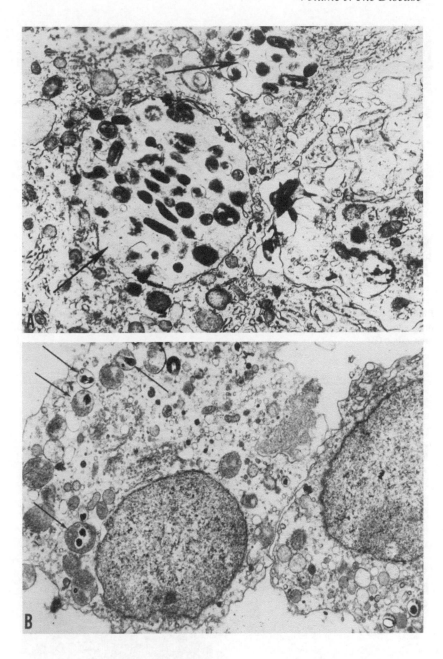

FIGURE 21. Murine alveolar macrophages with *Coxiella*-bearing vacuoles (A, arrows) and phagolysosomes containing degraded microorganisms (B, arrows). In A, most intravacuolar microorganisms have lost their structural integrity. Macrophages display numerous organelles, vesicles, and lysosomes indicating the activation state of the cell.

of *C. burnetii* described by McCaul and Williams[37] represents a dormant form of this microorganism, and whether the cell type harboring dormant coxiellae is a specific one, still needs to be established.

One can speculate that cells harboring dormant coxiellae are long-lived nonphagocytic cells rather than mononuclear phagocytes which have a short lifespan.[64] In this regard, stromal cells of the hematopoietic and lymphoid tissues deserve special attention. These cells also known under the common name limbocytes[71] comprise dendritic, interdigitating, veiled cells belonging

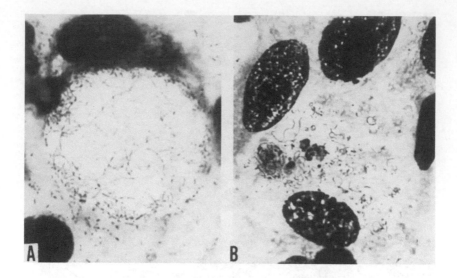

FIGURE 22. Filamentous forms of *C. burnetii* in the guinea pig exudate (A) and resident (B) macrophages, 12 d post challenge. Giemsa-Pappenheim stain of a smear (A) and a total omental preparation (B).

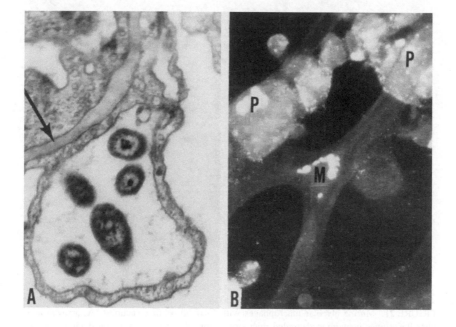

FIGURE 23. Electron microscopic (A) and immunofluorescent (B) images of murine nonphagocytic cells harboring *C. burnetii:* (A) type I pneumocyte; (B) mesothelial cell (M) in a total omental preparation. Arrow, pulmonary basal membrane separating pneumocytes and a vascular endothelium. P, peritoneal macrophages with specifically fluorescing organisms. (A from Khavkin, T. and Tabibzadeh, S., *Infect. Immun.*, 56, 1792, 1988; B, from Khavkin, T. N. and Amosenkova, N. I., *Proc. Pasteur Inst. Leningrad*, 25, 160, 1963. With permission.)

to the mononuclear phagocyte system. Actually, they are the nonphagocytic, nonhematopoietic cells with a comparatively long lifespan.[71,72] Ultrastructural features of some *Coxiella*-bearing stromal cells in lymph nodes and spleen (Figure 8) are consistent with those of limbocytes. It is worthwhile to remember that the existence of a latent typhus infection in man had been proved by Price[73] by the isolation of *R. prowazekii* from the lymph nodes of an individual with a history of epidemic typhus. Furthermore, *T. gondii* and *Mycobacterium leprae,* the parasites of mononuclear phagocytes, may also inhabit nonphagocytic cells with a long lifespan, such as muscles and neural and Schwann cells, when acute infection converts into chronic form and latency.

It is also to be established what cell type of the female genital system, such as those from myo- and endometrial tissues, are able to harbor dormant coxiellae and present a source of the infection of conceptus. Such a possibility is suggested by the isolation of *C. burnetii* from the feline uterus.[74] It has not been confirmed by morphological studies.

D. LOCAL AND SYSTEMIC PHENOMENA
1. Infectious Foci and Inflammatory Response

C. burnetii-induced infectious foci appear in tissues and organs when the pathogen starts multiplying within susceptible cells. At first, the infectious foci are not associated with an inflammatory response, since coxiellae confined to the intracellular vacuole do not exert chemotactic features. Transient influx of polymorphonuclear leukocytes (PMN) is observed only at the site of inoculation. The influx subsides in 1 or 2 d post challenge, when the inoculation site is cleared of extracellular microorganisms by a T-cell-independent mechanism, largely, by PMNs.

The inflammatory response develops upon release of microorganisms from destroyed host cells (Figure 24). In guinea pigs, this coincides with systemic phenomena, such as fever, and diffuse affection of the liver (Figure 3). The incubation period between challenge and onset of systemic events ranges from one to several days, depending on the inoculum dose and the virulence of a given *Coxiella* strain. A similar incubation-like period is also observed in guinea pigs inoculated with killed coxiellae.[75] This suggests that both local and systemic effects of *C. burnetii* require processing of microorganisms by inflammatory cells to release substances, such as LPS, which can trigger development of fever and other acute phenomena.

In most areas of the body, the inflammatory response to either living or killed coxiellae is dominated by macrophages,[20,41,75] with influx of lymphocytes, largely T cells. An admixture of PMNs is observed mostly in the vicinity of destroyed host cells. In areas that are rich in loose connective tissue and blood vessels, such as the dermis and hypodermis, the inflammatory response involves edema and hemorrhage.[57] This is similar, though less extensive, to the local effect of LPS from other Gram-negative bacteria, indicating that in some areas *C. burnetii* and its antigens are capable of triggering a cascade of local disorders involving interleukin 1, vasoactive and other substances associated with this effect (reviewed by Cybulski et al.)[76] From this point of view, dermal and hypodermal infectious foci in guinea pigs present a suitable model for studying a full spectrum of events in the acute inflammatory response elicited by *C. burnetti*.

The tissue damage in the infectious foci appears to be moderate and transient. It may involve noninfected parenchymal cells and vascular endothelium (Figure 25A), as well as phagocytic cells which degrade microorganisms (Figure 25B). The cell and tissue damage in various inflammatory foci is known to be triggered by a multitude of substances, such as LPS, oxygen radicals and lysosomal enzymes released from PMN, as well as by mediators of inflammation.[77,78] The specific mechanism of the cell damage in *C. burnetii* infectious foci is not known. One can speculate that *Coxiella* LPS is involved in indirect cell damage, since it occurs in the vicinity of free coxiellae and phagocytes which degrade microorganisms (Figure 25A). At the same time, macrophages and PMNs degrading coxiellae can be damaged directly. The presence of autophagosomes and other signs of cell damage in the vicinity of phagolysosomes with coxiellae (Figure 25B) suggest that these cells are affected by LPS released from degraded microorganisms. Similar autophagosome formation has also been found in granulocytes that degrade *R. prowazekii*[79] and *R. tsutsugamushi*.[80] The most extensive inflammatory response and

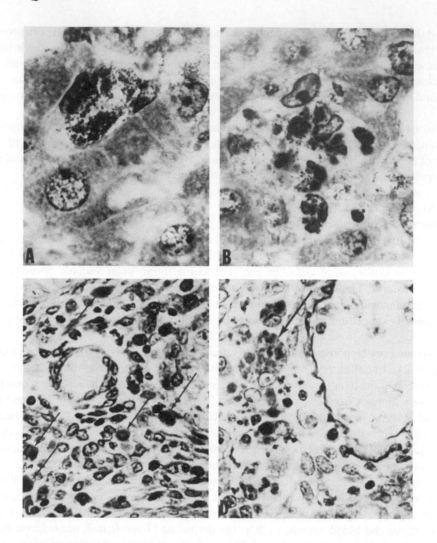

FIGURE 24.   Sequential steps of the development of infectious foci in the guinea pig liver (A, B,) and testicles (C, D), 8 (A—C) and 17 (D) d post intratesticular challenge. There are no inflammatory cells in the vicinity of nondamaged infected Kupffer cells (A), or only mononuclear exudate cells between infected macrophages (arrows) in the testicular interstitium (C). Extensive influx of leukocytes is seen in the vicinity of a destroyed Kupffer cell (B) and exudate macrophages (D). (A and B from Khavkin, T. N., *Arkh Patol. (Mosc.)*, 2, 75, 1977. With permission.)

tissue damage is observed in the heavily infected placenta. In placenta, which is rich in loose connective tissue, the inflammatory response is associated with severe edema and thrombo-hemorrhagic lesions.[81] These, most likely, are the cause of abortion in animals with intrauterine *C. burnetii* infection.

Since macrophages dominate and there is extensive influx of T-cells, the infectious and inflammatory foci resemble a delayed hypersensitivity response which is indicative of the involvement of cell-mediated immunity in the response. Cellular and humoral factors involved in this type of response, and its morphogenesis were reviewed by Dvorak et al.[82] and Kaufman et al.[83] We have observed that 4 and 5 d postchallenge, most lymphocytes in pulmonary infectious foci in mice have ultrastructural features of T-cells and granular lymphocytes.[21] This last type of cell comprises both cytotoxic T-and natural killer cells.[84] Some of granular and

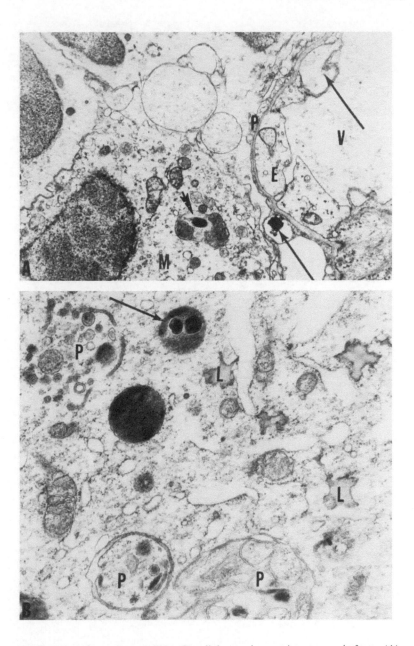

FIGURE 25. Indirect (A) and direct (B) cell damage in a murine pneumonic focus. (A) Damage to the vascular endothelium (E) and pneumocyte (P) in the vicinity of a macrophage (M) containing lysosome with a degraded coxiella (arrowhead). Edematous endothelial cell and a partially denuded pneumocyte display blister-like vesicles (arrows). V, blood capillary. B, alveolar macrophage with a degraded organism within phagolysosome (arrow), and autophagosomes (P) at different steps of their formation. Lipid dropletts (L), rarefied cytoplasmic matrix, and a widened endoplasmic reticulum are seen. Ultrathin sections. (From Khavkin, T. and Tabibzadeh, S., *Infect. Immun.*, 56, 1792, 1988. With permission.)

nongranular lymphocytes were closely attached to the *Coxiella*-bearing cells (Figure 26). This attachment pattern is similar to that of conjugated cytolytic lymphocytes and target cells *in vitro*.[85] Similarly, a close association of suppressor and cytotoxic lymphocytes with infected cells have been demonstrated by Kaplan et al.[86] in human lepromatous lesions in the process of

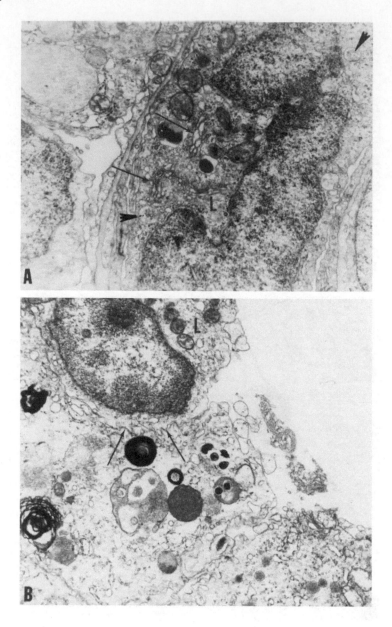

FIGURE 26. Lymphocytes (L) with ultrastructural features of granular lymphocytes (A) and T cells without granules (B) in a murine pneumonic focus. A, intravascular lymphocyte displays a reniform euchromatic nucleus, developed Golgi complexes (arrows), and microvesicular bodies (arrowheads). (B) a T cell attached to a macrophage containing phagolysosomes with degraded coxiellae. Arrows, interdigitating lymphocyte-macrophage contact area. (From Khavkin, T. and Tabibzadeh, S., *Infect. Immun.*, 56, 1792, 1988. With permission.)

healing, suggesting that *in vivo* these cells are directly involved in the tissue clearance of the pathogen.

In *C. burnetii* infectious foci, exact phenotype and kinetics of T-cell clones and their actual functional significance have not yet been identified. As Mullbacher and Ada[87] have pointed out, *in vivo* the same populations of T- and natural killer cells may exert diverse cytolytic, secretory, helper, and suppressor activities.

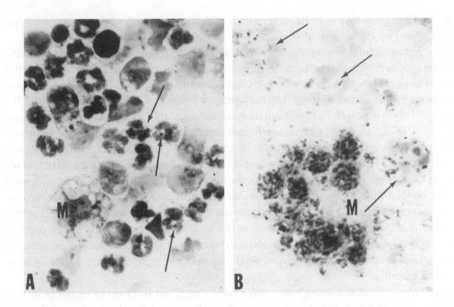

FIGURE 27. Phase I *C. burnetii* in polymorphonuclear leukocytes (PMN, arrows) and macrophages (M) in primary infectious foci 5 h (A) and 17 d (B) post challenge. (A) Giemsa-Pappenheim-stained omental exudate; single organisms are seen in PMNS and several ones in the macrophage. (B) basic fuchsin stained (Gimenez technique) testicular exudate; *Coxiella* colonies in the macrophage and an extensive phagocytosis of organisms by PMN are seen. (A from Khavkin, T. N. and Amosenkova, N. I., *Proc. Inst. Pasteur Leningrad*, 25, 160, 1963. With permission.)

The role of the PMN in *C. burnetii* infectious foci appears to be more understandable. The PMNs which involve both oxygen-dependent and independent antimicrobial mechanisms[88] effectively kill ingested coxiellae. Because unopsonized phase I *C. burnetii* are poorly taken up by phagocytes including PMNs,[46] these cells ingest and digest only a fraction of microorganisms at the site of inoculation (Figure 27A). Much more effectively, PMN ingest and degrade coxiellae, 2 and 3 weeks after challenge, during the mass destruction of parasitized host cells (Figure 27B). Coincidence of massive tissue clearance by PMNs with rising antibody titers (Figure 3) indirectly indicates that opsonization indeed promotes phagocytosis of *C. burnetii* by PMNs *in vivo*.

C. burnetii-induced inflammation has been designated a granulomatous one, without specification, however, of the granuloma type. Meanwhile, Racz and Tenner-Racz[89] counted at least 12 morphological types and subtypes of infectious granulomas, described in the literature: organized and unorganized, hypo- and hyperergic granulomas. Kinetics and composition of each granuloma type depends upon the pathogenic features of the agent and the immunologic status of the host. According to Kindler at al.,[90] cellular composition, especially the appearance of highly microbicidal epithelioid cells, reflects a degree of maturity of the infectious granulomas. Therefore, understanding the structural features of granuloma is important for the understanding of the pathogenesis of the particular disease.[78] Histologic studies of affected liver that is considered a suitable model of inflammatory lesions[91] have shown that in *C. burnetii* infection, granulomas appeared as unorganized cell collections which are largely composed of macrophages and lymphocytes.[20,92] In subcutaneous loose connective tissue, the inflammatory response evolves into the granulation tissue. Both liver granulomas and the subcutaneous granulation tissue eventually disappear without leaving a scar. These pathological features are comparable to transient unorganized granulomas caused by *Listeria* or *Salmonella*[93] rather than with organized epithelioid-cell granulomas typical of mycobacterial infections. The latter are considered the most mature granulomas.[90] They may form chronic lesions with alternating scarring and necrosis that is not characteristic of inflammatory lesions caused by *Listeria, Salmonella,* and by *C. burnetii* also.

Further ultrastructural, immunomorphological and kinetic studies are needed to specify the granuloma type in *C. burnetii* infection. These studies would also help comprehend whether *C. burnetii* granulomas evolve into chronic lesions. Chronic lesions are believed to underlie the pathogenesis of so-called chronic Q fever that is linked to Q fever endocarditis.[94] The development of chronic lesions implies that conditions exist that suppress immunologic control of the intracellular parasitism of *C. burnetii* and support its active multiplication in infectious foci. Q fever endocarditis has been shown to be, indeed, associated with a specific immunologic defect in patients.[67] However, without favorable local conditions, this is hardly sufficient for the formation of chronic lesions. One can hypothesize that preexisting chronic lesions, such as rheumatic endocarditis, atherosclerosis, valvular malformations, as well as valvular prostheses, rather than specific rickettsial granulomas, provide the local conditions required for the growth of *C. burnetii* in the heart valves. According to the review by Sawyer et al.,[94] either preexisting valvular lesions or prosthetic devices underlie most cases of Q fever endocarditis. Presence of the acute phase proteins in the patients' blood may contribute to the colonization of affected valves with coxiellae.

The so-called doughnut granulomas have been described in Q fever hepatitis in man as specific lesions. Actually, these granulomas, the collections of inflammatory cells oriented around lipid drops, represent nonspecific resorptive lipogranulomas rather than specific organized infectious granulomas. Lipogranulomas are observed in a variety of liver lesions unrelated to Q fever.[95] Animal experiments have shown that in acute Q fever, focal infectious lesions are superimposed by the diffuse lipidosis of the liver.[20] The resorption of lipid droplets released from hepatocytes rather than pathogenic features of *C. burnetii* underlie the pathogenesis of doughnut granulomas in Q fever.

2. Immunopathological Phenomena

Laboratory immunopathological phenomena of either passive cutaneous anaphylaxis[96] or delayed hypersensitivity[97] type have been produced in the guinea pig demonstrating the immunogenic and allergenic features of *C. burnetii* antigens. There are no indications, however, that these phenomena reflect the real process which might develop in the course of *C. burnetii* infection. Specific lesions of proved immunopathological nature have not yet been described in uncomplicated Q fever. The rickettsial etiology in cases of so-called Q fever glomerulonephritis in man has not been convincingly confirmed (reviewed by Leedom).[98] Although transient skin rash is occasionally observed in Q fever in man,[94] the destructive thrombovasculitis that is typical of some other rickettsioses[58] is uncommon in Q fever. Such a thrombovasculitis is believed to be associated with immune complexes deposited in the wall of the blood vessels.[99,100]

Both circulating and particulate immune complexes have been detected in the experimental *C. burnetii* infection. The circulating complex was observed in the guinea pig as a transient event.[101] As Hoibi et al.[102] and Couser[103] have pointed out, however, the mere appearance of circulating complex during any infection, and even its transient deposition in tissues, does not necessarily result in the immunopathological lesions. Evidence is needed for the interaction of circulating complexes with structural components of the kidney, especially with basal membranes, and with the phagocytic cells and the complements.[104] Current experimental data concerning the possible pathogenic effect of the circulating immune complex in tissue lesions in Q fever are conflicting. Yu et al.[105,106] have observed in guinea pig kidneys both transient depositions of circulating immune complexes and those containing complement fractions, which were associated with inflammatory glomerular lesions. In our experiments, nephritic glomerular lesions have not been found even in the presence of infectious foci in the kidney (Figure 9). Further studies are required to clarify whether tissue lesions associated with circulating immune complexes do really exist in *C. burnetii* infection.

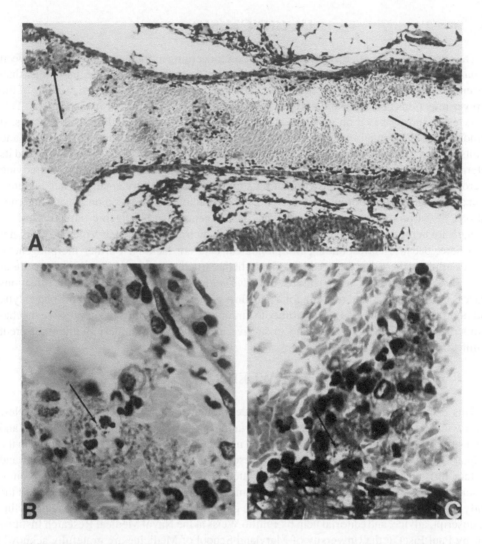

FIGURE 28.   Thrombocytic thrombi formed around agglutinated coxiellae (small arrows) within splenic vein of the guinea pig, 9 d postchallenge. (A) Overview; (B and C) details marked with large arrows. Paraffin section, Giemsa-Pappenheim stain.

The particulate immune complex has been observed within splenic and portal vessels in mice and guinea pigs. The complex resulted from the interaction of coxiellae and immunoglobulins which were simultaneously shed by respective spleen cells (Figure 2). This, in turn, led to the formation of microthrombi involving coxiellae and aggregated platelets without damage to the blood vessel (Figure 28). Such thrombi are typical of the blood clearance of circulating microorganisms[42] and, by themselves, cannot be considered an immunopathological event. However, in a severe *C. burnetii* infection, thrombi in the splenic and portal circulations underlie the pathogenesis for ischemic infarct-like necrotic foci that have been described in the liver and spleen of mice.[13,22] In this case, the particulate immune complex with the thrombosis represents an immunopathological event of a regional significance. It is not yet clear whether this particulate immune complex involves complement. Further studies are also needed to show whether the regional thrombosis in *C. burnetii* infection described here relates to the endotoxin-mediated allergic Schwartzman reaction.[107]

IV. CONCLUSION

The attempt to summarize available information concerning immediate interactions between *C. burnetii* and its arthropod, vertebrate, and "cellular" hosts provide fragmentary rather than systematic data. Still, the review, hopefully, attracts attention to morphogenetic studies of the Q fever infectious processes and outlines some areas that deserve further investigations. This overview is mostly based on experiments using phase I *C. burnetii* and nonbred animals. It demonstrates an essential species diversity in the response to *C. burnetii* and to its antigens and points out the necessity to "prove relevance of the biological property (of the pathogen) and its determinant to infection *in vivo*".[108] It does not meet contemporary requirements for studying microbial pathogenicity, such as comparing strains of different virulence genetically determined and animals of different responsiveness to the pathogen also based on genetic determinations.[108] Currently, essential observations on the genetic and antigenic variability of *C. burnetii* are being accumulated. These observations which are reviewed in Volume 2 were not used here, because corresponding morphological data are lacking. Further morphogenetic studies utilizing these data are needed to highlight the structural, genetic, chemical, and cellular bases for two major pathogenic features of *C. burnetii:* (1) its ability to initiate the infectious process by entering into and residing for a long time within phagolysosomes of both phagocytic and nonphagocytic cells, and (2) its ability to circumvent and/or adjust to suppressive mechanisms of cell-mediated immunity and to survive in the body indefinitely, without apparent manifestations.

ACKNOWLEDGMENTS

This chapter achieved its present form under support of Interferon Sciences, Inc., New Brunswick, NJ. A segment of my doctoral dissertation on the rickettsial infectious processes and the results of some related experiments that had not previously been published were used in the chapter. These studies were carried out at the Research Institute for Experimental Medicine and Pasteur Institute, Leningrad, U.S.S.R. I am grateful to the Russian pathologist M. V. Voino-Yasenetsky whose concepts on the combined morphogenetic and biological approach to the study of infectious processes inspired me while preparing this chapter. Critical reading of this manuscript, advice, and editorial help by Emilio Weiss at the Naval Medical Research Institute and by Paul Fiset at the University of Maryland School of Medicine are gratefully acknowledged. I am indebted to Edmund Evanovsky (Interferon Sciences, Inc.) for his library help.

REFERENCES

1. **Voino-Yasenetsky, M. V.,** General principles for the study of the pathogenesis of intestinal infections, in *Pathogenesis of Intestinal Infections. Microbiological and Pathological Principles,* Voino-Yasenetsky, M. V. and Bakacs, T., Eds., Akademiai Kiado, Budapest, 1977, 13.
2. **Baca, O. G. and Paretsky, D.,** Q fever and *Coxiella burnetii:* a model for host-parasite interaction, *Microbiol. Rev.,* 47, 127, 1983.
3. **Walker, D. H.,** Pathology of Q fever, in *Biology of Rickettsial Diseases,* Vol. 2, Walker, D. H., Ed., CRC Press, Boca Raton, FL, 1988, 17.
4. **Smith, H.,** The determinants of microbial pathogenicity, in Essays of Microbiology, Richmond, M. H. and Norris, T., Eds., Wiley & Sons, Chichester, England, 1978, 2.
5. **Wilson, G. S. and Miles, A.,** The mechanisms of bacterial infections, in *Topley and Wilson's Principles of Bacteriology, Virology and Immunity,* Vol. 2, Williams & Wilkins, Baltimore, 1975, 1273.
6. **Philip, C. B.,** Recent advances in knowledge of tick-associated rickettsia-like organisms, *J. Egypt Public Health Assoc.,* 38, 61, 1963.

7. **Balashov, Yu. S.,** The organism of ticks as a habitat of the agents of arthropod-borne diseases, in *Parazitologichesky Shornik* [Parasitol. Articles], Vol. 34, Nauka, Leningrad, 1987, 48.

8. **Rehácek, J.,** Some aspects of ecology of rickettsie, in *Ricketsia and Rickettsial Diseases,* Kazár, J., Ormsbee, R. A., and Tarasevich, I. N., Eds., VEDA, Bratislava, 1978, 475.

9. **Balashov, Yu. S., Daiter, A. B., and Khavkin, T. N.,** Distribution of Burnet's rickettsiae in the tick *Hyalomma asiaticum, Parazitologiya,* 6, 22, 1972.

10. **Khavkin, T. N., Daiter, A. B., and Berlin, L. B.,** Skin lesion in the area of bite by the tick infected with *Rickettsia burnetii,* in *Proc. Pasteur Inst. Epidemiol. Microbiol.,* Vol. 41, Pasteur Institute, Leningrad, 1972, 55.

11. **Raikhel, A. S.,** Intestine, in *An Atlas of Ixodid Tick Ultrastructure,* Balashov, Yu. S., Ed., Entomological Society of America, 1983, 59.

12. **Hall, W. C., White, J. D., Kishimoto, R. A., and Whitmire, R. R.,** Aerosol Q-fever infection of the nude mouse, *Vet. Pathol.,* 18, 672, 1981.

13. **Scott, G. H., Williams, J. C., and Stephenson, E. H.,** Animal models of Q-fever: pathological responses of inbred mice to phase I *Coxiella burnetii, J. Gen. Microbiol.,* 133, 691, 1987.

14. **Galanos, C., Freudenberg, M. A., and Reutter, W.,** Galactosamine-induced sensitization to the lethal effects of endotoxin, *Proc. Natl. Acad. Sci.,* 76., 5939, 1979.

15. **Amano, K.-I., Williams, J. C., Missler, S. R., and Reinhold, V. N.,** Structure and biological relationships of *Coxiella burnetii* lipopolysaccharides, *J. Biol. Chem.,* 262, 4740, 1987.

16. **Moore, R. N., Shackleford, G. M., and Berry, L. J.,** Glucocorticoid-antagonizing factor, in *Cellular Biology of Endotoxin,* Berry, L. G., Ed., Elsevier, Amsterdam, 1985, 123.

17. **Khavkin, T. N. and Amosenkova, N. I.,** Toxic effect of *Rickettsia burnetii* on adrenalectomized albino mice., *Bull. Exp. Biol. Med.,* 2, 73, 1974.

18. **Amosenkova, N. I., Khavkin, T. N., Kudrjavtseva, M. V., and Kudrjavtsev, B. N.,** Toxic features of *Rickettsia burnetii,* in *Proc. Pasteur Inst. Epidemiol. Microbiol.,* Vol. 41, Pasteur Institute, Leningrad, 1972, 38.

19. **Schramek, S., Brezina, R., Kazár, J., and Khavkin, T.,** Attempts at demonstration of lipopolysacharide in phase II *Coxiella burnetii, Acta Virol.,* 22, 509, 1978.

20. **Khavkin, T.,** Comparative Study of Infectious Processes Induced by *Rickettsia burnetii* and *Rickettsia prowazekii.* Doctoral Dissertation (Sci. Med.), Res. Inst. Exp. Med. and Pasteur Inst., Leningrad, U.S.S.R., 1972.

21. **Khavkin, T. and Tabibzadeh, S.,** Histologic, immunofluorescence, and electron microscopic study of infectious process in mouse lung after intranasal challenge with *Coxiella burnetii,* Infect. Immun., 56, 1792, 1988.

22. **Williams, J. C. and Cantrell, J. L.,** Biological and immunological properties of *Coxiella burnetii* vaccines in C57BL/10ScN endotoxin-nonresponder mice, *Infect. Immun.,* 35, 1091, 1982.

23. **Schramek, S., Kazár, J., Sekeyova, S., Freudenberg, M. A., and Galanos, C.,** Induction of hyperreactivity to endotoxin in mice by *Coxiella burnetii, Infect. Immun.,* 45, 713, 1984.

24. **Peavy, D. L., Baughn, R. E., and Musher, D. M.,** Mitogenic activity of bacterial lipopolysaccharides *in vivo:* morphological and functional characterization of responding cells, *Infect. Immun.,* 19, 71, 1978.

25. **Geldof, A. A., Rijnhart, P., Vanderende, M., Kors, N., and Langevoort, H. L.,** Morphology, kinetics and secretory activity of antibody–forming cells, *Immunobiology,* 166, 296, 1984.

26. **Melchers, F. and Andersson, J.,** B cell activation: three steps and their variations, *Cell,* 37, 715, 1984.

27. **Williams, J. C., Damrow, T. A., Waag, D. M., and Amano, K.-I.,** Characterization of a phase I *Coxiella burnetii* chlorophormmethanol residue vaccine that induces active immunity against Q fever in C57BL/10ScN mice, *Infect. Immun.,* 51, 851, 1986.

28. **Dinarello, C. A.,** Interleukin-1 and the pathogenesis of the acute phase response, *N. Engl. J. Med.,* 311, 1413, 1984.

29. **Powanda, C. M., Machotka S. Y., and Kishimoto, R. A.,** Metabolic sequelae of respiratory Q fever in the Guinea pig, *Proc. Soc. Exp. Biol. Med.,* 158, 1978, 626.

30. **Bradley, S. G.,** Interactions between endotoxin and protein synthesis, in *Cellular Biology of Endotoxin,* Berry, L.G., Ed., Elsevier, Amsterdam, 1985, 340.

31. **Wisseman, Ch. L., Jr,** Selected observations on rickettsiae and their host cells, *Acta. Virol.,* 30, 81, 1986..

32. **Moulder, J. W.,** Comparative biology of intracellular parasitism, *Microbiol. Rev.,* 49, 298, 1985.

33. **Baca, O. G. and Grissman, H. A.,** Correlation of DNA, RNA, and protein content by flow cytometry in normal and *Coxiella burnetii*-infected L929 cells, *Infect. Immun.,* 55, 1731, 1987.

34. **Weiss, E.,** History of rickettsiology, in *Biology of Rickettsial Diseases,* Vol. 1, Walker, D. H., Ed., CRC Press, Boca Raton, FL, 1988, 16.

35. **McDonald, T. L. and Mallavia, L. P.,** Host response to infection by *Coxiella burnetii, Can. J. Microbiol.,* 21, 675, 1975.

36. **Stoker, M. G. P and Fiset, P.,** Phase variation of the Nine Mile and other strains of *Rickettsia burnetii, Can. J. Microbiol.,* 2, 310, 1956.

37. **McCaul, T. E. and Williams, J. C.,** Developmental cycle of *Coxiella burnetii.* Structure and morphogenesis of vegetative and sporogenic differentiations, *J. Bacteriol.,* 147, 1063, 1981.

38. **Khavkin, T., Sukhinin, V., and Amosenkova, N.,** Host-parasite interactions and development of infraforms in chicken embryos infected with *Coxiella burnetii* via the yolk sac, *Infect. Immun.,* 32, 1281, 1981.

39. **Winkler, H. H.,** Early events in the interaction of the obligate intracellullar parasite *Rickettsia prowazekii* with eukaryotic cells: entry and lysis, *Ann. Inst. Pasteur Microbiol.,* 137A. 333, 1986.

40. **Gutzu, E., Sudran, C., and Timofte, V.,** La lymphadénite mésentérique primitive rickettsienne, *Rev. Roum. Inframicrobiol.,* 6, 159, 1969.

41. **Rychlo, A. and Pospisil, R.,** Zur Morphologie und Pathogenese des experimentellen Q-Fiebers beim Meerschweinchen, *Pathol. Microbiol.,* 23, 489, 1960.

42. **Donald, K.,** Ultrastructure of reticuloendothelial clearance, in *The Reticuloendothelial System. A Comprehensive Treatise,* Vol. 1, Carr, I. and Deans, W. T., Eds., Plenum Press, New York, 1980, 525.

43. **Hackstadt, T. and Williams, J. C.,** Metabolic Adaptation of *Coxiella burnetii* to intraphagolysosomal growth, in *Microbiology 84,* Leive, L. and Schlessinger, D., Eds., American Society for Microbiology, Washington, D.C., 1984, 266.

44. **Handley, J. D., Paretsky, D., and Stueckemann, J.,** Electron microscopic observations of *Coxiella burnetii* in the Guinea pig, *J. Bacteriol.,* 94, 263, 1967.

45. **Burton, P. L., Kordova, N., and Paretsky, D.,** Electron microscopic studies of the rickettsia *Coxiella burnetii*: entry, lysosomal response, and fate of rickettsial DNA in L-cells, *Can. J. Microbiol.,* 17, 143, 1971.

46. **Wisseman, C. L., Jr., Fiset, P., and Ormsbee, R. A.,** Interaction of rickettsiae and phagocytic host cells. V. Phagocytic and opsonic interactions of phase I and phase 2 *Coxiella burnetii* with normal and immune human leukocytes and antibodies, *J. Immunol.,* 99, 669, 1967.

47. **Williams, J. C., and Vodkin, M. H.,** Metabolism and genetics of chlamydias and rickettsias, *Onderstepoort J. Vet. Res.,* 54, 211, 1987.

48. **Khavkin, T. and Amosenkova, N.,** Release of *Coxiella burnetii* from the host cell, in *Rickettsiae and Rickettsial Diseases,* Burgdorfer, W. and Anacker, R. L., Eds., Academic Press, New York, 1981, 335.

49. **Anderson, R. G. and Pathak, R. K.,** Vesicles and cisternae in the trans Golgi apparatus of human fibroblasts are acidic compartments, *Cell,* 40, 635, 1983.

50. **Frehel, C., de Chastellier, C., Lang, T., and Rastogi, N.,** Evidence for inhibition of fusion of lysosomal and prelysosomal compartments with phagosomes in macrophages infected with pathogenic *Mycobacterium avium. Infect. Immun.,* 52, 252, 1986.

51. **Kishimoto, R. A., and Walker, J. S.,** Interaction between *Coxiella burnetii* and guinea pig peritoneal macrophages. *Infect. Immun.,* 14, 416. 1976.

52. **Moulder, J. M.,** Intracellular parasitism: life in an extreme environment, *J. Infect. Dis.,* 130, 300, 1974.

53. **Bogatchev, Yu., Khavkin, T. N., Shustrov, A. K., and Freidlin, I. S.,** Motion picture study of the response of cultured peritoneal macrophages to the invasion of endozoites of Toxoplasma gondii, RH strain, *Acta Microbiol. Acad. Sci. Hung.,* 27, 1, 1980.

54. **Murray, H. W.,** Survival of Intracellular pathogens within human mononuclear phagocytes, *Sem. Hematol.,* 25, 101, 1988.

55. **Ariel, B. M., Khavkin, T. N., and Amosenkova, N. I.,** Interaction between *Coxiella burnetii* and the cells in experimental Q-rickettsiosis. Histologic and electron microscope studies, *Pathol. Microbiol.,* 39, 412, 1973.

56. **Hassett, D. J. and Cohen, M. S.,** Bacterial adaptation to oxidative stress: implication for pathogenesis and interaction with phagocytic cell, *FASEB J.,* 3, 2574, 1989.

57. **Khavkin, T. N. and Amosenkova, N. I.,** On the Q-rickettsial inflammation, *Proc. Pasteur Inst. Epidemiol. Microbiol.* Vol. 29, Pasteur Institute, Leningrad, 1965, 212.

58. **Walker, D. H.,** Pathology and pathogenesis of the vasculotropic rickettsioses, in *Biology of Rickettsial Diseases,* Vol. 1, Walker, D. H., Ed., CRC Press, Boca Raton, FL., 1988, 116.

59. **Anitchkow, N. N.,** Einige Untersuchungsergebnisse über die Speicherung von Vitalfarbstoffen und Aufschwemmungen in Organismus, *Virchows Arch.,* 275, 93, 1930.

60. **van Oss, C. J.,** Phagocytosis: an overview, in *Methods in Enzymology,* Vol. 132, Di Sabato, G., and Everse, J., Eds., Academic Press, Orlando, FL, 1986, 3.

61. **Kishimoto, R. A., Veltri, B. J., Canonico, P. G., Shirey, F. G., and Walker, J. S.,** Electron microscope study on the interaction between normal guinea pig peritoneal macrophages and *Coxiella burnetii, Infect. Immun.,* 14, 1087, 1976.

62. **Turco, J., Thompson, H. A., and Winkler, H. H.,** Interferon-gamma inhibits growth of *Coxiella burnetii* in mouse fibroblasts, *Infect. Immun.,* 45, 781, 1984.

63. **Jerrelis, T. R.,** Mechanisms of immunity to Rickettsia species and *Coxiella bernetii,* in *Biology of Rickettsial Diseases,* Vol. 1, Walker, D. H., Ed., CRC Press, Boca Raton, FL, 1988, 79.

64. **van Furth, R., Diesselhoff-den Dulk, M. M. C., Sluiter, W., and van Dissel, J. T.,** New perspectives on the kinetics of mononuclear phagocytes, in *Mononuclear Phagocytes, Characteristics, Physiology and Function,* van Furth, R., Ed., *Martinus Nijhoff,* Dordrecht, 1985, 201.

65. **Radzun, H. J., Kriepe, H., Zavazava, N., Hansmann, M.-L., and Parwaresch, M. R.,** Diversity of the human monocyte/macrophage system as detected by monoclonal antibodies, *J. Leuk. Biol.,* 43, 41, 1988.

66. **Damrow, T. A., Williams, J. C., and Waag, D. M.,** Supression of *in vitro* lymphocyte proliferation in C57BL/10 ScN mice vaccinated with phase I *Coxiella burnetii, Infect. Immun.,* 47, 149, 1985.

67. **Koster, F. T., Williams, J. C., and Goodwin, J. S.,** Cellular immunity in Q fever: specific lymphocyte unresponsiveness in Q fever endocarditis, *J. Infect. Dis.,* 152, 1283, 1985.

68. **Huismann, O. and D'Ari, R.,** An inducible DNA replication—cell division coupling mechanism in *E. coli, Nature (London),* 290, 797, 1981.

69. **Kokorin, I. N.,** Biological peculiarities of the development of rickettsiae, *Acta Virol.,* 12, 31, 1968.

70. **Frenkel, J. K.,** Toxoplasmosis: parasite life cycle, pathology and immunology, in *The Coccidia,* Hammond, D. M. and Long, P. L., Eds., University Park Press, Baltimore, 1973, 343.

71. **Pugh, C. W. and MacPherson, G. C.,** The origin and turnover kinetics of lymbocytes, in *Mononuclear Phagocytes. Characteristics, Physiology, and Function.,* Van Furth., R., Ed., Martinus Nijhoff, Dordrecht, 1985, 211.

72. **Hume, D. A., Robinson, A. P., Macpherson, G. G., and Gordon, S.,** The mononuclear phagocyte system of the mouse defined by immunocytochemical localization of antigen F4/80. Relationship between macrophages, Langerhans cells, reticular cells and dendritic cells in lymphoid and hematopoietic organs, *J. Exp. Med.,* 158, 1522, 1983.

73. **Price, W. H.,** Studies on the interepidemic survival of louse borne epidemic typhus fever, *J. Bacteriol.,* 69, 106, 1955.

74. **Langley, J. M., Marrie, T. J., Covert, A., Waag, D. M., and Williams, J. C.,** Poker players pneumonia. An urban outbreak of Q fever following exposure to a parturient cat, *N. Engl. J. Med.,* 319, 354, 1988.

75. **Shifrin, I. and Khavkin, T.,** On the reactogenic properties of the Q-rickettsial vaccine, in *Problems of Infectious Pathology,* Anitchkov, N. and Yoffe, V., Eds., IEM, Leningrad, 1962, 92.

76. **Cybulsky, M. I., Chan, M. K. W., and Movat, H. Z.,** Acute inflammation and microthrombosis induced by endotoxin, interleukin 1, and tumor necrosis factor and their implication in Gram-negative infection, *Lab. Invest.,* 58, 365, 1988.

77. **Freeman, B. A. and Crapo, J. D.,** Biology of disease: free radicals and tissue injury, *Lab. Invest.,* 47, 412, 1982.

78. **Heymer, B.,** Causative agents, mediators and histomorphology of inflammation, *Pathol. Res. Pract.,* 180, 143, 1985.

79. **Khavkin, T. N., Ariel, B. M., Amosenkova, N. I., and Krasnik, F. I.,** Interaction of *Rickettsia prowazekii* with phagocytes in the course of infectious process, *Exp. Mol. Pathol.,* 22, 417, 1975

80. **Rikihisa, Y.,** Glycogen autophagosomes in polymorphonuclear leukocytes induced by rickettsiae, *Anat. Rec.,* 208, 319, 1984.

81. **Rády, M., Glávits, R., and Nagy, G.,** Demonstration of Q fever associated with abortions in cattle and sheep, *Acta Vet. Hung.,* 33, 169, 1985.

82. **Dvorak, H. F., Galli, S. J., and Dvorak, A. M.,** Cellular and vascular manifestations of cell-mediated immunity, *Human Pathol.,* 17, 122, 1986.

83. **Kaufmann, S. H. E., Flesch, I., Muller, I., Chiplunkar, S., and DeLibero, G.,** Macrophages, helper T-cells and cytolytic T-cells—possible contribution to host defense against mycobacteria, in *UCLA Symp. Molecular. Cellular Biology,* Vol. 64, Horovitz, M. A., Ed., Alan R. Liss, New York, 1988, 311.

84. **Dvorak, A. M., Galli, S. J., Marcum, J. A., Nabel, G., der Simonian, H., Goldin, J., Monahan, R. A., Pyne, K., Cantor, H., Rosenberg, R. D., and Dvorak, H. F.,** Cloned mouse cells with natural killer function and cloned supressor T cells express ultrastructural and biochemical features not shared by cloned inducer T cells, *J. Exp. Med.,* 157, 843, 1983.

85. **Berke, G.,** Cytotoxic T-lymophocytes. How do they function?, *Immunol. Rev.,* 72, 5, 1983.

86. **Kaplan, G., Laal, S., Sheftel, G., Nusrat, A., Nath, I., Mathur, N. K., Mishra, R. S., and Cohn, Z. A.,** The nature and kinetics of a delayed immune response to purified protein derivative of tuberculin in the skin of lepromatous leprosy patients, *J. Exp. Med.,* 168, 1811, 1988.

87. **Mullbacher, A. and Ada, G. L.,** How do cytotoxic T lymphocytes work *in vivo?, Microb. Pathogenesis,* 3, 315, 1987.

88. **Thomas, E. L., Lehrer, R. T., and Rest, R. F.,** Human neutrophil antimicrobial activity, *Rev. Infect. Dis.,* 10 (Suppl. 2), 450, 1988.

89. **Racz, P. and Tenner-Racz, K.,** Durch intrazellulare Parasiten verursachte Granulome, *Verh. Dtsch. Ges. Pathol.,* 64, 126, 1980.

90. **Kindler, V., Sappino, A. P., Grau, G. F., Piquet, P.-F., and Vassalli, P.,** The inducing role of tumor necrosis factor in the development of bacterial granulomas during BCG infection, *Cell,* 56, 731, 1989.

91. **Wahl, S. M.,** Hepatic granuloma as a model of inflammation and repair—an overview, in *Methods in Enzymology,* Vol. 163, Di Sabato, Ed., Academic. Press, San Diego, 1988, 605.

92. **Kokorin, I. N., Pushkareva, V. I., Kazár, J., and Schramek, S.,** Histological changes in mouse liver and spleen caused by different *Coxiella burnetii* antigenic preparations, *Acta Virol.,* 29, 410, 1985.

93. **Heymer, B., von Konig, W. C. H., Finger, H., Hof, H., and Emmerling, P.,** Histomorphology of experimental listeriosis, *Infection,* 16, S106, 1988.

94. **Sawyer, L. A., Fishbein, D. B., and McDade, J. E.,** Q-fever: current concepts., *Rev. Infect. Dis.,* 9, 935, 1987.
95. **Altmann, H., W.,** Die granulomatose Reactionen der Leber, *Verh. Dtsch. Ges. Pathol.,* 64, 132, 1980.
96. **Peacock, M., Munoz, J., Tallent, G. L., and Ormsbee, R. A.,** Passive cutaneous anaphylaxies with antigens from *Coxiella burnetii, J. Bacteriol.,* 95, 1580, 1978.
97. **Asher, M. S., Berman, M. A., Parker, D., and Turl, J.,** Experimental model for dermal granulomatous hypersensitivity in Q-fever, *Infect. Immun.,* 39, 388, 1983.
98. **Leedom, J. M.,** Q fever: an update, in *Current Clinical Topics in Infectious Diseases,* Vol. 1, Remington, J. S. and Swartz, M. N., Eds., McGraw-Hill, New York, 1980, 304.
99. **de Brito, T., Hoshino-Shimizu, S., Pereira, M. O., and Rigolon, N.,** The pathogenesis of the vascular lesions in experimental rickettsial disease of the guinea pig (Rocky Mountain spotted fever group). A light, immunofluorescent and electron microscopic study, *Virchows. Arch. A,* 358, 205, 1973.
100. **De Micco, C., Raoult, D., Benderitter, T., Gallais, H., and Toga, M.,** Immune complex vasculitis associated with Mediterranean spotted fever, *J. Infect.,* 14, 163, 1987.
101. **Wen, B. H. and Yu, S. R.,** Antigen-specific circulating immune complexes in *Coxiella burnetii* infected guinea pigs, *Exp. Mol. Pathol.,* 47, 175, 1987.
102. **Hoibi, N., Doring, G., and Schiotz, P. O.,** The role of immune complexes in the pathogenesis of bacterial infections, *Annu. Rev. Microbiol.,* 40, 29, 1986.
103. **Couser, W. G.,** Mechanisms of glomerular injury in immune complex disease, *Kidney Int.,* 28, 569, 1985.
104. **Penner, E. and Albini, B.,** Demonstration and analysis of immune complexes in tissues and circulation, with special reference to immunofluorescence techniques, in *Immunofluorescence Technology Selected Theoretical and Clinical Aspects,* Wick, G., Trail, K. N., and Schauenstein, K., Eds., Elsevier Biomedical Press, Amsterdam, 1982, 349.
105. **Yu, G. O., Zhang, X. U. E., Li, Q. I., and Wang, S. L.,** Experimental Q fever glomerulonephritis in guinea pig, in *Proc. 1st Chin. Natl. Symp. Rickettsiae, Chlamydiae and Toxoplasma Infections,* Fan, M. Y., Ed., Chinese Journal Epidemiology Press, Bejin, China, 1983, 80.
106. **Yu, S. R., Yu, G. O., Shi, I. C., Zhang, X., and Su, Y.P.,** Experimental Q fever glomerulonephritis in guinea pig, *Chin. J. Pathol.,* 2, 106, 1986.
107. **Movat, H. Z. and Burrowes, C. E.,** The local Schwartzman reaction: endotoxin-mediated inflammatory and thrombo-hemorrhagic lesions, in *Handbook of Endotoxin,* Berry, L. J., Ed., Elsvier, Amsterdam, 1985, 260.
108. **Smith, H.,** The state and future of studies on bacterial pathogenecity, in *Virulence Mechanisms of Bacterial Pathogenesis,* Roth, J. A., Ed., American Society for Microbiology, Washington, D.C., 1988, 365.

Chapter 5

IMMUNE RESPONSE TO *COXIELLA BURNETII* INFECTION

David M. Waag

TABLE OF CONTENTS

I. INTRODUCTION

Coxiella burnetii, the etiologic agent of Q fever, is an obligate intracellular bacterium which grows only in the phagolysosome of eukaryotic cells. Virulence of the microorganism and the immune state of the host influence the outcome of infection. Phase I virulent cells are characterized by a smooth-type lipopolysaccharide (LPS).[1,2] When grown in tissue culture or in an immune incompetent host, phase I *C. burnetii* undergoes a transition to avirulent phase II cells, characterized by a rough LPS.[2] The classic definition of virulence for C. burnetii has centered around the LPS chemotype.[3-5] However, genetic polymorphisms in chromosomal DNA were shown to correlate with phase differences,[6-7] while plasmid differences appear to correlate with disease expression.[8] The *C. burnetii* phase transition is analogous to the "smooth to rough" LPS mutational variation described for Gram-negative Enterobacteriaceae.[9] Strains structurally intermediate between phase I and phase II have been isolated and display partial virulence.[10]

The interaction of *C. burnetii* with the host immune system is complex, the intricacies of which are not yet understood. On the one hand, *C. burnetii* initiates a nonspecific enhancement of the immune system manifested in an ability to regress tumors and enhance resistance to bacterial and protozoan pathogens.[11-14] However, *C. burnetii* also is able to cause antigen–specific immunosuppression and adverse tissue reactions in the human and nonhuman host.[5,15-18] Whether these divergent activities are a function of antigen load, duration of exposure, lymphocyte subpopulations, or genetic makeup of the host remains unclear. In spite of the difficulty in understanding the global interrelationships in *C. burnetii* infection, small pieces of the puzzle are being worked out, and, hopefully, the interactions between the various elements will soon be known.

II. PREIMMUNE EVENTS

The interaction of a pathogen with a naive host sets in motion a defined series of events beginning with nonimmune control mechanisms and culminating in the development of a directed immune response involving humoral and cellular immunity. These events should result in control of the invading microorganism. The body's initial response to *C. burnetii* infection is presumed to be similar to responses against other microorganisms, both pathogens and nonpathogens. As a successful pathogen, *C. burnetii* may be able to evade certain aspects of the host immune system while using other endogenous host mechanisms to augment bacterial uptake into susceptible cells.

C. burnetii gains entrance to a susceptible host primarily by inhalation of infectious particles.[19] However, the ingestion of *C. burnetii* in milk products[20,21] or the injection of infectious materials during tick feeding[22,23] are alternate ports of entry for the acquisition of Q fever. *C. burnetii* is extremely infectious for the nonimmune host. One organism is sufficient to initiate infection in guinea pigs, as detected by serological response.[19] Although the initial attempts to prevent an infection are nonspecific, they are designed to sequester the infecting agent either at the site of entry or within regional lymph nodes. The entry of infecting agents by the aerosol route initiates the primary defensive mechanisms which clear particulate microorganisms by either mucociliary processes in conducting airways, resulting in the elimination of infectious microorganisms from the body, or phagocytosis and transport of the microorganisms from the lung to the regional lymph nodes. The latter mechanism may be effective in either the systemic spread of pathogenic microorganisms or the stimulation of inflammatory responses which limit access to vulnerable host tissue sites.

An important mechanism to control the proliferation of invading microorganisms is the development of an inflammatory response. Inflammatory processes represent a complex interaction between various interacting mediators and cells.[24] A complete discussion of the inflammatory process is beyond the scope of this article, and the reader is referred to Reference 25 for a general discussion of inflammatory mediators. Inflammation is nonspecific and usually beneficial, resulting in removal of the invading organisms and repair of the injured site. The initial development of the inflammatory response occurs as a response to tissue injury and/or foreign invaders and involves blood vessel permeability, neutrophil accumulation, and, finally, macrophage infiltration.[25]

Soluble mediators enhance blood flow, increase vessel permeability, and induce emigration of inflammatory cells from the blood. Mediators may be exogenous or endogenous. Bacterial products or toxins may play a role in this process. Endogenous mediators of inflammation include proteolytic fragments of complement; prostaglandin and leukotrienes; and metabolic products of arachidonic acid.

The first step in the inflammatory process is vasodilation and hyperemia. Prostaglandins may modulate blood flow and potentiate the effects of other mediators. As blood vessels become permeable to cellular trafficking, cell emigrants migrate from the capillaries to the tissue spaces and digest dead host cells and bacteria. Histamine, lipooxygenase metabolites of arachidonic acid and platelet activating factor may play a role in vasopermeability. Complement fragments (particularly C5a) produced by endopeptidases, bacterial products such as n-formulated methionyl peptides,[26] and certain lymphokines are chemotactic and attract other phagocytic cells to the site. Ligation of phagocyte receptors stimulates chemotaxis, secretion of lysosomal enzymes and the respiratory burst.[27]

Phagocytic cells internalize foreign particles into acid vesicles where chemical and enzymatic breakdown is initiated. Fusion of the phagosome with the lysosome subjects the internalized microorganisms to a battery of hydrolytic enzymes and oxygen metabolites which kill susceptible microorganisms. Surrounding tissue is destroyed by release of lysosomal contents from infected cells. The resolution of an inflammatory response is just as important as its initiation. The mechanisms, although not clearly delineated, probably involve a chemotactic factor inactivator to halt cell influx to the site.[28] Then, oxygen scavengers and antiproteases inactivate bactericidal products and inflammatory cells and tissue debris are removed from the affected site.

In contrast to its designed function, the inflammatory process may aid growth and dissemination of *C. burnetii* in a susceptible host. *C. burnetii* enlists naive host cells as facilitators of the systemic spread of the microorganism. The alveolar macrophages may participate in the systemic spread of *C. burnetii* by sequestering them in intracellular acidic vesicles during translocation to the regional tracheobronchial lymph nodes. Similarly, the mobile macrophages of the vascular and lymphatic systems of the skin are enlisted as translocators of microorganisms from the site of injection to the draining lymph node. Although the host is attempting to control proliferation of the infectious microorganisms, these processes may accelerate the onset of *C. burnetii* infections by providing the necessary intracellular environment for growth in phagolysosomes and a means of systemic spread.

Nonpathogenic microorganisms are killed during the inflammatory process and do not cause overt infections in the immunocompetent host. Pathogenic microorganisms, however, have mechanisms for evading the host immune system and surviving. Extracellular pathogens may interfere with uptake by phagocytes,[29,30] whereas intracellular pathogens may promote their uptake into target cells.[31]

As a successful obligate intracellular parasite, *C. burnetii* must gain entrance into a susceptible cell in order to multiply. Contact with the host membrane triggers phagocytic mechanisms. To be an efficient pathogen, *C. burnetii* must take advantage of certain host immune strategies,[32] thus promoting entry into susceptible host cells. *C. burnetii* has several cell surface receptors which can aid its uptake into susceptible cells. Acute phase proteins can act as facilitators of intracellular infection by enhancing uptake of *C. burnetii* before the initiation of a specific immune response. C-reactive protein (CRP) is induced during the inflammatory response and is synthesized by hepatocytes. CRP resembles immunoglobulins in its ability to bind C1q and can initiate precipitation, agglutination, and enhancement of phagocytosis.[33] Binding to phosphorylcholine and to polycationic substances has also been reported.[34,35] Recent studies have shown that *C. burnetii* binds CRP in a calcium dependent manner.[36] This interaction may facilitate uptake by phagocytosis during the early phases of infection with CRP serving as a bridge between the bacteria and the potential host cell. The structure and chemical composition of the *C. burnetii* CRP-binding sites is unknown but probably involves LPS. Cell surface polycationic substances have not been identified.

The activation of the classic complement pathway by either specific immunoglobulin or CRP binding may allow opsonization of the organism via C3b. Human C3b, C4 and C8 are deposited on the surface of *C. burnetii* after incubation in normal serum.[37-39] Complement receptors on the surface of *C. burnetii* have not been characterized. These receptors may play a role in aiding opsonization and internalization of the microbe.

Two possible sequelae of Q fever include granulomatous hepatitis[40,41] and endocarditis.[42,43] Tissue-tropic effects of *C. burnetii* can be explained by specific host receptors which either bind *C. burnetii* directly or share affinity for a common factor with *C. burnetii*. Ceruloplasmin is a glycoprotein involved in most of the plasma copper recycled in the liver. *C. burnetii* may gain entrance into hepatocytes, especially Kupffer cells, by the binding of host cell and bacterial receptors simultaneously via ceruloplasmin.[37,44] Similarly, ceruloplasmin receptors are found in the membranes of aortic and heart tissues[45] and may play a role in colonization of those tissues by *C. burnetii*.

C. burnetii also binds normal immunoglobulin.[46] After binding to *C. burnetii*, the antibody can still bind protein A, suggesting that a portion of the Fc region of the antibody molecule remains available to bind phagocyte Fc receptors. Therefore, attachment and possibly entry into the host cell could be facilitated via Fc receptor binding. The *C. burnetii* Fc binding site remains uncharacterized.

Many cell surface receptors are associated with clathran-coated pits.[47,48] Receptors that cannot associate with coated pits internalize ligand slowly, if at all. The destination of many receptor-ligand complexes is unknown, but the fate of most ligands is lysosomal hydrolysis.[49] An increase in hydrogen ion concentration and decrease in the calcium ion concentration in the lysosomal environment causes the ligand-receptor complex to dissociate. Whether binding ligand destined for the lysosome enhances the rate of *C. burnetii* transit to that organelle is unknown. Ligand binding may provide *C. burnetii* with a double advantage: attachment to a susceptible host cell may be facilitated and transit to the phagolysosome may be enhanced.

The putative existence of *C. burnetii* cell surface binding sites for acute phase reactants and nonspecific antibody allows one to postulate at least two mechanisms of *C. burnetii* entry into susceptible host cells: (1) nonspecific attachment of *C. burnetii* cell surface components to susceptible host cells followed by phagocytosis or (2) by adsorption of acute phase reactants or immunoglobulin to the bacterial surface and interaction of these ligands with specific receptors on the surface of host cells. *C. burnetii* may utilize specific receptor mechanisms to aid its internalization into host cells. Once inside the host cell, avoidance of the microbicidal mechanisms of the phagolysosome must then depend on the parasite strategy. The minimum qualifications a parasite must have to survive in this environment are acid resistance and an ability to cope with bactericidal conditions within the phagolysosome. Removal from the extracellular environment serves two important parasitic functions: (1) a suitable acidic environment triggers bacterial cell metabolism[50] and (2) the humoral arm of the host immune system is avoided.

III. AN IMMUNE RESPONSE

Nonspecific immune mechanisms of the naive host pose no major barriers to infection of susceptible cells with *C. burnetii*. However, the development of a specific immune response renders the intracellular environment inhospitable to *C. burnetii*, allowing the infection to be controlled. Events which occur during an immune response are well established[51] and are summarized in a scenario of *C. burnetii* infection. Internalization of invading microorganisms by nonimmune cells begins almost immediately after infection. Phagocytes internalize the bacteria into acidic vesicles. *C. burnetii* is resistant to intracellular degradation within quiescent host cells. However, soluble *C. burnetii* antigens[52] and specific acid-sensitive fragments of cell surface proteins may combine with a class II major histocompatibility complex gene product (Ia antigen) to provide a recognition complex for the T-cell receptor. This association may occur in acidic vesicles with the Ia-*C. burnetii* antigen complex reexpressed on the phagocyte cell surface.[53]

The recognition of the antigen-Ia complex by antigen-specific T-cells results in the activation of those cells to produce lymphokines, some of which activate macrophages (gamma interferon) while others stimulate T-cells to proliferate (interleukin 2). In addition, the initial interaction of antigen with macrophages causes the release of interleukin 1 (IL-1), one of numerous biological effects being the activation of T-cells.[54] This is the beginning of a specific immune response involving the production and release of soluble mediators which (1) activate macrophages during the initiation of a cellular immune response and (2) stimulate B-cells to differentiate and secrete antibody in a humoral immune response. The interaction of T-cells with specific antigen, expressed in the context of self Ia molecules, leads to the release of soluble T helper factors capable of interacting with B-lymphocytes. Binding of antigen to cell surface immunoglobulins (i.e., the B-cell receptor) and stimulation by T-cell factors cause transformation of resting B-cells to the blast cell stage. Upon further stimulation by T-cell products, the clone of antigen-specific B-cells increases as cells divide. Cell surface immunoglobulin disappears as B-cells become terminally differentiated antibody-secreting plasma cells.

A. ROLE OF HUMORAL ANTIBODY

The presence of specific antibody provides the best indication of prior exposure to *C. burnetii*. Q fever is an acute febrile disease with few characteristic symptoms to distinguish it from other febrile illnesses. Cultivation is hazardous, and the serological response is the primary method of diagnosis (see Volume II, Chapter 10).

The detection of anti-*C. burnetii* antibodies is measured by the complement fixation (CF) assay,[55,56] microagglutination assay (MAA),[57] indirect immunofluorescence assay (IFA),[58-60] and enzyme linked immunosorbent assay (ELISA).[61,62] Acute Q fever infections are characterized by an elevated phase II titer. The phase I titer may rise subsequent to the phase II titer, but rarely exceeds it. The phase II antibody is detectable before phase I antibody because the masked phase II antigen is more immunogenic than the phase I surface component.[57] Complement-fixing antibodies to phase II *C. burnetii* are detectable within 3 weeks of illness and reach a peak titer two weeks after onset.[63] Phase I complement-fixing antibodies are difficult to detect during the acute phase of the disease.

The IFA test is more sensitive than the CF test. Titers in the IgM class detected by IFA against phase II *C. burnetii* are observed within one week of illness. The onset of the anti-phase I IgM response follows that of the phase II response. During acute illness, IgG titers to phase II peak by week 8 of infection. However, titers in the IgG class detected by IFA against phase I organisms remain low for about one year. IFA titers of the IgA class are rarely elevated in patients suffering primary Q fever.[58]

The ELISA is the serological method of choice because of its sensitivity and reproducibility.[64] This test is superior to the CF and MAA tests because class specific IgM, IgG, and IgA can be evaluated. Guinea pigs infected with phase I *C. burnetii* had detectable titers of phase II antibody within one week, two d earlier than the CF, IFA, or MAA test.[61] Antibodies against phase I *C. burnetii* rise subsequent to the anti-phase II response and are detected at least one week earlier by the ELISA test than other serological assays. The greater sensitivity of the ELISA test has led to greater numbers of seropositive individuals, suggesting that cases of Q fever may be more prevalent than generally assumed.

Chronic Q fever can be manifested as granulomatous hepatitis or endocarditis. Persistent exposure to *C. burnetii* antigen results in a serological profile which is different from that measured in acute infection. When the titers of immunoglobulin to phase I and phase II *C. burnetii* are measured, a ratio of anti-phase II to anti-phase I of >1, $\geq$1, and $\leq$1, correlates with a diagnosis of primary Q fever, granulomatous hepatitis, and Q fever endocarditis, respectively.[58] Although anti-LPS reactivity can be found after a primary infection in guinea pigs,[61] high LPS titers are uncharacteristic of acute infection in humans.[65] However, chronically exposed patients may exhibit a significant antibody titer to the phase I *C. burnetii* LPS antigen.[66] Anti-phase I IgG may be the most important parameter for the diagnosis of chronic Q fever.[67] Titers are generally several times higher than those measured in acute Q fever. The IgA titer to phase I and phase II *C. burnetii* may also be elevated,[58,67] whereas IgM levels are not significantly different from those measured in patients suffering from primary Q fever.[58] High titers of IgG antibody lead to the formation of autoantibodies, called rheumatoid factor, directed against the heavy chain component. Frequently, clinical elevations in IgM titers are artificially high due to the presence of IgM rheumatoid factor which binds the Fc fragment of IgG and is subsequently detected by the antibody conjugate.[58] IgM titers fall when autoantibody is removed.

Although antibodies are prime effectors for the control of many infections, the role of antibodies in host defense against *C. burnetii* is equivocal. Early papers suggested that antibody played a key role in host resistance to *C. burnetii* infection.[68,69] Immune serum mixed with a suspension of viable phase I *C. burnetii* prior to injection suppressed fever in recipient animals. In addition, the number of microorganisms found in the smears of spleens harvested 7 d post infection was diminished when compared to controls.[70] The "mouse protective effect" could be removed by adsorption using phase I *C. burnetii*. These results, plus the coincidental appearance of protective effect and serum immunoglobulin provided strong initial evidence for protective antibody. In retrospect, lymphokines such as immune interferon may also have been transferred in serum to normal recipients which hastened the development of a cellular immune response and control of infection (see Section C).

Current evidence points to the presence of immune antibody facilitating the uptake of infectious microorganisms by macrophages and accelerating the appearance of a cellular immune response. Normal mice injected intravenously with immune serum were challenged 24 h later with viable phase I *C. burnetii*.[71] Spleen impression smears scored by direct microscopic examination 7 and 14 d post challenge contained fewer intracellular microorganisms than impression smears from mice pretreated with normal serum. One might predict that the number of phagocytosed microorganisms due to opsonization would be high soon after infection. If the onset of a cellular immune response is accelerated due to opsonization of microorganisms, macrophages will have fewer infectious particles as time passes. Passive transfer of serum to nude mice did not affect the clearance of microorganisms from the spleen. Therefore, the development of acquired resistance to infection by *C. burnetii* appears to require a T-cell mediated immune response. Specific antibody may effectively promote a rapid development of activated macrophages and, ultimately, control of the parasite.

A correlate experiment utilized peritoneal exudate cells from guinea pigs. Cells were infected *in vitro* with phase I *C. burnetii*. Hinrichs and Jerrells determined that specific antiserum added to *normal* macrophage cultures before or after infection, or reacted directly with *C. burnetii*, failed to control subsequent intracellular replication.[72] There was a noticeable enhancement of bacterial uptake when organisms were pretreated with immune serum. Immune antibody had no apparent effect on bacterial replication. In contrast, when lymphokines or immune lymphocytes were added to normal macrophages, the number of intracellular microorganisms was markedly reduced. As phagocytes became activated *in vitro*, their bacteriocidal abilities were enhanced.

The importance of activated macrophages in controlling the intracellular growth of *C. burnetii* has also been shown in work by Kishimoto et. al.[73,74] The presence of immune serum potentiated phagocytosis of microorganisms by normal and activated guinea pig macrophages *in vitro*. The rate of destruction of microorganisms by activated macrophages from phase I *C. burnetii*-injected animals was also enhanced. A major difference between this protocol and the previous one centered on whether the peritoneal macrophages had been previously activated by *in vivo* exposure to phase I *C. burnetii*. Activated macrophages provide a much less hospitable environment for the growth of *C. burnetii*. Therefore, the induction of a classic cellular immune response by mechanisms including lymphokine production is the primary mechanism in control of the microorganism. Any participation by immune antibody is ancillary to control of *C. burnetii* replication by activated macrophages.

The presence of high titers of specific antibody does not uniformly correlate with resistance to infection. Inbred strains of mice may be partitioned into 3 groups: sensitive, intermediately sensitive, and resistant, based on susceptibility to *C. burnetii* infection.[75] Mice in all categories developed similar antibody titers against *C. burnetii* phase I, phase II, and LPS antigen. Critical differences in susceptibility may correlate with the ability to mount a strong cellular immune response. The most sensitive strain, A/J, is a low interferon producer and has deficiencies in the complement pathway[76] and macrophage functions.[77]

The presence or absence of functional B-cells may have little bearing on the survivability of *C. burnetii* infection. This conclusion has been derived from experiments involving cyclophosphamide, an immunosuppressive agent which cripples all immune responses but preferentially eliminates the humoral immune response in a time- and dose-dependent manner.[78] Although the use of this agent makes strict interpretation difficult, published reports have suggested that pretreatment of mice with cyclophosphamide did not cause increased growth of *C. burnetii in vivo*, even though antibody formation was inhibited.[79,80] This may indicate that antibodies are not required for control of *C. burnetii* infection. To circumvent many problems in interpreting results gathered using toxic reagents *in vivo*, advantage should be taken of mutant strains of inbred mice having defined genetic lesions. The CBA/N mouse is deficient in B-cell function and could be used in future experiments delineating the role of antibody in *C. burnetii* infection.

Although not a primary mechanism of host resistance to *C. burnetii* infection, there is evidence for direct lytic effects by complement. Human serum serologically negative for anti-*C. burnetii* antibodies by the MAA is able to kill nonvirulent, but not virulent, *C. burnetii* by the alternate complement pathway.[39] Phase II nonvirulent microorganisms may be less able to cause infections in the nonimmune host due to their susceptibility to complement-mediated serum killing. The mechanism of resistance of phase I *C. burnetii* to complement-mediated killing may be due to weak binding or nonbinding of C3b. Alternatively, C3b may be rapidly converted to the C3d fragment which does not lead to a stable membrane attack complex. Although the alternate pathway of complement activation may play a role in lysing *C. burnetii*, experiments to demonstrate that specific antibody from infected mice has a direct bacteriocidal effect on phase I *C. burnetii* through activation of the classic complement pathway have been unsuccessful.[69]

Antibody-dependent cell-mediated cytotoxicity is a cytolytic mechanism whereby macrophage and natural killer cell Fc receptors bind target cells coated with antibody.[81,82] Target cells are then killed by a mechanism not involving complement and without a requirement for a previous exposure to target antigen. This mechanism may be operative in the lysis of *C. burnetii*-infected target cells.[83] A mouse macrophage tumor cell line, J774, persistently infected with phase I *C. burnetii* was incubated in the presence of nonimmune human peripheral blood cells. These blood cells had, in turn, been incubated for 16 h in the presence of immune sera from Q fever hepatitis or endocarditis patients and were able to lyse target cells, possibly allowing the spread of infection. Monocytes were shown to be more efficient effector cells than lymphocytes. Whether similar mechanisms occur and are effective in controlling growth of *C. burnetii in vivo* are questions remaining to be answered.

B. ROLE OF CELL-MEDIATED IMMUNITY

Development of a cellular immune response is essential for the control of parasites that live within the eukaryotic cell. Although early reports stressed the importance of antibody in Q fever immunity,[69,84] later papers showed that the development of cell-mediated immunity was also critical. The contribution of cellular immunity to host resistance against infection has been gauged by comparing susceptibilities to infection of athymic (nude) mice and their heterozygous (euthymic) littermates.[85,86] Infected euthymic mice cleared the organisms from peripheral circulation and the spleen within 14 d. In contrast, *C. burnetii* could be detected and isolated from the spleen and blood of athymic mice through 60 d. All animals produced similar levels of antibody. These results support those of Scott et. al.[75] and suggest that antibody, at best, plays a peripheral role in the control of *C. burnetii* infection. In a similar experiment, nude mice were infected via aerosol and underwent examination of various tissues for the presence of *C. burnetii*.[86] The number of bacteria within infected cells and the gross appearance of organs was similar in nude and euthymic mice on days 7 and 14 post injection. However, the infection was controlled in euthymic mice within 30 d, whereas nude mice displayed splenomegaly and a progressive macrophage infiltration of most tissues, especially spleen and liver. Indirect fluorescent antibody titers were similar in both sets of mice. The phagocytic ability of macrophages from nude mice was not impaired. However, once phagocytized, these macrophages had a decreased ability to kill, suggesting a lesion in bacteriocidal mechanisms. The inability of nude mice to control *C. burnetii* points to a role of T-cells in host resistance to infection. Without functional T-cells, these animals have a decreased ability to activate macrophages and kill *C. burnetii*. Although experiments involving nude mice have been used to support conclusions regarding the importance of cell-mediated immunity in the control of *C. burnetii* infection, it has not been demonstrated that nude mice reconstituted with T-cells from euthymic littermates control infection.

In order to delineate the host factors responsible for controlling the growth of *C. burnetii*, one must examine differences between nonimmune (i.e., permissive) macrophages and immune (i.e., activated) macrophages. Quiescent macrophages superficially resemble monocytes morphologically, in phagocytic competence and activity and in decreased capacity to respond to lymphokines. Striking changes in size, content of acid hydrolases, secretion of acid hydrolases, and secretion of neutral proteases accompany macrophage activation.[87] Activation of macrophages results in functional changes in three broad areas: (1) those intracellular constituents important to cellular metabolism and degradation of ingested materials; (2) the ability of macrophages to recognize and interact with their environment, mediated by specific cellular receptors; and, (3) the synthesis and secretion of a variety of products, acting both intra- and extracellularly.[81]

Over 60 hydrolytic enzymes can be found in lysosomes. Activation of macrophages by lymphokines and phagocytic stimuli results in cells gaining increased numbers of lysosomes and their associated enzymes.[88] Phagocytic mechanisms and/or lymphokine preparations can stimulate production and secretion of a wide variety of lysosomal enzymes, including lysosomal hydrolases, neutral proteases, and complement. In contrast, lysozyme is produced constitutively. As a phagolysosomal parasite, *C. burnetii* must cope with lysozyme in the living environment. Although the peptidoglycan of *C. burnetii* is susceptible to lysozyme degradation, the sacculus is stabilized by protein-protein interactions, and the cell does not lyse.[89] Macrophage stimulation also leads to an activation of the mitochondrial system and the associated respiratory chain, the Embden-Meyerhoff-Parnas pathway and hexose monophosphate shunt. Phagocytosis can stimulate a respiratory burst and activate NADPH oxidase, resulting in the generation of a toxic superoxide ion.[90] Presumably, the induction of oxidative degradative pathways would restrict or eliminate intracellular growth of *C. burnetii.*

There are at least three categories of receptors found on the external surfaces of macrophages: Fc receptors, receptors for complement, and receptors for mannose-fucose. A quantitative increase in the number of Fc receptors has been documented during macrophage activation.[91] Macrophages in all stages of development are able to bind the third component of complement. However, resident macrophages bind but do not ingest C3b-coated particles, while inflammatory macrophages bind and internalize these particles.[92] In contrast, a quantitative decrease in the numbers of receptors for mannose-fucose are observed during activation.[93] Receptors for N-formylated peptides are found on mononuclear phagocytes of many species. Receptor binding initiates oxidative metabolism and the discharge of lysosomal enzymes.[94] Although binding of cell receptors may enhance the uptake and growth of *C. burnetii* in nonimmune macrophages, these events may lead to phagocyte activation and restriction of *C. burnetii* growth.

C. burnetii is an efficient inducer of cell-mediated immunity. When guinea pigs were infected with the phase I Henzerling strain via aerosol, peritoneal exudate cells showed macrophage migration inhibition in 3 d when cultured in the presence of phase I, but not phase II, formalin-killed whole cells. By day 14, macrophage migration inhibition could be shown using phase I or phase II killed whole cells as recall antigen. An antibody response to phase II antigen was not detectable by IFA until day 14 and rose through day 42 post infection. Phase I antibodies appeared on day 21 and also rose through day 42. The data indicate that the development of a cell-mediated immune response is an early event following *C. burnetii* infection and is functional before IFA-detectable humoral antibody is present.[95] Formalin-killed phase I *C. burnetii* also induced measurable cell-mediated immunity when injected into guinea pigs as shown by a positive skin test reaction and positive macrophage migration inhibition test.[96]

Moreover, human exposure to *C. burnetii* either by vaccination or through natural infection by bite or aerosol, induced lymphocyte responsiveness to *in vitro C. burnetii* antigen in a lymphocyte transformation assay.[15,97] Previously sensitized individuals responded to phase I and phase II whole cell antigen and to the protein component of trichloroacetic acid-soluble phase I, but not phase II, antigen. No antibody titer by the MAA was detectable at the time lymphocyte transformation assays were performed.[97]

The development of a strong cellular immunity may be verified *in vivo* with a skin test reaction. Intradermal injection of a small amount dose of antigen to which the subject has been sensitized results in erythema and induration characteristic of delayed type hypersensitivity (DTH).[98] Histological examination of DTH reactions in *C. burnetii* phase I whole cell-immunized guinea pigs skin tested with homologous antigen revealed epithelioid cell infiltration and the presence of large numbers of multinucleated giant cells in granulomatous tissue.[99] The immunologic significance of granulomatous hypersensitivity in Q fever remains uncertain but probably reflects a high degree of macrophage activation.

Although much evidence exists showing that phase I *C. burnetii* causes pathogenesis in susceptible animals,[17,18] the immune response may simultaneously be augmented nonspecifically. Normal macrophages have limited antimicrobial and antitumor activity. One milligram of formalin-killed phase I *C. burnetii* activates guinea pig peritoneal exudate cells to exhibit listericidal activity *in vitro.*[11] In addition, peritoneal exudate cell culture supernatants from macrophages activated *in vivo* were able to kill *Listeria monocytogenes in vitro.*

C. burnetii injection also prevents ascites tumor formation and growth of the intracellular protozoan parasites, *Babesia* and *Plasmodium,* in mice.[12,13] An "extract" containing an unknown amount of *C. burnetii* cellular material gave complete protection against *B. microti* when given 7 d before infection. The shorter the interval between injection of the *C. burnetii* extract and subsequent protozoan infection, the less effectively the parasite was suppressed. Although a mechanism of protection was not determined, the author postulated that interferon, natural killer cells, and/or tumor necrosis factor may be involved. Using a guinea pig tumor system, Kelly et. al., were able to show that injections of killed phase I *C. burnetii* into Line 10 hepatocellular carcinomas led to tumor regression.[14] These parameters of cell activation are important demonstrations of *C. burnetii*'s ability to specifically and nonspecifically stimulate cells of the immune system.

The injection of killed phase I or phase II *C. burnetii* was reported to protect mice from the ascites tumor, sarcoma 180. The extent of tumor protection was a function of the relative dose of *C. burnetii*, tumor cells, and time of injection.[13] Although *C. burnetii* can undoubtedly stimulate nonspecific protection, the 3-mg dose used in these experiments seems excessive. In our hands, 750 µg of formalin-killed phase I *C. burnetii* constitutes an LD_{50} in resistant strains of mice.[18,100] Maximum doses in the 100-µg range may be more appropriate for experiments of this nature.

Some of the data on the interaction of *C. burnetii* with the immune system have been gathered after injection of experimental animals with killed phase I *C. burnetii*. This tack seems appropriate since there have been no inconsistencies when comparing abilities of killed and viable microorganisms to stimulate a cellular or humoral immune response, to induce immunosuppression, or to be phagocytosed by eukaryotic cells.

C. STIMULATION OF LYMPHOKINES

In an attempt to evaluate the ability of *C. burnetii* to activate the host immune system, we have assayed the levels of lymphokines produced in tissue culture by spleen or peritoneal exudate cells.[101] Tumor necrosis factor (TNF) is a macrophage-derived cytotoxin that has been implicated in tumor necrosis[102,103] and endotoxic shock.[104] TNF alone or in combination with other lymphokines may also be important in macrophage activation.[105] Spleen and peritoneal exudate cell culture supernatants from mice injected with a killed *C. burnetii* phase I whole cell preparation were assayed for the presence of TNF.[101] Maximum detectable levels (64 U/well) of TNF were released into the culture supernatant within 48 h and were significantly higher than levels measured in control cell supernatants. There was no release of TNF unless cells were restimulated *in vitro* with *C. burnetti* phase I whole cells. When levels of TNF from phase I *C. burnetii*-injected mice were compared to those from mice injected with Bacillus Calmette Guerin (BCG), they were consistently higher. This suggests that mechanisms operative in the induction of TNF activity were active and just as efficient in *C. burnetii*-injected mice as in a standard agent of TNF induction, BCG.[102] Although anti-TNF antibody neutralized a majority of the TNF activity in the assayed supernatants, a residual cytotoxic activity remained. The source of this activity is unknown, but may be lymphotoxin.

Early studies suggested that *C. burnetii* may stimulate detectable levels of serum interferon.[106] Although the presence of interferon was not shown conclusively, immune serum was shown to inhibit plaque formation of encephalomyocarditis virus *in vitro*. Gamma-interferon is a lymphokine produced by antigen- or mitogen-stimulated T-lymphocytes. Biological effects include the induction of an antiviral state[107] and stimulation of macrophages to display antitumor[108] and antibacterial activity.[109] Phase I *C. burnetii* is an efficient inducer of interferon. Spleen cell supernatants prepared similarly to those for TNF analysis were tested in a viral inhibition assay for the presence of interferon.[101] Whereas the maximum release of TNF occurred within 48 h and was maximal in cell supernatants from *C. burnetii*-injected mice, interferon release was measurable within 4 h of culture but only when cells were incubated in the presence of phase I *C. burnetii* antigen. In addition, cells from control mice produced quantities of interferon similar to that of antigen-injected mice. Interferon levels were maximal at 24 h of culture and remained constant through 72 h. Although spleen cells incubated in the presence of phase I *C. burnetii* produced measurable amounts of interferon, release was maximal when cells were incubated in the presence of phase I *C. burnetii* and concanavalin A or interleukin 2. Nonspecific stimulation of T-cells may make them more receptive to stimulation by phase I *C. burnetii*. Interferon activity was acid labile and presumed to be due to gamma interferon. Although lymphocyte proliferative responses of spleen cells from mice injected with phase I *C. burnetii* can be depressed by 90% in a lymphocyte transformation assay,[5,17] there is no decrease in the ability of spleen cells from these animals to produce interferon. In fact, polyinosine cytosine, used to stimulate interferon production,[110] showed no difference in the ability to elicit interferon from control cells and cells from phase I *C. burnetii*-injected mice. Although protein synthesis by spleen cells in infected mice is inhibited,[17] release or synthesis of mediators is proceeding in suppressed cells.

Gamma-interferon production may be one mechanism of host defense against *C. burnetii*. This lymphokine has been noted to inhibit the growth of *C. burnetii* in mouse fibroblast macrophage-free cultures.[111] Whether gamma-interferon directly inhibits the microorganism or stimulates microbicidal or microbistatic mechanisms within host cells is uncertain. Professional phagocytes may also be activated by gamma interferon following the interaction of immune T-lymphocytes with specific antigen[112,113] and aid in the control of *C. burnetii* infection.

IV. IMMUNOSUPPRESSION

Microorganisms and their products are able to modulate the immune response of the host in various ways to include immune enhancement through adjuvant effects,[114-116] immune suppressive effects (i.e., specific and nonspecific), and through immunologic recognition of self induced by cross-reactive bacterial epitopes.[117,118] Several diseases may result in a depression of cellular immune processes, including measles,[119,120] rubella,[121] influenza,[122] viral hepatitis,[123] leprosy,[124] leishmaniasis,[125] syphilis,[126] candidiasis,[127] trypanosomiasis,[128] and Q fever.[15]

Toxins, enzymes, cell wall components, and microbial metabolites are examples of microbial products that can suppress immune responses. Such immunosuppressive factors have been shown to be produced by several microorganisms including *Vibrio cholera*,[129] Group A streptococci,[130] *Pseudomonas aeruginosa*,[131] and spirochetes.[132] These immunomodulatory agents may act via several different mechanisms that interfere with either the induction or expression of immunity. Phase I *C. burnetii* possesses an immune suppressive complex which is attached to the cell matrix by disulphide bonds.[5] Biochemical analysis has shown that components of the immunosuppressive complex are amphipathic and sensitive to alkali, acid, periodate, lysozyme, and neuraminidase. Therefore, peptidoglycan and sialic acid may be structural components. The mechanism of action is unknown. The suppressive effects of the immunosuppressive complex on *in vitro* lymphocyte responses are well known.[15,17,18,133] Lymphocyte responses of mice and monkeys are dramatically suppressed when animals are exposed to viable or killed phase I *C. burnetii*. Suppression of the uptake of thymidine radiolabel can be measured 3 to 28 d post injection.

Q fever is regarded as an acute febrile disease with the majority of patients developing immunity to reinfection. However, a minority of persons who contract the disease develop sequelae, including endocarditis, which until recently has been uniformly fatal. Peripheral blood lymphocytes from Q fever endocarditis patients display profound lymphocyte unresponsiveness to *C. burnetii* antigens *in vitro* while proliferating normally to control antigens.[15] The mechanisms of endocarditis induction are presently unknown but may involve suppressor cells allowing latent *C. burnetii* infection. Specific lymphocyte unresponsiveness may be an important factor in persistent infection with *C. burnetii*.

The ambiguity between immune-enhancing and immune-suppressing characteristics of *C. burnetii* remains unresolved. Important factors may be duration of infection, dose, strain of *C. burnetii*, and host. Lymphocytes from mice injected with small doses of phase I *C. burnetii* do not exhibit suppression, but with increasing amounts of antigen, suppression becomes more pronounced.[65] Host and strain differences are apparent when *C. burnetii* isolates from human heart valves are injected into mice.[5] Although peripheral blood lymphocytes from human endocarditis patients can be suppressed after *in vitro* exposure to phase I *C. burnetii* by over 60%, cardiotropic strains fail to cause suppression of *in vitro* lymphocyte proliferation 14 d post murine injection. A relatively short *in vivo* exposure time in comparison with a time course frequently lasting several years in human chronic Q fever patients may be a contributing factor.

Cellular sequestration and suppressor cell induction may provide a tandem mechanism of evasion of the immune system. Noninterference with normal cell metabolism and function is an important strategy for microorganisms able to persistently infect the host. *C. burnetii* has been shown to inhabit nonphagocytic cells, such as pneumocytes, without causing obvious cell damage.[134] In addition, short- and long term infection of continuous cell lines with *C. burnetii* did not alter generation times or cell cycles.[135] A model to explain persistent infection of tissue culture cells may involve asymmetric division of daughter cells.[136] The parasite vacuole segregates into only one daughter cell, leaving the companion cell daughter cell parasite free. This generation of uninfected cells from an infected cell population may provide a key explanation to persistent infection.

The mechanisms of long term persistence *in vivo* are relatively unexplored, but the consequences allow the parasite to maintain an infection over the lifetime of the patient. Individuals may be persistently infected without having symptoms of chronic Q fever exposure.[137] Phase I *C. burnetii* has been isolated from the placentas of women who contracted Q fever up to 3 years previously and were asymptomatic. Infection may be reactivated in experimental animals after parturition,[138] cortisone injection,[139] or x-irradiation.[140] Suppressor cell involvement in latency and chronicity is an important area of future research.

Aspects of leprosy bear striking resemblance to Q fever. Etiological agents, *Mycobacterium leprae* and *C. burnetii,* are intracellular microorganisms primarily infecting mononuclear phagocytes. Diseases range in severity from tuberculoid leprosy and acute Q fever, which are least severe and are characterized by high levels of cell-mediated immunity, to lepromatous leprosy and chronic Q fever at the opposite spectrum. Antibody is found in all forms of each disease, but the highest levels occur in the lepromatous form of leprosy and the chronic form of Q fever.[58,141] Antibodies have little to do with protecting the host from either disease. Lepromatous and chronic Q fever patients have good *in vitro* lymphocyte responsiveness to a variety of recall antigens, but are suppressed in the presence of homologous antigen.[15,124] A complex phenolic glycolipid is an antigen unique to *M. leprae,* which stimulates suppressor T-cell activity.[142] Activated suppressor cells may have the capability to block the responsiveness of T helper cells to other specific or cross-reactive determinants. The immunosuppressive complex of phase I *C. burnetii* may have similar composition.[5] The mechanism of suppression in lepromatous leprosy may be related to an inability to produce IL-2.[124] T-cells from lepromatous patients have IL-2 receptors and respond to IL-2. An identical mechanism of lymphocyte hyporesponsiveness is unlikely in the mouse Q fever model as lymphocytes from mice injected with phase I *C. burnetii* do not proliferate when incubated in the presence of IL-2.[101] The expression of IL-2 receptors on lymphocytes from phase I *C. burnetii* whole cell-injected mice is being investigated.

V. CONCLUSIONS

The encounter of *C. burnetii* with the nonimmune host sets in motion a series of events involving nonspecific immune mechanisms. These mechanisms do not result in bacterial killing and may, in fact, aid entry into host cells. As a specific immune response is generated, activated macrophages become less hospitable to the intracellular growth of *C. burnetii* and the infection can be controlled. *C. burnetii* is an excellent inducer of cellular immunity and other infectious agents may be killed nonspecifically. Suppressor cells are induced by the phase I *C. burnetii* immune suppressive complex and may play a role in chronic Q fever infections. There remain several questions to be answered regarding interactions between the immune suppressive complex and suppressor cells, cells involved in the induction and expression of immunosuppression, biological response modifiers induced by *C. burnetii,* and the cell surface epitopes responsible for stimulating helper and suppressor cells.

REFERENCES

1. **Schramek, S. and Meyer, H.,** Different sugar compositions of lipopolysaccharides isolated from phase I and phase II cells of *Coxiella burnetii, Infect. Immun.,* 38, 53, 1982.
2. **Amano, K.-I. and Williams, J. C.,** Chemical and immunological characterization of lipopolysaccharides from phase I and phase II *Coxiella burnetii, J. Bacteriol.,* 160, 994, 1984.
3. **Baca, O. G. and Paretsky, D.,** Some physiological and biochemical effects of a *Coxiella burnetii* lipololysaccharide preparation on guinea pigs, *Infect. Immun.,* 9, 939, 1974.
4. **Kazar, J., Brezina, R., Schramek, S., Urvolgyi, J., Pospisil, V., and Kovacova, E.,** Virulence, antigenic properties and physicochemical characteristics of *Coxiella burnetii* strains with different chick embryo yolk sac passage history, *Acta. Virol.,* 18, 434, 1974.
5. **Waag, D. M. and Williams J. C.,** Immune modulation by *Coxiella burnetii:* characterization of a phase I immunosuppressive complex differentially expressed among strains, *Immunopharmacol. Immunotoxicol.,* 10, 231, 1988.
6. **Vodkin, M. H., Williams, J. C., and Stephenson, E. H.,** Genetic heterogeneity among isolates of *Coxiella burnetii, J. Gen. Microbiol.,* 132, 455, 1986.
7. **Vodkin, M. H. and Williams, J. C.,** Overlapping deletion in two spontaneous phase varients of *Coxiella burnetii, J. Gen. Microbiol.,* 132, 2587, 1986.
8. **Samuel, J. E., Frazier, M. E., and Mallavia, L. P.,** Correlation of plasmid type and disease caused by *Coxiella burnetii, Infect. Immun.,* 49, 775, 1985.
9. **Luderitz, O., Staub, A. M., and Wetphal, O.,** Immunochemistry of O and R antigens of *Salmonella* and related *Enterobacteriaceae, Bacteriol. Rev.,* 30, 192, 1966.
10. **Hackstadt, T., Peacock, M. G., Hitchcock, P. J., and Cole, R. L.,** Lipopolysaccharide variation in *Coxiella burnetii:* intrastrain heterogeneity in structure and antigenicity, *Infect. Immun.,* 48, 359, 1985.
11. **Kelly, M. T.,** Activation of guinea pig macrophages by Q fever rickettsia, *Cell. Immunol.,* 28, 198, 1977.
12. **Clark, I. A.,** Resistance to *Babesia* sp. and *Plasmodium* sp. in mice pretreated with an extract of *Coxiella burnetii, Infect. Immun.,* 24, 319, 1979.
13. **Kazar, J. and Schramek, S.,** Inhibition by *Coxiella burnetii* of ascites tumour formation in mice, *Acta. Virol.,* 23, 267, 1979.
14. **Kelly, M. T., Granger, D. L., Ribi, E., Milner, K. C., Strain, S. M., and Stoenner, H. G.,** Tumor regression with Q fever rickettsiae and a mycobacterial glycolipid, *Cancer Immunol. Immunother.,* 1, 187, 1976.
15. **Koster, F. T., Williams, J. C., and Goodwin, J. S.,** Cellular immunity in Q fever; modulation of responsiveness by a suppressor T cell-monocyte circuit, *J. Immunol.,* 135, 1067, 1985.
16. **Williams, J. C., Damrow, T. A., Waag, D. M., and Amano, K.-I.,** Characterization of a phase I *Coxiella burnetii* chloroform-methanol residue vaccine that induces active immunity against Q fever in C57BL/10 ScN mice, *Infect. Immun.,* 51, 851, 1986.
17. **Damrow, T. A., Williams, J. C., and Waag, D. M.,** Suppression of *in vitro* proliferation in C57BL/10 ScN mice vaccinated with phase I *Coxiella burnetii, Infect. Immun.,* 47, 149, 1985.
18. **Williams, J. C. and Cantrell, J. L.,** Biological and immunological properties of *Coxiella burnetii* vaccine in C57BL/10 ScN endotoxin-nonresponder mice, *Infect. Immun.,* 35, 1091, 1982.
19. **Tigertt, W. D., Benenson, A. S., and Gochenour, W. S.,** Airborne Q fever, *Bacteriol. Rev.,* 25, 285, 1961.
20. **Benson, W. W., Brock, D. W., and Mather, J.,** Serologic analysis of a penitentiary group using raw milk from a Q fever infected herd, *Public Health Rep.,* 78, 707, 1963.
21. **Biberstein, E. L., Behymer, D. E., Bushnell, R., Crenshaw, G., Ripmann, H. P., and Franti, C. E.,** A survey of Q fever *(Coxiella burnetii)* in California dairy cows, *Am. J. Vet. Res.,* 35, 1577, 1974.
22. **Davis, G. E. and Cox, H. R.,** A filter-passing infectious agent isolated from ticks. I. Isolation from *Dermacentor andersoni,* reactions in animals, and filtration experiments, *Public Health Rep.,* 53, 2259, 1938.
23. **Cox, H. R.,** Studies of a filter-passing infectious agent isolated from ticks. V. Further attempts to cultivate in cell-free media. Suggested Classification, *Public Health Rep.,* 54, 1822, 1939.
24. **Ratnoff, O. D.,** A tangled web. The interdependence of mechanisms of blood clotting, fibrinolysis, immunity and inflammation, *Thromb. Diath. Haemorrh.,* 45, 109, 1971.
25. **Larsen, G. L. and Henson, P. M.,** Mediators of inflammation, in *Annual Review of Immunology,* Paul, W. E., Fathman, C. G., and Metzger, H., Eds., Annual Reviews Inc., 1983, 335.

26. **Schiffmann, E., et al.,** Some characteristics of the neutrophil receptor for chemotactic peptides, *FEBS Lett.,* 117, 1, 1980.
27. **Adams, D. O.,** Molecular interactions in macrophage activation, *Immunol. Today,* 10, 33, 1989.
28. **Berenberg, J. L. and Ward, P. A.,** Chemotactic factor inactivator in normal human serum, *J. Clin. Invest.,* 52, 1200, 1973.
29. **Moulder, J. W.,** Comparative biology of intracellular parasitism, *Microbiol. Rev.,* 49, 298, 1985.
30. **Jonson, P., Lindberg, M., Haraldsson, I., and Wadstrom, T.,** Virulence of *Staphylococcus aureus* in a mouse mastitis model: studies of alpha hemolysin, coagulase, and protein A as possible virulence determinants with protoplast fusion and gene cloning, *Infect. Immun.,* 49, 765, 1985.
31. **Cohn, Z. A., Bozeman, F. M., Campbell, J. M., Humphries, J. W., and Sawyer, T. K.,** Study on growth of rickettsia. V. Penetration of *Rickettsia tsutsugamushi* into mammalian cells *in vitro, J. Exp. Med.,* 109, 271, 1959.
32. **Williams, J. C., McCaul, T. F., Thompson, H. A., and Waag, D. M.,** Molecular strategies for uptake and phagolysosomal growth of *Coxiella burnetii* in nonimmune and immune hosts, in *Intracellular Parasites,* Moulder, J. W., Ed., CRC Press, Boca Raton, FL, in press.
33. **Joiner, K. A., Brown, E. J., and Frank, M. M.,** Complement and bacteria: chemistry and biology in host defense, *Annu. Rev. Immunol.,* 2, 461, 1984.
34. **Volanakis, J. E. and Kaplan, M. H.,** Specificity of C-reactive protein for choline phosphate residues of pneumococcal C-polysaccharide, *Proc. Soc. Exp. Biol. Med.,* 136, 612, 1971.
35. **DiCamelli, R., Potempa, L. A., Siegel, J., Suyehira, L., Petras, K., and Gewuz, H.,** Binding reactivity of C-reactive protein for polycations, *J. Immunol.,* 125, 1933, 1980.
36. **Kindmark, C. O. and Williams, J. C.,** Interaction between *Coxiella burnetii* and C-reactive protein, in press.
37. **Kindmark, C. O. and Williams, J. C.,** unpublished results.
38. **Williams, J. C. and Joiner, K.,** unpublished observations.
39. **Vishwanath, S. and Hackstadt, T.,** Lipopolysaccharide phase variation determines the complement-mediated serum susceptibility of *Coxiella burnetii, Infect. Immun.,* 56, 40, 1988.
40. **Westlake, P., Price, L. M., Russell, M., and Kelly, J. K.,** The pathology of Q fever hepatitis. A case diagnosed by liver biopsy, *J. Clin. Gastroenterol.,* 9, 357, 1987.
41. **Pelligrin, M., Deisol, D., Auvergnat, J. C., Familiades, J., Faure, H., Guiu, M., and Voigt, J. J.,** Granulomatous hepatitis in Q fever, *Hum. Pathol.,* 11, 51, 1980.
42. **Tobin, M. J., Cahill, N., Gearty, G., Maurer, B., Blake, S., Daly, K., and Hone, R.,** Q fever endocarditis, *Am. J. Med.,* 72, 396, 1982.
43. **Haldane, E. V., Marrie, T. J., Faulkner, R. S., Lee, S. H., Cooper, J. H., MacPherson, D. D., and Montague, T. J.,** Endocarditis due to Q fever in Nova Scotia: experience with five patients in 1981-1982, *J. Inf. Dis.,* 148, 978, 1983.
44. **Steer, G. J. and Ashwell, G.,** Studies on a mammalian hepatic binding protein specific for asialoglycoproteins. Evidence for receptor recycling in isolated rat hepatocytes, *J. Biol. Chem.,* 255, 3008, 1980.
45. **Stevens, M. D., DiSilvestro, R. A., and Harris, E. D.,** Specific receptor for ceruloplasmin in membrane fragments from aortic and heart tissues, *Biochemistry,* 23, 261, 1984.
46. **Williams, J. C.,** unpublished results.
47. **Fearon, D. T., Kaneko, I., and Thompson, G. G.,** Membrane distribution and adsorptive endocytosis by C3b receptors on human polymorphonuclear leukocytes, *J. Exp. Med.,* 153, 1615, 1981.
48. **Anderson, R. G. W. and Kaplan, J.,** Receptor-mediated endocytosis, *Modern Cell Biol.,* 1, 1, 1983.
49. **Kaplan, J.,** Patterns in receptor behavior and function, in *Mechanisms of Receptor Regulation,* Poste, G. and Crooke, S. T., Eds., Plenum Press, New York, 1985, 13.
50. **Hackstadt, T. and Williams, J. C.,** Biochemical stratagem for obligate parasitism of eukaryotic cells by *Coxiella burnetii, Proc. Natl. Acad. Sci. U.S.A.,* 78, 3240, 1981.
51. **Klien, J.,** Immunology: the science of self-nonself discrimination, Wiley-Interscience, New York, 1982.
52. **Williams, J. C., Peacock, M. G., and Kindmark, C. L.,** Detection of *Coxiella burnetii* soluble antigens by immunoelectrophoresis: demonstration of antigen in the sera of guinea pigs during experimental Q fever, in *Rickettsiae and Rickettsial Diseases,* Burgdorfer, W. and Anacker, R. L., Eds., Academic Press, New York, 1981, 103.
53. **Harding, C. V., Leyva-Corbian, F., and Unanue, U. R.,** Mechanisms of antigen processing, *Immunol. Rev.,* 106, 77, 1988.
54. **Durum, S. K., Schmidt, J. A., and Oppenheim, J. J.,** Interleukin 1: an immunological perspective, *Annu. Rev. Immunol.,* 3, 263, 1985.
55. **Lennette, E. H., Clark, W. H., Jensen, F. W., and Toomb, C. J.,** Q fever studies. XV. Development and persistence in man of complement-fixing and agglutinating antibodies to *Coxiella burnetii, J. Immunol.,* 68, 591, 1952.
56. **Stoker, M. G. P. and Fiset, P.,** Phase variation of the Nine Mile and other strains of *Rickettsia burnetii, Can. J. Microbiol.,* 2, 310, 1956.
57. **Fiset, P., Ormsbee, R. A., Silberman, R., Peacock, M., and Spielman, S. H.,** The microagglutination technique for detection and measurement of rickettsial antibodies, *Acta. Virol.,* 13, 60, 1969.

58. **Peacock, M. G., Philip, R. N., Williams, J. C., and Faulkner, R. S.,** Serological evaluation of Q fever in humans: enhanced phase I titers of immunoglobulins G and A are diagnostic for Q fever endocarditis, *Infect. Immun.,* 41, 1089, 1983.

59. **Field, P. R., Hunt, J. G., and Murphy, A. M.,** Detection and persistence of specific IgM antibody to *Coxiella burnetii* by enzyme-linked immunosorbent assay: a comparison with immunofluorescence and complement fixation tests, *J. Inf. Dis.,* 148, 477, 1983.

60. **Hunt, J. G., Field, P. R., and Murphy, A. M.,** Immunoglobulin responses to *Coxiella burnetii* (Q fever): single-serum diagnosis of acute infection, using an immunofluorescence technique, *Infect. Immun.,* 39, 977, 1983.

61. **Williams, J. C., Thomas, L. A., and Peacock, M. G.,** Humoral immune response to Q fever: enzyme-linked immunosorbent assay antibody response to *Coxiella burnetii* in experimentally infected guinea pigs, *J. Clin. Microbiol.,* 24, 935, 1986.

62. **Behymer, D. E., Ruppanner, R., Brooks, D., Williams, J. C., and Franti, C. E.,** Enzyme immunoassay for surveillance of Q fever, *Am. J. Vet. Res.,* 46, 2413, 1985.

63. **Dupuis, G., Peter, O., Peacock, M., Burgdorfer, W., and Haller, E.,** Immunoglobulin responses in acute Q fever, *J. Clin. Microbiol.,* 22, 484, 1985.

64. **Peter, O., Dupuis, G., Peacock, M. G., and Burgdorfer, W.,** Comparison of enzyme-linked immunosorbent assay and complement fixation and indirect fluorescent-antibody tests for detection of *Coxiella burnetii* antibody, *J. Clin. Microbiol.,* 25, 1063, 1987.

65. **Williams, J. C. and Waag, D. M.,** unpublished results.

66. **Williams, J. C., Waag, D. M., and Marrie, T.,** manuscript in preparation.

67. **Peter, O., Dupuis, G., Bee, D., Luthy, R., Nicolet, J., and Burgdorfer, W.,** Enzyme-linked immunosorbent assay for diagnosis of chronic Q fever, *J. Clin. Microbiol.,* 26, 1978, 1988.

68. **Derrick, E. H., Smith, D. J. W., Brown, H. E., and Freeman, M.,** The role of the bandicoot in the epidemiology of "Q" fever: a preliminary study, *Med. J. Aust.,* 1, 150, 1939.

69. **Abinanti, F. R. and Marmion, B. P.,** Protective or neutralizing antibody in Q fever, *Am. J. Hyg.,* 66, 173, 1957.

70. **Burnet, F. M. and Freeman, M.,** "Q" fever: factors affecting the appearance of rickettsiae in mice, *Med. J. Aust.,* 2, 1114, 1938.

71. **Humphries, R. C. and Hinrichs, D. J.,** Role of antibody in *Coxiella burnetii* infection, *Infect. Immun.,* 31, 641, 1981.

72. **Hinrichs, D. J. and Jerrells, T. R.,** *In vitro* evaluation of immunity to *Coxiella burnetii, J. Immunol.,* 117, 996, 1976.

73. **Kishimoto, R. A., Veltri, B. J., Shirey, F. G., Canonico, P. G., and Walker, J. S.,** Fate of *Coxiella burnetii* in macrophages from immune guinea pigs, *Infect. Immun.,* 15, 601, 1977.

74. **Kishimoto, R. A. and Walker, J. S.,** Interaction between *Coxiella burnetii* and guinea pig peritoneal macrophages, *Infect. Immun.,* 14, 416, 1976.

75. **Scott, G. H., Williams, J. C., and Stephenson, E. H.,** Animal models in Q fever: pathological responses of inbred mice to phase I *Coxiella burnetii, J. Gen. Microbiol.,* 133, 691, 1987.

76. **Cerquetti, M. C., Sordelli, D. O., Ortegon, R. A., and Bellanti, J. A.,** Impaired lung defenses against *Staphylococcus aureus* in mice with hereditary deficiency of the fifth component of complement, *Infect. Immun.,* 41, 1072, 1983.

77. **Boraschi, D. and Meltzer, M. S.,** Defective tumoricidal capacity of macrophages from A/J mice. II. Comparison of the macrophage cytotoxic defect of A/J mice with that of lipid A-unresponsive C3H/HeJ mice, *J. Immunol.,* 122, 1592, 1979.

78. **Stockman, G. D., Heim, L. R., South, M. A., and Trentin, J. J.,** Differential effects of cyclophosphamide on the B and T cell compartments of adult mice, *J. Immunol.,* 110, 277, 1973.

79. **Kazar, J., Rajcani, J., and Schramek, S.,** Differential effects of cyclophosphamide on *Coxiella burnetii* infection in mice, *Acta. Virol.,* 26, 174, 1982.

80. **Ascher, M. S., Jahrling, P. B., Harrington, D. G., Kishimoto, R. A., and McGann, V. G.,** Mechanisms of protective immunogenicity of microbial vaccines. Effects of cyclophosphamide pretreatment in Venezuelan encephalitis, Q fever, and tularemia, *Clin. Exp. Immunol.,* 41, 225, 1980.

81. **Adams, D. O. and Hamilton, T. A.,** The cell biology of macrophage activation, *Annu. Rev. Immunol.,* 2, 283, 1984.

82. **Ortaldo, J. R. and Herberman, R. B.,** Hererogeneity of natural killer cells, *Annu. Rev. Immunol.,* 2, 359, 1984.

83. **Koster, F. T., Kirkpatrick, T. L., Rowatt, J. D., and Baca, O. G.,** Antibody-dependent cellular cytotoxicity of *Coxiella burnetii*-infected J774 macrophage target cells, *Infect. Immun.,* 43, 253, 1984.

84. **Burnet, F. M. and Freeman, M.,** Studies of the X strain (Dyer) of *Rickettsia burnetii.* II. Guinea pig infections, with special reference to immunological phenomena, *J. Immunol.,* 40, 421, 1940.

85. **Kishimoto, R. A., Rozmiarek, H., and Larson, E. W.,** Experimental Q fever infection in congenitally athymic nude mice, *Infect. Immun.,* 22, 69, 1978.

86. **Hall, W. C., White, J. D., Kishimoto, R. A., and Whitmire, R. E.,** Aerosol Q fever infection of the nude mouse, *Vet. Pathol.,* 18, 672, 1981.

87. **Cohn, Z. A.,** The activation of mononuclear phagocytes: fact, fancy, and future, *J. Immunol.,* 121, 813, 1978.
88. **Page, R. C., Davies, P., and Allison, A. C.,** The macrophages as a secretory cell, *Int. Rev. Cytol.,* 52, 119, 1978.
89. **Amano, K.-I. and Williams, J. C.,** Sensitivity of *Coxiella burnetii* peptidoglycan to lysozyme hydrolysis and correlation of sacculus rigidity with peptidoglycan-associated proteins, *J. Bacteriol.,* 160, 989, 1984.
90. **Johnston, R. B.,** Enhancement of phagocytosis-associated oxidase metabolism as a manifestation of macrophage activation, *Lymphokines,* 3, 33, 1981.
91. **Unkeless, J. C., Scigliano, E., and Freedman, V. H.,** Structure and function of human and murine receptors for IgG, *Annu. Rev. of Immunol.,* 6, 251, 1988.
92. **Griffin, F. M.,** Activation of macrophage complement receptors for phagocytosis, *Contemp. Top. Immunbiol.,* 14, 57, 1984.
93. **Ezekowitz, R. A. B. and Gordon, S.,** Down-regulation of mannosyl receptor-mediated endocytosis and antigen F4/80 in bacillus Calmette-Guerin-activated mouse macrophages. Role of T lymphocytes and lymphokines, *J. Exp. Med.,* 155, 1623, 1982.
94. **Snyderman, R. and Goetzl, E.,** Chemoattractant receptors on phagocytic cells, *Annu. Rev. Immunol.,* 1, 257, 1981.
95. **Kishimoto, R. A. and Burger, G. T.,** Appearance of cellular and humoral immunity in guinea pigs after infection with *Coxiella burnetii* administered in small-particle aerosols, *Infect. Immun.,* 16, 518, 1977.
96. **Heggers, J. P., Mallavia, L. P., and Hinrichs, D. J.,** The cellular immune response to antigens of *Coxiella burnetii, Can. J. Microbiol.,* 20, 657, 1974.
97. **Jerrells, T. R., Mallavia, L. P., and Hinrichs, D. J.,** Detection of long-term cellular immunity to *Coxiella burnetii* as assayed by lymphocyte transformation, *Infect Immun.,* 11, 280, 1975.
98. **Turk, J. L.,** *Delayed Hypersensitivity,* 3rd ed., Elsevier/North-Holland, Amsterdam, 1980.
99. **Ascher, M. S., Berman, M. A., Parker, D., and Turk, J. L.,** Experimental model for dermal granulomatous hypersensitivity in Q fever, *Infect. Immun.,* 39, 388, 1983.
100. **Damrow, T. A., Williams, J. C., and Waag, D. M.,** unpublished results.
101. **Waag, D. M. and Williams, J. C.,** manuscript in preparation.
102. **Carswell, E. A., Old, L. J., Kassel, R. L., Green, S., Fiore, N., and Williamson, B.,** An endotoxin-induced serum factor that causes necrosis of tumors, *Proc. Natl. Acad. Sci. U.S.A.,* 72, 3666, 1975.
103. **Old, L. J.,** Tumor necrosis factor (TNF), *Science,* 230, 630, 1985.
104. **Beutler, B., Milsark, I. W., and Cerami, A. C.,** Passive immunization against cachectin/tumor necrosis factor protects mice from lethal effect of endotoxin, *Science,* 229, 869, 1985.
105. **Heidenreich, S., Weyers, M., Gong, J.-H., Sprenger, H., Nain, M., and Gemsa, D.,** Potentiation of lymphokine-induced macrophage activation by tumor necrosis factor, *J. Immunol.,* 140, 1511, 1988.
106. **Kazar, J.,** Interferon-like inhibitor in mouse sera induced by rickettsiae, *Acta. Virol.,* 10, 277, 1966.
107. **Kuwata, T., Fuse, A, and Moringa, N.,** Effect of cycloheximide and puromycin on the antiviral and anticellular activities of human interferon, *J. Gen. Virol.,* 37, 195, 1977.
108. **Pace, J. L., Russell, S. W., Torres, B. A., Johnson, H. M., and Gray, P. W.,** Recombinant mouse gamma interferon induces the priming step in macrophage activation for tumor cell killing, *J. Immunol.,* 130, 2011, 1983.
109. **Buchmeier, N. A. and Schreiber, R. D.,** Requirement of endogenous gamma interferon production for resolution of *Listeria* monocytogenes infection, *Proc. Natl. Acad. Sci. U.S.A.,* 82, 7404, 1985.
110. **Tamura, M., Saito, S., and Sasakawa, S.,** Induction of human interferon gamma in nylon-column nonadherent human lymphocyte cells by heat-treated poly I:poly C, *J. Interferon Res.,* 5, 77, 1985.
111. **Turco, J., Thompson, H. A., and Winkler, H. H.,** Interferon gamma inhibits growth of *Coxiella burnetii* in mouse fibroblasts, *Infect. Immun.,* 45, 781, 1984.
112. **Cole, P.,** Activation of mouse peritoneal cells to kill *Listeria monocytogenes* by T-lymphocyte products, *Infect. Immun.,* 12, 36, 1975.
113. **Patterson, R. J. and Youmans, G. P.,** Demonstration in tissue culture of lymphocyte-mediated immunity to tuberculosis, *Infect. Immun.,* 1, 600, 1970.
114. **Ziegler, H. K., Staffileno, L. K., and Wentworth, P.,** Modulation of macrophage Ia-expression by lipopoly-saccharide. I. Induction of Ia expression *in vivo, J. Immunol.,* 133, 1825, 1984.
115. **Sewell, W. A., Munoz, J. J., and Vadas, M. A.,** Enhancement of the intensity, persistence and passive transfer of delayed type hypersensitivity lesions by pertussigen in mice, *J. Exp. Med.,* 157, 2087, 1983.
116. **Merser, C., Sinay, P., and Adam, A.,** Total synthesis and adjuvant activity of bacterial peptidoglycan derivatives, *Biochem. Biophys. Res. Commun.,* 66, 1316, 1975.
117. **Zabriskie, J.,** Mimetic relationships between group A streptococci and mammalian tissues, *Adv. Immunol.,* 7, 147, 1967.
118. **Cunningham, M. W. and Swerlick, R. A.,** Polyspecificity of antistreptococcal murine monoclonal antibodies and their implications in autoimmunity, *J. Exp. Med.,* 164, 998, 1986.
119. **Arneborn, P. and Biberfeld, G.,** T-lymphocyte subpopulations in relation to immunosuppression in measles and varicella, *Infect. Immun.,* 39, 29, 1983.

120. **Starr, S. and Berkovich, S.,** Effects of measles, gamma globulin modified measles and vaccine measles on the tuberculin test, *N. Engl. J. Med.,* 270, 386, 1964.
121. **Kauffman, C. F., Phair, J. P., Linnemann, C. C., Jr., and Schiff, G. M.,** Cell-mediated immunity in humans during viral infection. I. Effect of rubella on dermal hypersensitivity, phytohemagglutinin response and T lymphocyte numbers, *Infect. Immun.,* 10, 212, 1974.
122. **Kantzler, G. B., Lauteria, S. F., Cusumano, C. L., Lee, J. D., Ganguly, R., and Waldman, R. H.,** Immunosuppression during influenza virus infection, *Infect. Immun.,* 10, 996, 1974.
123. **Williams, F. T. C., Melnick, J. L., and Rawls, W. E.,** Viral inhibition of the phytohemagglutinin response of human lymphocytes and application to viral hepatitis, *Proc. Soc. Exp. Biol. Med.,* 130, 652, 1969.
124. **Bloom, B. R. and Mehra, V.,** Immunological unresponsiveness in leprosy, *Immunol. Rev.,* 80, 5, 1984.
125. **Reiner, N. E. and Finke, J. H.,** Interleukin 2 deficiency in murine *Leishmaniasis donovani* and its relationship to depressed spleen cell responses to phytohemagglutinin, *J. Immunol.,* 131, 1487, 1983.
126. **Albright, J. W., Albright, J. F., and Dusanic, D. G.,** Mechanism of trypanosome-mediated suppression of humoral immunity in mice, *Proc. Natl. Acad. Sci. U.S.A.,* 75, 3923, 1978.
127. **Rivas, V. and Rogers, T. J.,** Studies on the cellular nature of Candida albicans-induced suppression, *J. Immunol.,* 130, 376, 1983.
128. **Eardley, D. D. and Jayawardena, A. N.,** Suppressor cells in mice infected with *Trypanosoma brucei, J. Immunol.,* 119, 1029, 1977.
129. **Holmgren, J., Lindholm, L., and Lonnröth, I.,** Interaction of cholera toxin and toxin derivatives with lymphocytes. I. Binding properties and interference with lectin-induced cellular stimulation, *J. Exp. Med.,* 139, 801, 1974.
130. **Malakian, A. H. and Schwab, J. H.,** Immunosuppressant from group A streptococci, *Science,* 159, 880, 1968.
131. **Floerscheim, G. L., Hopff, W. H., Gasser, M., and Bucher, K.,** Impairment of cell mediated immune response by *Pseudomonas aeruginosa, Clin. Exp. Immunol.,* 9, 241, 1971.
132. **Shenker, B. J., Listgarten, M. A., and Tichman, N. S.,** Suppression of human lymphocyte responses by oral spirochetes: a monocyte-dependent phenomenon, *J. Immunol.,* 132, 2039, 1984.
133. **Kishimoto, R. A. and Gonder, J. C.,** Suppression of PHA-stimulated lymphocyte transformation in cynomolgus monkeys following infection with *Coxiella burnetii, Can. J. Microbiol.,* 25, 949, 1979.
134. **Khavkin, T. and Tabibzadeh, S. S.,** Histologic, immunofluorescence, and electron microscopic study of infectious process in mouse lung after intranasal challenge with *Coxiella burnetii, Infect. Immun.,* 56, 1792, 1988.
135. **Baca, O. G., Scott, T. O., Akporiaye, E. T., DeBlassie, K. R., and Crissman, H. A.,** Cell cycle distribution patterns and generation times of L929 fibroblast cells persistently infected with *Coxiella burnetii, Infect. Immun.,* 47, 366, 1985.
136. **Roman, M. J., Coriz, P. D., and Baca, O. G.,** A proposed model to explain persistent infection of host cells with *Coxiella burnetii, J. Gen. Microbiol.,* 132, 1415, 1986.
137. **Syrucek, L., Sobeslavsky, O., and Gutvirth, I.,** Isolation of *Coxiella burnetii* from human placentas, *J. Hyg. Epidemiol. Microbiol. Immunobiol.,* 2, 29, 1958.
138. **Sidwell, R. W. and Gebhardt, L. P.,** Studies of latent Q fever infections. III. Effects of partiturition upon latently infected guinea pigs and white mice, *Am. J. Epidemiol.,* 84, 132, 1966.
139. **Sidwell, R. W., Thorpe, B. D., and Gebhardt, L. P.,** Studies of latent Q fever infections. II. Effects of multiple cortisone injections, *Am. J. Hyg.,* 79, 113, 1964.
140. **Sidwell, R. W., Thorpe, B. D., and Gebhardt, L. P.,** Studies of latent Q fever infections. I. Effects of whole body X-irradiation upon latently infected guinea pigs, white mice and deer mice, *Am. J. Hyg.,* 79, 113, 1964.
141. **Young, D. B. and Buchanan, T. M.,** A serological test for leprosy with a glycolipid specific for *Mycobacterium leprae, Science,* 221, 1057, 1983.
142. **Mehra, V., Brennan, P. J., Rada, E., Convit, J., and Bloom, B. R.,** Lymphocyte suppression in leprosy induced by a unique *Mycobacterium leprae* glycolipid, *Nature (London),* 308, 194, 1984.

Chapter 6

ACUTE Q FEVER

Thomas. J. Marrie

TABLE OF CONTENTS

I. INTRODUCTION

The clinical manifestations of Q fever are readily divided into acute and chronic forms. The latter almost invariably means endocarditis,[1] but vertebral osteomyelitis and granulomatous hepatitis may be other manifestations of chronic Q fever.[2] In this chapter, the manifestations of acute Q fever will be described. Also, Q fever pneumonia will be compared with pneumonia due to *Mycoplasma pneumoniae* and to *Legionella pneumophila*.

II. ACUTE Q FEVER SYNDROME

A. SELF-LIMITED FEBRILE ILLNESS

This is probably the most frequent manifestation of Q fever. In many areas 11 to 12% of the population has serological evidence of previous infection with *Coxiella burnetii*[3] — most do not recall pneumonia or other serious illness.[3] It is likely that some infections are totally asymptomatic.[4]

The symptoms of acute Q fever vary somewhat from country to country. Table 1 is a compilation of the symptoms reported by investigators who studied patients with Q fever in five different countries.[5-9]

The onset of acute Q fever is nearly always abrupt. Fever, fatigue, chills, and headache are the most common symptoms. The spectrum of Q fever is illustrated by cases 1 to 6 in Sections V through VIII and by Figures 1 to 9. These cases should be reviewed before reading the remainder of this chapter.

B. SYMPTOMS OF Q FEVER
1. Fever

The duration of fever in untreated patients ranged from 5 to 57 days.[10] In the 138 patients studied by Derrick, the duration of fever was 13 d or less for 67% of the cases.[10] Temperature peaks of 39 to 40°C are common. In our experience the temperature remains elevated all day but does show considerable fluctuation during the day. In untreated Q fever the most common pattern was a rapid ascent for 2 to 4 d, a plateau at 39 to 40°C, defervescence by rapid lysis, and a duration of 5 to 14 d.[10] About one quarter of the patients had biphasic fever. The first phase was as described above, but after falling to normal or nearly normal, the temperature rose again. This phase lasted 1 to 19 d, the temperature was usually lower than in the first phase and intermittent. Both Derrick[10] and Clark and co-workers[5] observed that the duration of fever increased with increasing age. In the California study[5] fever lasted over 14 d in 60% of those 40 years of age and older, compared with 29% of those under 40 years of age. The longest duration of fever in untreated Q fever was 57 d.

TABLE 1

Symptoms of Acute Q Fever in Patients from Five Different Countries

	Northern California, 1948—1949 (5)	Australia, 1962—1981 (6)	Switzerland, 1983 (7)	Uruguay, 1975—1985 (8)	Nova Scotia, 1983—1986 (9)
			Place and time of study (ref.)		
			No. of patients studied		
	180	111	91	1358[a]	51
			Percent with indicated symptom		
Fever	100	100	88	98	94
Fatigue	100	NS[b]	97	98	98
Chills	74	68	NS	NS	88
Headache	65	86	77	98	73
Myalgia	47	60	64	NS	69
Sweats	31	NS	NS	98	84
Cough	24	32	70	90	28
Nausea	22	25	25	NS	49
Vomiting	13	42	25	NS	25
Chest pain	10	NS	34	NS	28
Diarrhea	5	7	NS	NS	22
Sore throat	5	NS	27	NS	14
Rash	4	8	5	NS	18

[a] All cases occurred in workers at meat-processing plants.

[b] NS = not stated.

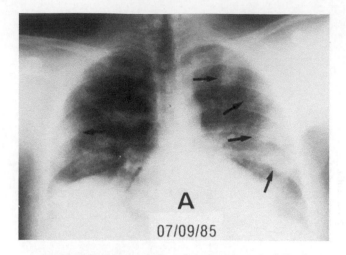

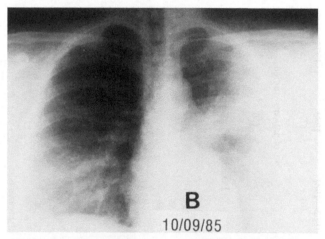

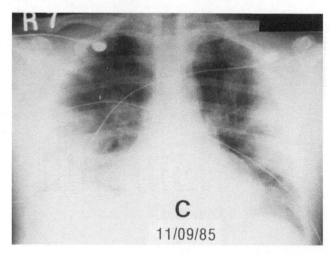

FIGURE 1.   Serial chest radiographs of a young man who developed Q fever pneumonia following exposure to an infected cat placenta. He cleaned up the products of conception from a sofa on which his cat gave birth to kittens. Note the multiple rounded opacities in both lung fields on 07.09.85. Also note the coalescence and progressive increase in size of these opacities on 10.09.85 and 11.09.85.

2. Fatigue

This is the second most common symptom in all the series summarized in Table 1. It occurs significantly more often in Q fever than in other lower respiratory tract infections (Table 2). It is described as a feeling of extreme lassitude. Rarely the feeling of fatigue persists for weeks following resolution of all other symptoms.

3. Chills

Many patients with Q fever complain of a chilly sensation. This usually lasts 3 to 4 d. Of our patients with Q fever, 20% had rigors. True rigors were uncommon among allied troops who acquired Q fever in the Mediterranean area. Clark et al.[5] found that on occasion rigors occurred repeatedly at daily intervals for as long as 3 weeks.

4. Headache

Headache is usually the most distressing complaint to the patient. Many state it is "the worst headache that I have ever had". The complaint of severe pain which on occasion is combined with nuchal rigidity, prompts a lumbar puncture to rule out meningitis. The headache is usually frontal and throbbing in nature. Retroorbital pain accompanies the headache in about 10 to 15% of patients. Headache occurs significantly more often in Q fever than it does in other lower respiratory tract infections (Table 2). The headache usually resolves when the patient becomes afebrile. It is likely that headache represents infection of the central nervous system, as evidenced by isolation of *C. burnetii* from the spinal fluid of a soldier with Q fever.[11] His only manifestation of central nervous system infection was headache. Denlinger[12] performed lumbar punctures on 21 patients with Q fever. The cerebrospinal fluid pressure was increased in some patients; the highest value was 250 mm of water. The protein concentration was normal, and the white blood cell count was normal in all but one. The range was 0 to 10 lymphocytes per cubic millimeter, and the mean white blood cell count was 2/mm.[3]

5. Myalgia

About half the patients with Q fever complain of myalgia (Table 1); however, these pains are usually mild and do not occur significantly more often in Q fever than in other lower respiratory tract infections (Table 2).

6. Sweats

Sweats usually cease promptly following institution of appropriate antibiotic therapy. Rarely, drenching sweats continue for 2 to 4 d following admission to hospital.

7. Cough

Cough is a surprisingly infrequent manifestation for an infection that enters the body through the respiratory tract. The number of patients who experienced cough varied considerably among the five series of patients shown in Table 1. Patients who developed Q fever in Switzerland and Uruguay frequently had a cough, 70 and 90%, respectively, whereas in the other series it ranged from 4 to 32% of patients. Cough occurred significantly less frequently in patients with Q fever in Nova Scotia than it did among patients with other lower respiratory tract infections (Table 2). The cough is almost always nonproductive.

8. Chest Pain

Chest pain is also an infrequent complaint (Tables 1 and 2); however, in contrast to cough when chest pain occurs, it can be very severe. It is usually pleuritic in nature. The sudden onset of severe pleuritic chest pain in combination with a pleural-based, wedge-shaped pulmonary opacity suggests a diagnosis of pulmonary infarction. We have seen patients with Q fever who have been misdiagnosed as pulmonary embolism with pulmonary infarction.

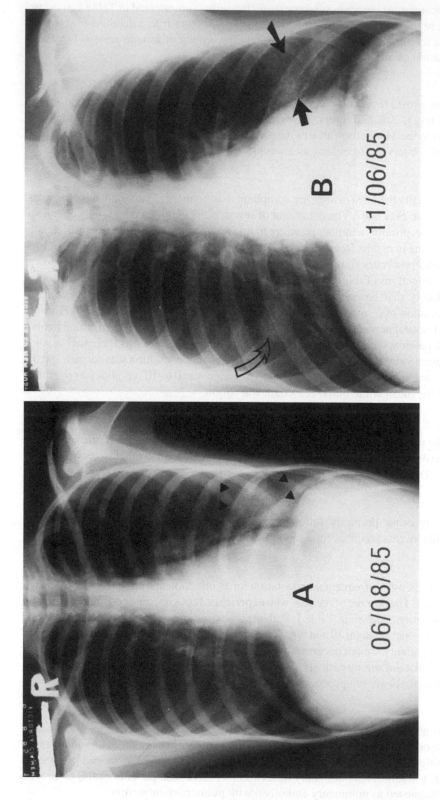

FIGURE 2.   Composite of the chest radiographs of four patients with Q fever pneumonia. (A) Chest radiograph of a 13-year-old female. There is a rounded opacity in the left lower lobe (arrowheads). (B) Chest radiograph of a 30-year-old male. There is a rounded opacity in the right lower lobe and a subsegmental opacity at the left base (arrows). (C) Chest radiograph of a 19-year-old female. A small left upper zone opacity is evident (arrows). (D) Chest radiograph of a 17-year-old female. A small opacity is seen in the right upper lobe. These four patients were moderately ill. All were part of an outbreak of Q fever.

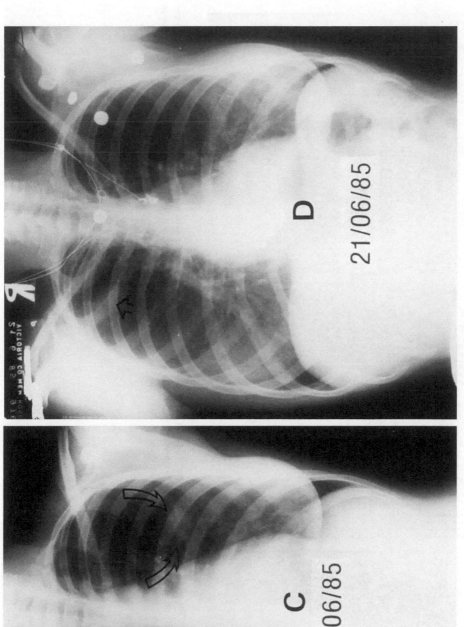

FIGURE 2 continued.

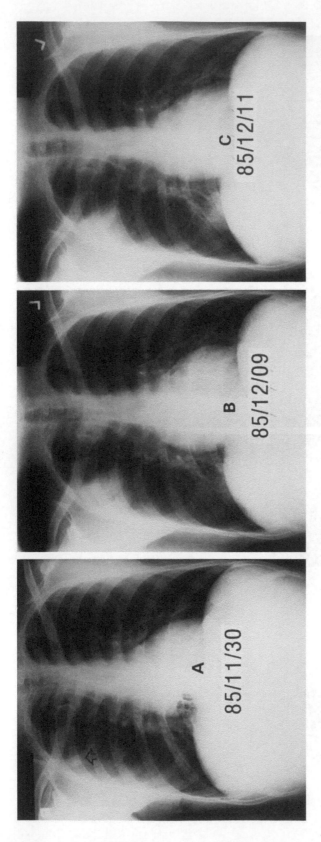

FIGURE 3.   Serial chest radiographs of Case 1. The chest radiograph on November 30, 1985 (A) was obtained a few hours following the onset of pleuritic chest pain. Note the small right upper lobe rounded opacity (arrow). Nine days later (B), this opacity had increased considerably in size, had become wedge-shaped in appearance, and was pleural based. Tetracycline therapy was begun on December 16, 1985. One month later, January 22, 1986 (F), the pulmonary opacity had almost completely resolved.

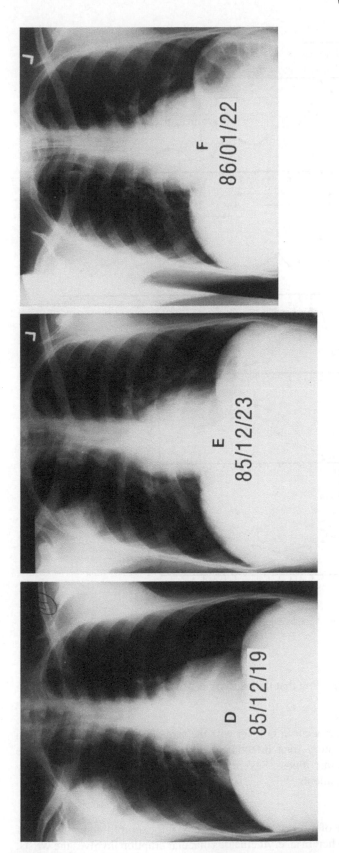

FIGURE 3 continued.

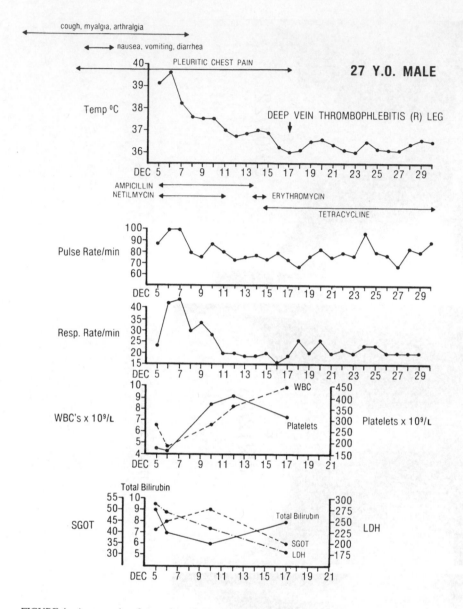

FIGURE 4.   A composite of several graphs depicting various clinical and laboratory parameters of Case 1.

9. Sore Throat

Sore throat occurs rarely, suggesting that the pharynx is usually not infected.

10. Gastrointestinal Symptoms

Nausea, vomiting, and diarrhea occur in Q fever with the same frequency as they occur in patients with other lower respiratory tract infections (Table 2), suggesting that these are nonspecific symptoms of a systemic illness. Severe diarrhea has been described as the only manifestation of Q fever in two patients.[13]

11. Rash

Rash is a characteristic feature of most rickettsial infections. It is exemplified by the rash of Rocky Mountain spotted fever, wherein an erythematous macular eruption involves the wrists,

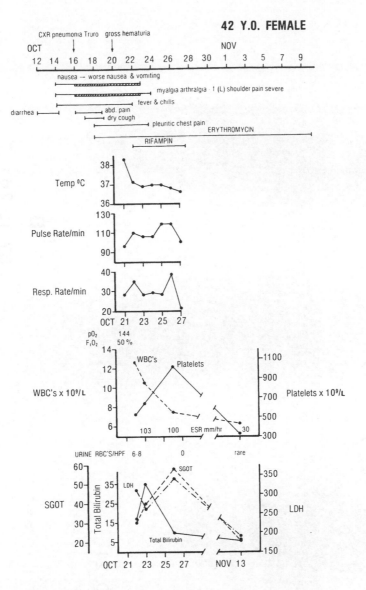

FIGURE 5. Clinical and laboratory data Case 2. Note the marked thrombocytosis and prompt resolution of the fever following the addition of rifampin.

ankles, soles, palms, and forearms. It then progresses to become macular and 3 to 4 d later petechial. These lesions correlate with infection of the endothelial cells of the arterioles and capillaries, resulting in an infectious vasculitis.[14]

A small number of patients with Q fever, 4 to 18%, complain of rash. In contrast to other rickettsial infections, the rash in Q fever is nonspecific. Clark et al.[5] observed 7 patients with Q fever and rash:

> Two displayed pruritic red papules over the upper part of the body on days 8 and 12 of the illness respectively; two patients exhibited faint pink macular lesions on the abdomen, shoulders and thighs during the second week of illness; these rashes were evanescent, disappearing within a few hours. Numerous petechiae were observed in two patients, shown by laboratory studies to have considerable impairment of hepatic function.

I have not observed a rash in the 70 patients with acute Q fever that I have examined, despite the

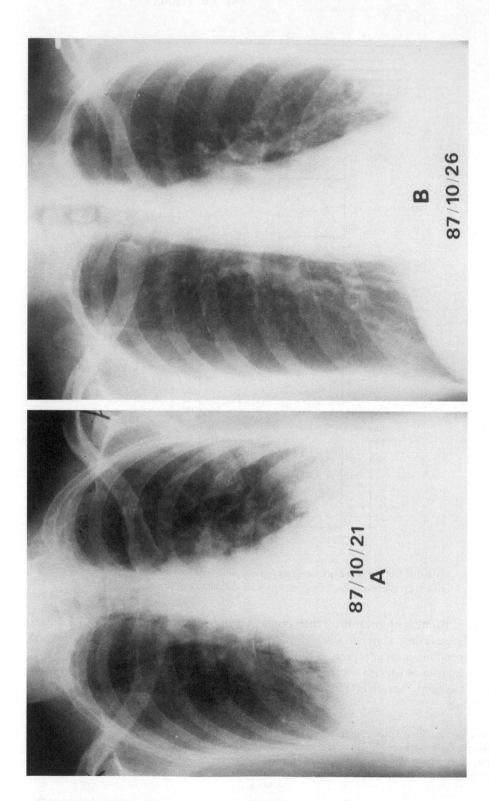

FIGURE 6.   Chest radiographs Case 2. The radiograph shown in A was obtained 9 d following the onset of symptoms. Bilateral lower lobe opacities are evident. Five days later (B) the right side has cleared; the left has improved, but there is still an opacity and a pleural reaction.

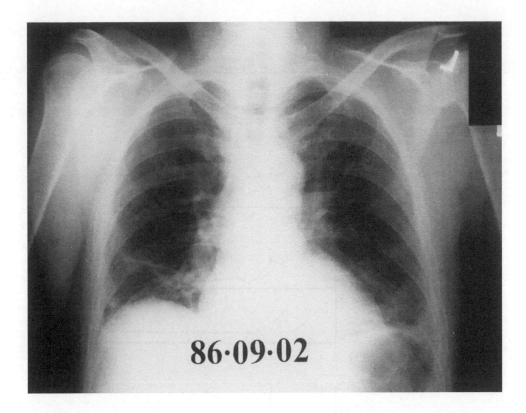

FIGURE 7.   Chest radiograph, Case 3. Note the small but definite lower lobe subsegmental opacities.

fact that some of these patients complained of rash prior to admission to hospital (Table 1). It is likely, then, that rash in Q fever represents the flushing that accompanies fever, nonspecific eruption, or drug-related rash.

C. PHYSICAL EXAMINATION

The general appearance of the patient varies according to the degree of illness. Some patients are severely ill; most are mildly to moderately ill. Jaundice is uncommon but has been reported in 1[6] to 5%[5] of patients. We have not observed jaundice in our patients with Q fever.

Hepatomegaly and splenomegaly have been reported in 11 and 4%, respectively, in a study from northern California in 1948—1949.[5] A study from Australia[6] of 111 patients, carried out from 1962—1981, found that 51% had hepatomegaly and 30% had splenomegaly. Three patients in this later study had acute hepatitis. It is likely that hepatosplenomegaly depends on the strain of *C. burnetii* causing the infection. I have not found hepatosplenomegaly in any of the patients who have acquired Q fever following exposure to parturient cats. In contrast, both patients who developed Q fever following exposure to an infected cow had hepatomegaly. Two of five patients who developed Q fever following exposure to infected wild rabbits had splenomegaly.

Relative bradycardia has been reported by some investigators and not by others. Powell[61] found that 96% of his patients had this finding: with a temperature of 103°F the pulse rate was usually about 100/min or less. Denlinger[12] noted that the average pulse rate for his group of 80 patients was 98/minute when an average temperature of 102.8°F was present. Clark et al.[5] stated that relative bradycardia "was not a prominent sign in this series of patients"; however, no data were given to support this statement.

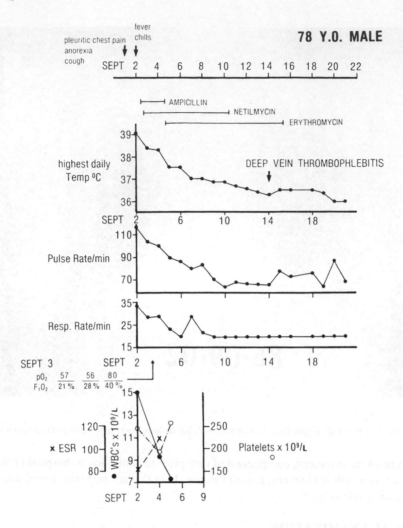

FIGURE 8.   Clinical and laboratory events Case 3.

Pleural friction rubs are heard in about 10% of patients. Abdominal distension and abdominal tenderness are infrequent findings.

D. PNEUMONIA

Pneumonia is an important clinical manifestation of acute Q fever. The frequency with which pneumonia is present clinically and radiographically varies from country to country. It is uncommon in Australia, occurring in 4% of 72 patients[61] in one series and 7% of 111 cases in another series,[6] whereas in Nova Scotia 80% of those who have had chest radiographs performed have had pneumonia.

We studied all patients admitted to the Victoria General Hospital (an urban hospital) with community-acquired pneumonia over a 5-year period from November 1981 to March 1987 (there was a 4½-month gap in the study due to illness in one of the investigators). Seven hundred and nineteen persons met the study criteria. Twenty-one (2.9%) had *C. burnetii* pneumonia. In contrast, in rural Nova Scotia 20% of patients admitted to hospital with pneumonia had Q fever.[15] In other studies of   community-acquired pneumonia,[16-31] only 3 of the 16 reports[24,27,28] noted cases of Q fever, a total of 8 cases among these 499 patients (1.6%).

Q fever pneumonia is an "atypical" pneumonia. This is a term used to describe pneumonia characterized by a dry nonproductive cough and blood and sputum cultures negative for

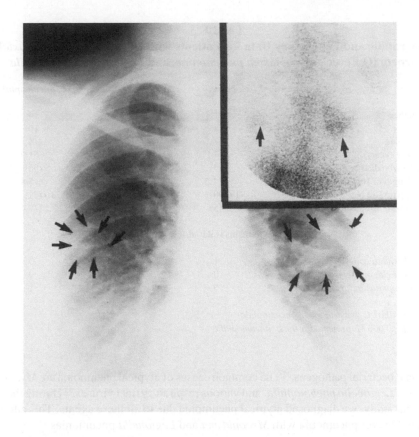

FIGURE 9. Chest radiograph Case 6, and [67]Gallium scan (inset). Note the small plate-like atelectatic lesions (arrows). The [67]Gallium scan shows uptake of the isotope at sites that correspond to the location of these opacities.

TABLE 2
A Comparison of the Symptoms Experienced by 51 Patients with Q Fever with Those Experienced by 102 Age- and Sex-Matched Patients with Pneumonia Due to Agents Other than *Coxiella burnetii*

Symptom	Q fever	Pneumonia patients	Probability
Fatigue	50 (98)	81 (79.4)	<0.0004
Fever	48 (94.1)	70 (68.6)	<0.001
Chills	45 (88.2)	74 (72.5)	<0.04
Sweats	43 (84)	70 (68.6)	<0.001
Headache	37 (72.5)	54 (52.9)	<0.01
Myalgia	35 (68.6)	58 (56.8)	
Nausea	25 (49)	42 (41.1)	
Cough	29 (28.4)	81 (79.4)	<0.006
Chest pain	29 (28.4)	60 (58.5)	
Vomiting	13 (25.4)	31 (30.3)	
Diarrhea	11 (21.5)	28 (27.4)	
Rash	9 (17.6)	10 (9.8)	
Earache	9 (17.6)	13 (12.7)	
Sore throat	7 (13.7)	34 (33)	<0.01

From Marrie, T. J., Durant, H., Williams, J. C., Mintz, E., and Waag, D. M., *J. Infect. Dis.*, 158, 101, 198. With permission.

TABLE 3

Demographic and Laboratory Data for Patients with Pneumonia Due to *Coxiella burnetii* (Q Fever), *Mycoplasma pneumoniae*, and *Legionella pneumophila*

	C. burnetii	*M. pneumoniae*	*L. pneumophila*
No. studied	21	40	14
No. males	14	25	10
No. females	7	15	4
Mean age (years)	47.1	42.3	61.9
No. (%) who died	1 (5)	1 (3)[a]	5 (36)[b]
Mean length of stay (d)	11.5	13.6	15.8
No.(%) who had blood cultures done	19 (90)	29 (73)	10 (71)
No.(%) who had positive blood cultures	1 (5)	1 (3)	1 (10)

Microorganisms isolated from blood

Escherichia coli	1	0	0
Salmonella montevideo	0	1	0
Haemophilus influenzae	0	0	1

[a] *p* <0.02 *C. burnetii* vs. *L. pneumophila*.
[b] *p* <0.003 *M. pneumoniae* vs. *L. pneumophila*.

conventional bacterial pathogens.[32] The common causes of atypical pneumonia are *Mycoplasma pneumoniae*, *Legionella pneumophila*, and various respiratory tract viruses.[33] During the course of our 5-year study, we diagnosed atypical pneumonia due to all these agents. This allowed us to compare Q fever pneumonia with *Mycoplasma* and *Legionella* pneumonias.

III. A COMPARISON OF PNEUMONIA DUE TO *COXIELLA BURNETII*, *MYCOPLASMA PNEUMONIAE*, AND *LEGIONELLA PNEUMOPHILA*

The various features of patients hospitalized with Q fever pneumonia, *M. pneumoniae*, and *L. pneumophila* pneumonia are compared in Tables 3 to 8. The patients with Q fever and *Mycoplasma* were younger than those with *Legionella* pneumonia. Mortality was uncommon in Q fever and *Mycoplasma* patients; however, 36% of the patients with Legionnaires' disease died. The one patient with Q fever who died had an unrecognized cardiomyopathy. He became ill with Q fever and was found in an unconscious state in his home. He had aspirated gastric contents and at the time of admission was bacteremic with *Escherichia coli*. At this point he was in established septic shock and had renal and cardiac failure. In Derrick's study, 4 of 273 (1.5%) patients with Q fever died.[10] Huebner,[34] in a review of Q fever cases reported up to 1949, stated that 5 of 625 (0.8%) patients with Q fever died.

A comparison of the symptoms experienced by patients with *C. burnetii*, *Mycoplasma*, and *Legionella* infections is given in Table 4. Headache was more common among Q fever patients. This group of patients with Q fever had a high incidence of cough: 71% of them complained of this symptom. This was a select group of patients. All had severe pneumonia and were hospitalized. Confusion was more common among Q fever patients compared with the other two groups, even though they were younger than the patients with *Mycoplasma* and *Legionella*. Other investigators have reported that abnormalities of mental status are common in Legionnaires' disease and are found in about one fourth of patients,[35] a figure similar to the 29% found in our study.

Rales and rhonchi were found less frequently on physical examination of the Q fever patients,

TABLE 4
Comparison of Symptoms and Signs for Patients with Pneumonia Due to the Indicated Pathogens

Symptoms	Etiology of pneumonia		
	C. burnetii (%) n = 21	*M. pneumoniae* (%) n = 40	*L. pneumophila* (%) n = 14
Fever	95	90	71
Chills	71	75	29
Rigors	38	20	7
Pleuritic chest pain	57	43	50
Headache	71	60	21[a]
Nausea	38	30	36
Vomiting	24	25	36
Abdominal pain	5	10	14
Diarrhea	10	5	14
Sore throat	14	33	0[b]
Anorexia	81	80	43
Myalgia	43	53	36
Arthralgia	33	30	7
Cough	71	95	71
Productive cough	33	83	43
Confusion	43	18	29

Signs

Temperature >37°C when admitted	90	92	100
Mean temperature when admitted °C	38.4	38.3	38.7
Rales	52	88	79
Rhonchi	5	20	29
Consolidation	29	23	43
Mean respiratory rate when admitted	24.7	26.5	26.4
Mean number of days febrile	2.2	2.45	4

[a] *L. pneumophila* is significantly different from *C. burnetii*.
[b] *L. pneumophila* is significantly different from *M. pneumoniae*.

52 and 5%, respectively, compared with the other two groups. There were no significant differences in the mean temperatures of the three groups of patients at the time of admission (Table 4). The mean respiratory rate of 24.7/min for those with Q fever was lower than that of 26.5 and 26.4 for *Mycoplasma* and *Legionella* patients. The patients with Q fever were febrile for a mean of 2.2 d following admission, shorter than the 2.45 d for *Mycoplasma* patients and 4 d for those with *Legionella*.

Patients with Q fever were less likely to have an underlying disease. Fifty-seven percent were in this category compared with 40% of the patients with *Mycoplasma* and only 14% of the patients with *Legionella* (Table 5).

The therapeutic approach of physicians to these three groups of patients was similar. Over half the Q fever and *Mycoplasma* patients received antibiotics prior to admission, while 36% of those with *Legionella* received such medication. Following admission most of the patients received multiple antibiotics. Erythromycin was the most commonly used antibiotic. It is noteworthy that

TABLE 5

Comparison of Patients with Pneumonia Due to *Coxiella burnetii, Mycoplasma pneumoniae*, and *Legionella pneumophila* with Respect to Underlying Diseases and Antibiotic Therapy

	C. burnetii n = 21	*M. pneumoniae* n = 40	*L. pneumophila* n = 14
No. (%) with no underlying disease	12 (57)	16 (40)	2 (14)[a]
Mean No. underlying diseases	0.76	1.05	1.86
No. (%) who received antibiotics prior to admission	11 (52)	24 (60)	5 (36)
Mean number of antibiotics used to treat pneumonia	1.86	1.6	2.57
Percent of patients who received antibiotic			
Erythromycin	71	68	57
Rifampin	19	3	21
Tetracycline	14	3	7
Aminoglycosides	34	18	57
Ampicillin	14	18	29
Cloxacillin	0	5	21

[a]   *L. pneumophila* significantly different from *C. burnetii* and *M. pneumoniae*.

TABLE 6

Radiographic Features of Pneumonia Due to *Coxiella burnetii, Mycoplasma pneumoniae*, and *Legionella pneumophila*

	C. burnetii n = 21	*M. pneumoniae* n = 40	*L. pneumophila* n = 14
Mean No. of lobes involved at time of admission	1.62	1.72	1.29
No. (%) whose pneumonia progressed following admission to hospital	3 (14)	2 (5)	1 (14)
No. (%) whose pneumonia resolved prior to discharge	7 (33)	24 (60)	5 (36)

only 57% of the *Legionella* group received erythromycin, the drug of choice for this infection.[36] This was due to the fact that for most patients a diagnosis was made by serological means, and the information was not available during the acute phase of the illness.[37]

Only 33% of the patients with Q fever received optimal antimicrobial therapy, i.e., tetracycline or rifampin. Some investigators[38,39] have reported that Q fever pneumonia has responded to treatment with erythromycin. This has not been so in our experience. Indeed, patients with severe pneumonia have continued to show progression of their illness despite treatment with 4 g of erythromycin per day. It is likely that the apparent favorable response to erythromycin therapy reflects the natural history of Q fever; i.e., patients with mild to moderate degrees of pneumonia will probably recover without antibiotic therapy. Rifampin is very effective against *C. burnetii in vitro*.[40] In one of the few randomized studies of antibiotic therapy

TABLE 7

Complications of Pneumonia Due to *Coxiella burnetii*, *Mycoplasma pneumoniae*, and *Legionella pneumophila*

	C. burnetii n = 21	*M. pneumoniae* n = 40	*L. pneumophila* n = 14
Mean No. of complications	0.38	0.53	1.21
No. with indicated complication			
Renal failure	1[a]	3	3
Congestive heart failure	1[a]	0	3
Hemolysis	0	4	0
Pulmonary embolism	0	1	0
Ataxia	0	0	1
Deep vein thrombophlebitis	1	0	0
Respiratory failure	1[a]	3	3
Stroke	0	1	1
Urinary tract infection	1	2	0
Gastrointestinal hemorrhage	0	1	0
Diarrhea	0	1	0
Pneumothorax	0	1	0
Seizures	0	1	1
Hypothermia	1	0	1
Myocardial infarction	1	0	0
Disseminated intravascular coagulation	0	1	0
Relapse of pneumonia	0	1	0
Death	1[b]	1[b]	5[b]

[a] All these complications occurred in one patient.
[b] *L. pneumophila* significantly different from *M. pneumoniae* (*p* <0.003) and *C. burnetii* (*p* <0.02).

TABLE 8

Mean Number of Indicated Laboratory Investigations Carried Out for Patients with Pneumonia Due to *Coxiella burnetii*, *Mycoplasma pneumoniae*, and *Legionella pneumophila*

Mean no.	*C. burnetii* n = 21	*M. pneumoniae* n = 40	*L. pneumophila* n = 14
Hemograms	3.5	5.59	9.62
Chest radiographs	3.1	3.67	4.23
Chemistries	2.79	2.90	3.50
Blood gases	2.71	7.38	5.75
Sputum cultures	0.36	0.78	1.00

in Q fever, tetracycline[42] reduced the duration of fever by 50% if administered during the first 3 d of the illness. Spelman[6] compared (nonrandomized) tetracycline 500 mg four times daily with doxycycline 100 mg twice daily for the treatment of acute infection. He found that the mean duration of fever in the untreated group was 3.3 d, while for those who received tetracycline it was 2.0 d, and 1.7 d for the doxycycline-treated group. Studies of patients with chronic Q fever suggest that various combinations of antimicrobials are effective, including tetracycline in combination with trimethoprim-sulfamethoxazole,[42-44] and rifampin in combination with doxycycline.[45] It is likely that these combinations would prove effective for the treatment of acute Q fever as well. However for the patients with severe pneumonia the best combination is erythromycin and rifampin. This combination will effectively treat *M. pneumoniae, L. pneumophila,* and *C. burnetii.* In addition, it will treat *Streptococcus pneumoniae.* We have observed dramatic responses of patients seriously ill with Q fever pneumonia to this combination of antibiotics. It is noteworthy that tetracycline had no effect on *C. burnetii* in infected L1929 fibroblast cells.[40] Likewise, gentamicin and streptomycin had no effect.[40] Huebner et al.[34] found that all six patients with Q fever who were treated with streptomycin for 5 d relapsed following a 5-d course of treatment.

The mean number of lobes involved by the pneumonic process is shown in Table 6. There were no significant differences among the patients with Q fever, *Mycoplasma,* or *Legionella* infections. Of the Q fever and *Legionella* patients, 14% showed radiographic progression of the pneumonia following admission to hospital, whereas only 5% of the patients with *Mycoplasma* infection demonstrated such progression. Of the patients with *Mycoplasma* infection, 60% had total radiographic resolution of pneumonia by discharge. The mean length of stay for these patients was 13.6 d. In contrast, only one third of the Q fever and *Legionella* patients showed resolution of the pneumonia prior to discharge.

In a previous study,[45] we reviewed the chest radiographs of 25 patients with Q fever pneumonia, 11 of whom were part of an outbreak of Q fever that followed exposure to an infected parturient cat.[46] When the cat-associated cases were compared with sporadic cases of Q fever, rounded pulmonary opacities and multiple rounded lesions were found more often among the cat-related cases. These lesions are illustrated in Figure 1. Pleural effusions were noted in 35% of the sporadic cases but in none of the epidemic cases; however, subsequently, we observed pleural effusions in patients with cat-related Q fever. Increased reticular markings and atelectasis were frequently present. Millar[47] in a review of 35 patients with Q fever also noted multiple round segmental consolidations, 5 to 10 cm in diameter, of ground-glass density, and usually occurring in the lower lobes. Jacobson and co-workers[48] found that 17 of 65 (26%) patients had small pleural effusions. Rarely, cavitation of pulmonary nodular opacities has been described.[49] The radiographic features of Q fever are illustrated in Figures 1 to 3.

Complications were infrequent with *C. burnetii* pneumonia (Table 7). There was a mean of 0.38 complications per patient, but one patient accounted for 5 of the 8 complications. This patient developed Q fever and was admitted in septic shock following aspiration and subsequent *E. coli* bacteremia. He had a previously unrecognized cardiomyopathy. One patient developed a deep vein thrombophlebitis, and another developed a myocardial infarction. Four patients with *Mycoplasma* pneumonia had cold agglutinin-induced hemolytic anemia.[50,51] Respiratory failure occurred in three patients. This complication has been previously reported in *M. pneumoniae* and has resulted in death.[52] The adult respiratory distress syndrome has also been described in Q fever.[54] Patients with community-acquired Legionnaires' disease had the highest number of complications and the highest mortality rate — 36%. Only some of the complications are shown in Table 7, but renal failure and congestive heart failure were the most common complications.

The mean number of laboratory investigations is given in Table 8. The patients with Q fever pneumonia had a lower mean number of all the investigations compared with the other two groups. Those with Legionnaires' disease had a higher number of investigations compared with the other two groups, with the exception of blood gases. The *M. pneumoniae* patients had a mean of 7.38 compared with a mean of 5.75 for the *Legionella* patients and 2.71 for the Q fever patients.

The mean values for selected laboratory tests are shown in Table 9, while in Table 10 the results of these and other tests are expressed as the number and percent that are above the normal range.

Twenty-nine percent of the patients with Q fever had an abnormally high serum creatinine; 10% of those with *Mycoplasma* and 50% of the *Legionella* patients had such values. None of the patients with Q fever showed a significant rise in creatinine during their hospital stay. The creatine phosphokinase (CPK) was elevated in 29, 17.5, and 42.8%, respectively, of the patients

TABLE 9
Results of Various Laboratory Tests in Three Groups of Patients with Atypical Pneumonia (Normal Values Given in Parentheses)

Test	C. burnetii n = 21	M. pneumoniae n = 40	L. pneumophila n = 14
	Mean value of indicated test		
White blood cell count × 10⁹/l (5—10)	8.95	12.52	10.40
SGOT U/l (8—29)	54.35	71.85	72.38
SGPT U/l (1—41)	48.1	142.0	20
Bilirubin μmol/l (0—16)	14.1	10.7	4.0
Creatine phosphokinase U/l (18—199)	912	154	527
Platelet count × 10⁹/l (150—350)	259.95	330.5	261.42

TABLE 10
Laboratory Data[a] Expressed as the Number and Percentage above the Upper Limit of Normal for Patients with Pneumonia Due to *C. burnetii, M. pneumoniae,* and *Legionella pneumophila*

	C. burnetii n = 21	M. pneumoniae n = 40	L. pneumophila n = 14
No.(%) with creatinine >110 mmol/l	6 (29)	4 (10)	7 (50)[b]
No.(%) in whom creatinine increased by >100 mmol/l from day 1 to day 7	0	1 (2.5)	2 (14)
No.(%) with WBC >10 × 10⁹/l	6 (29)	20 (50)	6 (42.8)
No.(%) with CPK >200 U/l	6 (29)	7 (17.5)	6 (42.8)
No.(%) with alkaline phosphatase >104 U/l	12 (57)	19 (47.5)	9 (64)
No.(%) with SGOT >29 U/l	9 (42.8)	15 (37.5)	8 (57)
No.(%) with SGPT >41 U/l	10/19 (52.6)	11 (27.5)	4/13 (30.7)
No.(%) with total bilirubin >16 μmol/l	3 (14.3)	8 (20)	3 (21)
No.(%) with ESR >50 mm/h	7/8 (87.5)	15/19 (78.9)	3/6 (50)
No.(%) with >20 g/l drop in hemoglobin from hospital day 1 to 5—7	6/13 (46)	11/29 (37.9)	6/10 (60)
No.(%) with PO 2 <60 torr on admission	4/15 (26.6)	12/32 (37.5)	7/13 (53.8)

[a] At time of admission unless otherwise indicated.
[b] *L. pneumophila* significantly different from *M. pneumoniae.*

(Table 10). Elevations of CPK have been reported in Legionnaires' disease, and on occasion rhabdomyolysis has been a manifestation of this disease.[55] One of our patients with Q fever had a very high CPK level — the patient in septic shock due to *E. coli* bacteremia. It is likely that the rhabdomyolysis was due to the septic shock. The actual CPK values for the three groups of patients are given in Figure 10.

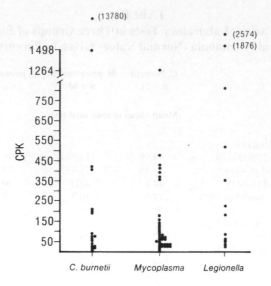

FIGURE 10. Creatine phosphokinase values for the patients with pneumonia due to *C. burnetii*, *M. pneumoniae*, and *L. pneumophila*. The normal value is 18 to 199 U/l.

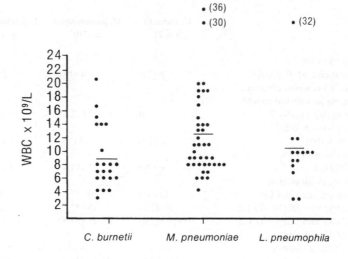

FIGURE 11. White blood cell counts for patients with pneumonia due to the indicated agents. The bar indicates the mean value. The normal range is 4.5 to 10×10^9/l.

The white blood cell count was $\geq 10 \times 10^9$/l for 29% of the Q fever patients (Figure 11). Spelman[6] found that 2.7% of his patients with Q fever had white blood cell counts in this range; Denlinger[12] reported both leukocytosis and leukopenia. Eight of Powell's 72 patients (13.8%) had a total white blood cell count $\geq 10 \times 10^9$/l. Woodhead and MacFarlane[56] found that 60% of patients with Legionnaires' disease and 39% of those with *Mycoplasma* infection had a WBC $\geq 10 \times 10^9$/l. A relative lymphocytosis has been reported in Q fever by some investigators[6] but not by others.[12] The lymphocyte counts for our three groups of patients are given in Figure 12.

Eighty-five percent (85/100) of patients with Q fever in an Australian study had abnormal liver function tests.[6] The serum bilirubin was elevated in 9%, the alkaline phosphatase in 30%,

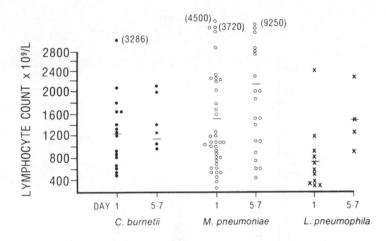

FIGURE 12. Lymphocyte counts on hospital days 1 and 5—7 for patients with Q
fever *(C. burnetii)*, *M. pneumoniae*, and *L. pneumophila*. The bars indicate the mean
values. A value >2.5 × 10⁹/l indicates lymphocytosis, and a value <1 × 10⁹/l indicates
lymphopenia.

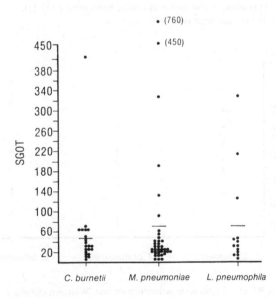

FIGURE 13.   Values for SGOT (serum glutamic oxaloacetic
transaminase, also known as asparate transaminase
[AST]). The normal range is 8 to 29 U/l.

and the serum glutamic pyruvic transaminase in 72%. Corresponding findings from our study
were 14% for the bilirubin, 57% for the alkaline phosphatase, and 52% for the SGPT. The
proportion of patients who had elevated liver function tests was the same for all three groups of
pneumonia patients (Table 10 and Figures 13 and 14). Domingo et al.[57] studied 18 patients with
Q fever pneumonia who were hospitalized in Barcelona, Spain. Seven (38.8%) had hepato- and/
or splenomegaly. The SGPT was elevated in 7 of the 18 patients (38.3%), the SGPT in 6 (33.3%),
the alkaline phosphatase in 5 (27.7%), and the lactic dehydrogenase in 6 (33.3%). The authors
noted that the level of the transaminases was significantly lower for patients with Q fever
pneumonia compared with that of those who had hepatitis due to *C. burnetii*.

The erythrocyte sedimentation rate was elevated in most of the patients with *M. pneumoniae*,

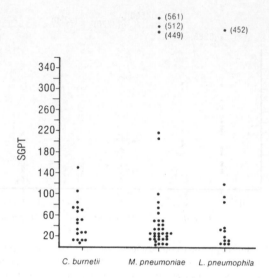

FIGURE 14.   Values for SGPT (serum glutamic pyruvic transaminase, also known as alanine transaminase [ALT]). The normal range is 1—41.

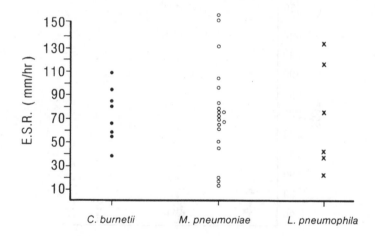

FIGURE 15.   Erythrocyte sedimentation rate, Westgren technique.

C. burnetii, and *L. pneumophila* pneumonia (Table 10 and Figure 15). It was not uncommon to find an erythrocyte sedimentation rate over 50 mm/h, and some were 100 mm/h or greater.

A drop in hemoglobin concentration of >20 g/l from hospital day 1 to hospital days 5 to 7 was common: 46% of those with Q fever pneumonia, 38% of those with *Mycoplasma* pneumonia, and 60% of those with *Legionella* pneumonia. Three of the patients with *Mycoplasma* pneumonia had severe cold agglutinin-induced hemolytic anemia.[51] Cardellach and co-workers[58] described a 26-year-old male Spaniard with Q fever pneumonia who was admitted to hospital with a hemoglobin of 101 g/l. Twenty-four hours later it had declined to 79 g/l. Coombs tests (direct and indirect) were negative, and the cold agglutinin titer was 1:20. The researchers concluded that this was a hemolytic anemia due to Q fever.

We do not know how many of our patients with Q fever who had a 20-g/l drop in hemoglobin had hemolysis. We suspect that in many patients the hemoglobin drop was due to blood loss (for various tests) and underproduction due to the acute illness. Spelman[6] described hemolysis in a

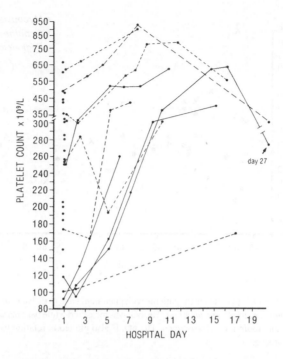

FIGURE 16. Distribution and time course of platelet counts for 32 patients with Q fever. The normal range is 150 to 350 × 10⁹/l. A value of >400 × 10⁹/l is indicative of thrombocytosis.

53-year-old meat worker with Q fever whose hemoglobin dropped from 119 to 99 g/l. Hemolytic anemia has been described as an unusual complication of Legionnaires' disease.[59,60]

Atypical lymphocytes have been observed in Q fever by some investigators.[61] We did not observe atypical lymphocytes in the blood films of any of our patients or in any of another 30 patients with acute Q fever for whom we examined the peripheral blood films. Lymphocytosis has been observed by some during the recovery phase, the third or fourth week.[61] Our observations are limited in this regard (Figure 12). The patients with *Legionella* infection were lymphopenic at the time of admission.

The mean platelet counts for the patients with Q fever, *Mycoplasma*, and *Legionella* disease are shown in Table 9. In a separate study of 32 patients with Q fever, we noted 4 patients (12.5%) were thrombocytopenic when admitted and 17 (53%) demonstrated thrombocytosis (platelet count ≥ 400 × 10⁹/l) at some point during the hospital stay. Figure 16 shows the initial platelet counts and the time course of the platelet counts of selected patients.

Of 30 patients with Q fever, 15 (50%) had microscopic hematuria. Kosatsky[46] reported that 6 of 13 patients (46%) with Q fever had hematuria, three of whom had gross hematuria. Powell[61] found that 7 (9.7%) of his patients had hematuria. In 5 of the 7, the abnormality was slight with 10 to 15 erythrocytes per high-power field, in 2, several hundred red cells were noted. One of these patients had thrombocytopenia. Strains of *C. burnetii* isolated from cats in Nova Scotia did not cause enlargement of the kidneys when injected into mice,[62] but as-yet-undefined unique features of the feline strains of *C. burnetii* may explain the high rate of hematuria among Nova Scotia patients with Q fever.

Four of the 15 patients (26.6%) with Q fever who had blood gases measured were hypoxemic (Table 10 and Figure 17). This compares with 37.5% of the *Mycoplasma* patients and 53.8% of the patients with Legionnaires' disease (Figure 17).

The serum sodium concentration for the three groups of patients is given in Figure 18.

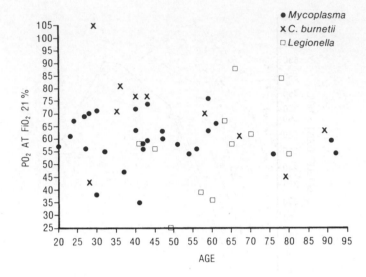

FIGURE 17.   pO$_2$ values for patients with pneumonia due to *C. burnetii, M. pneumoniae*, and *L. pneumophila*. The pO$_2$ was measured while patients were breathing room air ~21% oxygen. The pO$_2$ is also plotted in relation to the patient's age.

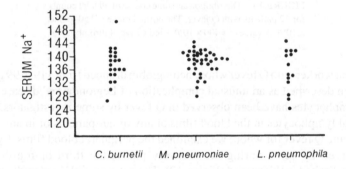

FIGURE 18.   Serum sodium concentration (mmol/l) for patients with Q fever, *Mycoplasma*, and *Legionella* pneumonia. The normal range is 138 to 145 mmol/l.

IV. UNCOMMON MANIFESTATIONS OF ACUTE Q FEVER

A. NEUROLOGICAL MANIFESTATIONS

Reference works state that encephalitis and encephalomyelitis may arise late in the course of an illness caused by Q fever[63]. Dementia, toxic confusional states, and extrapyramidal disease are other neurological complications.[63] Dementia and extrapyramidal manifestations are very difficult to substantiate as being due to Q fever.

Derrick reported that 1 of his 273 patients (0.3%) with Q fever had encephalomyelitis.[10] One of the 72 patients (1.3%) described by Powell[61] had encephalitis, and 2 of 814 patients (0.2%) from Uruguay[8] had neurological findings: 1 had encephalitis and the other behavioral disturbance. Eight of Powell's patients (11%) had neck stiffness which was moderately severe in 6 and marked in 2 patients.

One of our 170 patients (0.5%) with Q fever had meningoencephalitis.[64] This 35-year-old male had Q fever pneumonia and a change in his behavior. His normally placid behavior was replaced by hostility. The cerebrospinal fluid contained $18 \times 10^7/l$ white blood cells, of which 87% were mononuclear cells; the protein concentration was elevated to 2000 mg/l, and the glucose concentration was normal. The cerebrospinal fluid is usually normal in Q fever.[65] Ladurner et al.[66] described an 18-year-old female who became ill with fever, headache, vomiting, psychomotor disturbance, disorientation, right-sided weakness, and speech disturbance. Papilledema was present, and the cerebrospinal fluid contained $1.392 \times 10^9/l$ lymphocytes, the serum glucose was 4.3 mmol/l, while the CSF glucose was 0.88 mmol/l. Recovery occurred in 8 d.

Brooks and co-workers[67] described a 34-year-old man with aortic stenosis who presented with disorientation and confusion following cough, low grade fever and myalgia 6 months earlier. The CSF contained $72 \times 10^7/l$ lymphocytes, and the electroencephalogram showed moderately severe generalized cortical dysfunction. Ten days later the CSF was normal; however 3 weeks later photophobia, headache, and low-grade fever returned. The CSF now contained $79 \times 10^7/l$ white blood cells. The serological studies showed very high antibody titers to phase I and phase II *C. burnetii* antigens by complement fixation and microagglutination techniques. Indeed, the antibody levels were in the range usually associated with endocarditis.[68] Thus, the possibility that these neurological manifestations represented embolic manifestations in a patient with Q fever endocarditis was very likely, even though computed tomographic (CT) scans were normal. Gomez-Aranda et al.[69] described the case of a 7-year-old female who developed coma as part of a febrile illness accompanied by nausea and vomiting. The cerebrospinal fluid was normal, and the CT scan showed decreased absorption coefficient in both hemispheres. Two days later the CT scan was normal except for a hypodense area in the right frontobasal region. Treatment was begun with dexamethasone and adenine arabinoside for presumed herpes encephalitis with complete resolution of signs and symptoms in 24 h. A fourfold rise in antibody to phase II *C. burnetii* antigen from 1:512 to 1:2048 occurred.

Wallace and Zealley[70] in study of 78 febrile children (53 with seizures and 25 without) performed neurological and electroencephalographic examinations on 2 children with Q fever. One had prior left hemiplegia, and the other was neurologically normal. The neurologically normal child had a normal EEG.

Manic psychosis has been associated with Q fever.[71] A 40-year-old prisoner developed a psychotic illness 13 d following the onset of pneumonia which was later shown serologically to be due to *C. burnetii*. The manic-psychosis persisted for 9 weeks.

The presumed extrapyramidal manifestations of Q fever are exemplified by a 46-year-old male who 1 month following an illness characterized by fever, rigors, nausea, and vomiting developed a recurrence of the fever and a coarse tremor of the head and upper extremities.[72] He had consumed excessive alcohol until 6 months prior to this illness. When examined 1 month following the onset of tremor, there was a fixed facial expression, unsustained left ankle clonus, left extensor plantar response, cogwheel rigidity, and a gross 4-per-second intention tremor of the upper extremities. The CSF contained $29 \times 10^7/l$ white blood cells. The Q fever titer was 1:256 and rose to 1:4096. A liver biopsy revealed granulomatous hepatitis and scattered liver cell necrosis. The neurological abnormalities responded to treatment with trihexphenidyl hydrochloride (Artane™).

The only report of neuropathological findings in Q fever is very tenuous, with the diagnosis of Q fever resting on the visualization of cytoplasmic granules that looked like rickettsia.[73]

We studied a 54-year-old man who became confused on the second day of hospitalization for Q fever pneumonia. At this time he was hypertensive, blood pressure 230/140 mmHg. Over the next 48 h he became more confused and spent most of his time sleeping. During the next 9 d an expressive dysphasia became evident. It is likely that this patient suffered a cerebrovascular accident as a result of his hypertension, and the neurological component of his illness was not due to Q fever.

The following neurological deficits have also occurred in association with Q fever: neuritis with atrophy of the proximal musculature, polyradiculopathy, and isolated neuritis of the cranial nerves.[74]

B. OPTIC NEURITIS

Optic neuritis has been described as complicating the course of six patients with Q fever.[75-79] In all instances, there was bilateral optic disk abnormalities, and three had other neurological defects, encephalitis in two, and confusion in one. Five of the six patients had a lasting decrease in visual acuity and optic atrophy.

C. THYROIDITIS

One case of acute thyroiditis precipitated by and possibly due to *C. burnetii* infection has been reported.[80] This 42-year-old female was admitted because of a painful swelling in the right lower anterior part of her neck. Five weeks previously she had a lower respiratory tract infection with a pleural friction rub. *C. burnetii* was isolated from her urine and throat washings. There was no fourfold rise in serum complement fixation titers against *C. burnetii* antigens — acute phase sample 1:64, convalescent phase titer 1:64. This patient was treated with chlortetracycline. She was discharged on the 18th hospital day, and her thyroid was normal when examined 1 month following the institution of therapy.

D. MYOCARDITIS AND PERICARDITIS

Endocarditis is a well-recognized manifestation of chronic Q fever (Chapter 9). Clinical myocarditis is reported infrequently although subclinical myocarditis may be more common than we appreciate. Powell[61] performed electrocardiograms on 50 patients with Q fever. Of these tracings, 10 (20%) were abnormal. The most common abnormality was T wave change. Flat, diphasic, or inverted T waves occurred in seven patients. Sheridan and co-workers[81] described two patients with frank myocarditis due to Q fever, both of whom recovered. The following case history of our patient with myocarditis illustrates the clinical features of this entity.

A 38-year-old coal miner became ill 11 d prior to hospital admission with vomiting followed by pleuritic chest pain, fever, chills, and pain and tenderness in the calves of both legs. He was admitted to his local hospital where a chest radiograph revealed bilateral upper lobe (lobar) opacities. Despite treatment with ampicillin and gentamicin, his condition continued to deteriorate, and he became hypoxemic (pO$_2$ 43 torr while breathing 40% O$_2$) and required intubation and ventilatory support. Therapy was begun with erythromycin and rifampin for presumed Legionnaires' disease. Eight days later, chest radiographs were suggestive of a pericardial effusion, and electrocardiograms revealed diffuse ST-T wave changes (Figure 19) consistent with myocarditis. Within 1 week of starting erythromycin and rifampin therapy, his heart had decreased in size and the pulmonary opacities cleared. There was a fourfold rise in antibody to *C. burnetii* by complement fixation, but no rise to *L. pneumophila*.

Two of 80 patients with Q fever studied over a 5-year period in Iran had pericarditis.[82] One of 80 patients studied by Denlinger had pericarditis.[12]

E. PLEURAL FLUID

There is very little information regarding the composition of the pleural fluid that on occasion complicates *C. burnetii* pneumonia. In two cases reported to date,[82,106] eosinophils accounted for 80 and 70%, respectively, of the cells in the fluid.

F. BONE MARROW LESIONS

Bone marrow necrosis is rarely diagnosed during life. Brada and Bellingham[83] described a 42-year-old male who had pneumonia 3 weeks prior to his admission for back pain and tenderness. At that time his hemoglobin was 7.5 g/dl, and he had a leukoerythroblastic picture

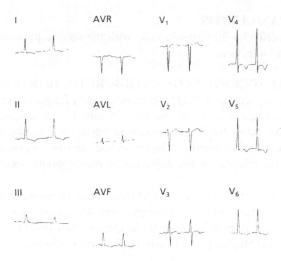

FIGURE 19.　Electrocardiogram of a 38-year-old male with myocarditis due to *C. burnetii* infection. Note the widespread ST-T wave changes.

in his blood smear. Examination revealed bone tenderness. Bone marrow aspirates (three) and biopsy showed bone marrow necrosis. Nuclei were basophilic, recognizable features of all cell types were obscured, and cells were surrounded by amorphous acidophilic material. He died 3 weeks following admission from cerebral and gastrointestinal hemorrhage. The Q fever titer rose from 1:10 to 1:640.

The granulomas in the bone marrow in Q fever are of the doughnut type; i.e., the granulomas often have an empty space in the center (probably representing dissolved lipid), surrounded by a mixture of polymorphonuclear cells and concentric lamination of fibrinoid material. These granulomas frequently appear in one perivascular region.[84-87] Such granulomas have also been described in the bone marrow of animals infected with *C. burnetii*.[88] The bone marrow granulomas in Q fever are similar to those that occur in the liver[89,90] in this illness. Doughnut granulomas are not diagnostic of Q fever as such. Granulomas have been seen in typhoid fever,[91] Hodgkin's disease, and infectious mononucleosis.[90]

C. burnetii involvement of the bone marrow has resulted in transient severe hypoplastic anemia in one patient.[86] This patient had severe thrombocytopenia 10 to $20 \times 10^9/l$ and anemia; his white blood cell count was $4.4 \times 10^9/l$. A bone marrow biopsy showed severe hypoplasia. Five days later areas, of regenerating marrow and granulomas were observed. Histiocytic hemophagocytosis is another manifestation of Q fever involvement of the bone marrow.[93]

G. LYMPHADENOPATHY

Denlinger[12] noted that none of his 80 patients had marked lymphadenopathy. Ten had small palpable anterior cervical nodes. It is likely that this degree of lymphadenopathy was nonspecific. We have seen one patient with a large 4 × 3-cm node in his left axilla. This patient had cut his left index finger while skinning wild rabbits, and he subsequently developed Q fever. Ramos and colleagues[93] described a 19-year-old male with fever, hepatosplenomegaly, and diffuse lymphadenopathy — groins and axilla — and a circular density in the left upper lobe. Biopsy of axillary nodes showed nonspecific lymphadenitis. They considered the diagnosis to be lymphoma until a biopsy was obtained and seroconversion occurred to *C. burnetii*.

H. PANCREATITIS

Case 2 described in Section VI had mild pancreatitis.

I. MESENTERIC PANICULITIS

Uriarte et al.[94] described a 39-year-old male with mesenteric paniculitis and serological evidence of *C. burnetii* infection.

J. INAPPROPRIATE SECRETION OF ANTIDIURETIC HORMONE

A 56-year-old brewery worker with Q fever manifested by fatigue, sweating, headache, and cough developed hyponatremia, serum sodium 125 mmol/l, on hospital day six. Despite treatment with oxytetracycline for 32 d, he remained febrile for 9 weeks. At that point he was treated with rifampin and trimethoprim. The serum sodium rose to normal after 4 d of fluid restriction, fulfilling the criteria for the diagnosis of inappropriate secretion of antidiuretic hormone.[95]

The serum sodium levels of our patients with Q fever pneumonia are shown in Figure 18 and are compared with the values obtained in patients with *M. pneumoniae* and *L. pneumophila* pneumonia. Hyponatremia is relatively common in Q fever, 28% (6/21) compared with 20% of *Mycoplasma* patients and 50% of patients with Legionnaires' disease.

K. ASSOCIATION WITH KAWASAKI DISEASE

Several reports have appeared[96,97] suggesting that *C. burnetii* may have a role in the etiology of Kawasaki disease. However, Lambert et al.[98] in subsequent studies were unable to demonstrate antibodies to *C. burnetii* in five patients with Kawasaki disease, and even the patient they previously reported[97] was negative. In this study[97] they used antigens of *C. burnetii* obtained from the Rocky Mountain Laboratory, Hamilton, MT. They went on to suggest that the Colindale antigen that they had previously used may have undergone an antigenic shift or had become contaminated with an agent causing Kawasaki disease or one related to such an agent.

L. ERYTHEMA NODOSUM

Conget et al.[99] described a 20-year-old pregnant female who presented with pneumonia and small tender cervical lymph nodes. A throat swab was negative for Group A Streptococci. Seroconversion occurred to *C. burnetii*. On day 15 in the hospital, erythema nodosum developed on the knees and shins.

M. GENITAL DISORDERS

Epididymitis and orchitis were reported by Gallaher[72] to complicate Q fever. Unfortunately, no references were provided. Huebner et al.[34] reviewed the literature to 1949 and stated that three cases of orchitis and one case of epididymitis had occurred among these patients. We have not observed either orchitis or epididymitis in 30 patients hospitalized with Q fever. Likewise, this complication has not occurred in the 170 cases of Q fever seen in Nova Scotia to date. We have, however, observed one case of priapism in a 27-year-old male with Q fever pneumonia who underwent an open lung biopsy for severe pneumonia, later diagnosed as Q fever. While in the intensive care unit, he had an indwelling urinary catheter. The priapism resolved without specific therapy and did not impair his potency.

N. Q FEVER IN THE IMMUNOCOMPROMISED HOST

Infection of the immunocompromised host seems to be an uncommon manifestation of Q fever.[100] Heard et al.[100] reported on five such patients, four of whom had *C. burnetii* pneumonia. The fifth had pyrexia and lymphadenopathy. The underlying diseases in these patients were Hodgkin's disease, acute myeloid leukemia, Ewing's sarcoma with acute lymphoblastic leukemia, Crohn's disease (patient receiving treatment with azathioprine and prednisolone), and alcohol abuse. Irradiation[101] and treatment with cortisone[102] have resulted in reactivation of *C. burnetii* in experimental animals. Exacerbation of quiescent Q fever endocarditis in a 40-year-old woman followed aortic valve replacement and corticosteroid therapy for presumed postpericardiotomy syndrome.[103] One bone marrow recipient, who was suffering from graft vs. host disease, developed fever, and a productive cough on day 8 posttransplantation. Coarse crepitations were evident on auscultation of the chest, but the radiograph was normal.[104] A diagnosis of Q fever was made serologically. A 5-year-old girl who had received a fetal liver-thymic transplant for severe combined immunodeficiency developed acute lymphoblastic leukemia at age 21 months. At 4 years of age co-trimoxazole prophylaxis was discontinued, and 4 weeks later she became ill with fever, headache, photophobia, myalgia, sore throat, dry cough, irritability, and a skin rash. The Q fever titer peaked at 1:512.[105]

In conclusion, then, the most common manifestation of acute Q fever is that of a self-limited mild febrile illness, and some infections are even asymptomatic. However, the range of manifestations of this illness is great and includes severe atypical pneumonia and involvement of almost every organ in the body. Usually, dysfunction of one organ predominates the clinical picture such as in encephalitis, myocarditis, or bone marrow infarction. It is likely that differences in strains of *C. burnetii* explain the differences in the manifestations of Q fever in different parts of the world.

V. CASE 1

This 27-year-old male was admitted to hospital on December 5, 1985, with a chief complaint of chest pain of 5-d duration. On November 28, 1985, he noted sudden onset of fever, cough, myalgia, and arthralgia. Two days later, severe right-side pleuritic chest pain developed, and he coughed up a small amount of blood. A few hours following the onset of the chest pain, he was seen in the emergency room where a chest radiograph showed a small right upper lobe opacity (Figure 3A). He remained febrile and was admitted to hospital where inspiratory crackles and rhonchi were heard on auscultation over the right upper lobe.

His past health was significant in that he had abused alcohol for the previous 9 years. Three weeks prior to admission he was arrested for stealing a deer (from a hunter's car) while he was drunk. Also while he was on this spree he slept in a field and drank unpasteurized milk that he obtained from a cow in the field.

His course in hospital is illustrated graphically in Figure 4. The duration of fever was 11 d, his pleuritic chest pain persisted for 16 d, and his respiratory rate was above normal for 11 d. His white blood cell count was $6.8 \times 10^9/l$ at the time of admission and rose to $10 \times 10^9/l$ by hospital day 12. His platelet count also rose during his hospital stay. The serum glutamic oxaloacetic transaminase (SGOT) was slightly elevated at the time of admission. The total bilirubin and lactic dehydrogenase levels were normal.

A diagnosis of Q fever was not made until hospital day 10. He had become afebrile by then, so that the duration of his fever represents the natural history of his disease.

He developed deep vein thrombophlebitis of his right calf on hospital day 12. This was treated with heparin and then with warfarin. He was discharged on hospital day 25 and was readmitted 22 d later with a gastrointestinal hemorrhage secondary to an excessively prolonged prothrombin time from his warfarin. At that time a chest radiograph showed that his pneumonia had almost completely resolved (Figure 3F).

Comment—This patient had a moderately severe case of Q fever pneumonia, complicated by deep vein thrombophlebitis. The source of his Q fever is unknown but possibilities include the deer or exposure to cows in the field.

VI. CASE 2

This 42-year-old female became ill on October 12, 1987, with diarrhea. Two days later (Figure 5) she developed nausea, myalgia, arthralgia, fever and chills. On day 4 of her illness the nausea became worse, and she noted left shoulder and abdominal pain. One day later she complained of a dry cough which persisted for 48 h. The next day (day 6 of her illness) left pleuritic chest pain developed. This seemed to be an extension of her abdominal pain, which was located in the left upper quadrant and in the left flank. The abdominal pain was severe, constant, and aggravated by movement. A chest radiograph done at her local hospital on day 4 showed a subsegmental opacity of the left lower lobe. Gross hematuria was noted on day 8, and she was admitted to our hospital. At that time she was very ill. Her temperature was 38.3°C, pulse rate 93, respiratory rate 28/min, blood pressure 112/70 mmHg. A pleural friction rub and crackles were heard over the base of the left lung both laterally and posteriorly. Her abdomen was distended, bowel sounds were decreased, and guarding and tenderness were noted on palpation of the left side of her abdomen.

A chest radiograph showed bilateral alveolar opacities (Figure 6A) at the bases, with the process being more extensive on the left. A pleural reaction and pleural effusion were also evident on the left. A radiograph of the abdomen showed slight gaseous distension of the colon from rectum to cecum. These findings were believed to be nonspecific but "could be indicative of colitis".

Serial white blood cell and platelet counts and selected biochemical parameters are shown in Figure 5. The white blood cell count fell during her hospital stay, and the platelet count rose to $1018 \times 10^9/l$. The SGPT was 30 U/l on admission but rose to 58 U/l before discharge (normal range, 8 to 29 U/l). The LDH was slightly elevated at 309 U/l (normal, 117 to 259 U/l). The total bilirubin was mildly elevated at 17 µmol(normal, 0 to 16 µmol) and rose to 35 µmol. All these parameters had returned to normal by the time of her follow-up visit on November 13, 1987, 1 month following the onset of her illness. The serum amylase was elevated to 708 U/l (normal, 13 to 83).

Our working diagnosis at the time of admission was Legionnaires' disease. Two days prior to admission erythromycin was prescribed. The dose was increased to 1 g every 6 h and was given intravenously. Rifampin was added in a dose of 600 mg once daily orally. There was a dramatic response to therapy (Figure 5). Within 3 d she was afebrile, and her abdominal distension had resolved. As she improved, her pulse rate increased, and her respiratory rate did not return to normal until the sixth hospital day. Her chest radiograph showed improvement within 3 d (Figure 6b). Blood, urine and stool were negative on culture for bacteria and viruses.

Acute and 3-week convalescent serum samples were tested for antibodies to *L. pneumophila* serogroups I-IV; *Coxiella burnetii;* influenza viruses A and B; parainfluenza viruses 1, 2, 3; respiratory syncytial virus; cytomegalovirus; adenovirus; *M. pneumoniae.* The only rise in antibody titer was to *C. burnetii,* from 1:128 in the acute phase specimen to 1:1024 in the convalescent phase sample.

Comment—This person was part of an outbreak of Q fever in a truck repair plant. Two of the 16 affected were seriously ill. The abdominal pain and distension were probably due to the pancreatitis, an uncommon complication of Q fever.

VII. CASE 3

This 78-year-old former coal miner was en route to the hospital for a scheduled reevaluation of transitional cell carcinoma of the bladder when he developed fever and pleuritic chest pain. Examination revealed a pleural friction rub over the left lower lung field anteriorly and decreased air entry at the bases. A chest radiograph (Figure 7) showed bilateral lower lobe subsegmental opacities and atelectasis. Therapy was begun with ampicillin and netilmicin, and later erythromycin was added (Figure 8). His fever, tachycardia, and tachypnea resolved over a period of 4 to 6 d. He developed a deep vein thrombophlebitis involving the right leg on hospital day 12.

Q fever was never suspected and was diagnosed only because the appropriate serological tests were performed as part of a study of community-acquired pneumonia. The acute phase serum had a titer to *C. burnetii* phase II antigen of <1:8. The convalescent phase titer was 1:128.

VIII. ASYMPTOMATIC (POSSIBLY UNRECOGNIZED) Q FEVER

A. CASE 4
This 45-year-old lady was part of a prospective study of Q fever. A serum sample obtained on August 16, 1987, revealed an antibody titer to phase II antigen of 1:16. By September 30 the titer was 1:64, and in February 1988 it was 1:128. The only illness that she could remember was a "slight cold with a headache" at the end of August.

B. CASE 5
This 72-year-old lady had antibody titers of 1:8 on May 22, 1987, and 1:64 on September 4, 1987. The only symptoms that she had during this time was a flare-up of her osteoarthritis.

IX. Q FEVER PNEUMONIA WITH POSITIVE [67]GALLIUM SCAN

A. CASE 6
This 52-year-old female was admitted for recurrence of pulmonary opacities associated with fever. In late July 1984, she had fever, cough, and a right upper lobe opacity. At that time she seroconverted to *M. pneumoniae,* acute phase titer <1:8, convalescent phase titer 1:64. On October 29, 1984, she developed fever and headache. Two days later when seen in the emergency room of her local hospital her temperature was 39°C. A chest radiograph revealed a rounded opacity in the right lower lobe. When seen in consultation on November 7, 1987, small but definite opacities were present in both lower lobes (Figure 9). A [67]Gallium scan showed uptake of the isotope in these areas (Figure 9 inset). Her antibody titers to *C. burnetii* rose from <1:8 to 1:64 (complement fixation test). She was treated with tetracycline and made an uneventful recovery.

Comment—A patient mildly ill with Q fever pneumonia. This patient and one other in our series illustrate that [67]Gallium can be taken up by the inflammatory response that occurs in Q fever pneumonia.

REFERENCES

1. **Turck, W. P. G., Howitt, G., Turnberg, L. A., Fox, H., Longson, M., Matthews, M. B., and Das Gupta, R.,** Chronic Q fever, *Q. J. Med.,* 178, 193, 1976.
2. **Ellis, M. E., Smith, C. C., and Moffat, M. A. J.,** Chronic or fatal Q fever infection: a review of 16 patients seen in north-east Scotland, *Q. J. Med.,* 52, 54, 1983.
3. **Clark, W. H., Romer, M. S., Holmes, M. A., Welsh, H. H., Lennette, E. H., and Abinanti, F. R.,** Q fever in California. VIII. An epidemic of Q fever in a small rural community in Northern California, *Am. J. Hyg.,* 54, 25, 1951.
4. **Luoto, L., Casey, M. L., and Pickens, E. G.,** Q fever studies in Montana. Detection of asymptomatic infection among residents of infected dairy premises, *Am. J. Epidemiol.,* 3, 81, 356, 1965.
5. **Clark, W. H., Lennette, E. H., Railsback, O. C., and Romer, M. S.,** Q fever in California. VII, Clinical features in one hundred eighty cases, *Arch. Intern. Med.,* 88, 155, 1951.
6. **Spelman, D. W.,** Q fever: a study of 111 consecutive cases, *Med. J. Aust.,* 1, 547, 1982.
7. **Dupuis, G., Péter, O., Pedroni, D., and Petite, J.,** Aspects cliniques observés lors d'une épidémie de 415 cas de fièvre Q, *Schweiz. Med. Wochenschr.,* 115, 814, 1985.
8. **Somma-Moreira, R. E., Caffarena, R. M., Somma, S., Pérez, G., and Monteiro, M.,** Analysis of Q fever in Uruguay, *Rev. Infect. Dis.,* 9, 386, 1987.

9. **Marrie, T. J., Durant, H., Williams, J. C., Mintz, E., and Waag, D. M.,** Exposure to parturient cats is a risk factor for acquisition of Q fever in Maritime Canada, *J. Infect. Dis.,* 158, 101, 1988.

10. **Derrick, E. H.,** The course of infection with *Coxiella burnetii, Med. J. Aust.,* 1, 1051, 1973.

11. **Robbins, F. C. and Ragan, C.,** Q fever in the Mediterranean area: report of its occurrence in allied troops; clinical features of the disease, *Am. J. Hyg.,* 44, 6, 1946.

12. **Denlinger, R. B.,** Clinical aspects of Q fever in southern California; study of 80 hospitalized cases, *Ann. Intern. Med.,* 30, 510, 1949.

13. **Lim, K. C. L. and Kang, J. Y.,** Q fever presenting with gastroenteritis, *Med. J. Aust.,* 1, 327, 1980.

14. **Bisno, A. L.,** Rocky Mountain Spotted Fever in *Medical Microbiology and Infectious Diseases,* Braude, A. I., Davis, C. E., and Fierer, J., Eds., W. B. Saunders, Philadelphia, 1981, 1458.

15. **Marrie, T. J., Haldane, E. V., Faulkner, R. S., Kwan, C., Grant, B., and Cook, F.,** The importance of *Coxiella burnetii* as a cause of pneumonia in Nova Scotia, *Can. J. Public Health,* 76, 233, 1985.

16. **Bath, J. C. J. L., Boissard, G. P. B., Calder, M. A., and Moffatt, M. J.,** Pneumonia in hospital practice in Edinburgh 1960—1962, *Br. J. Dis. Chest,* 58, 1, 1964.

17. **Mufson, M. A., Chang, V., Gill, V., Wood, S. C., Romansky, M. J., and Chanock, R. M.,** The role of viruses, mycoplasmas and bacteria in acute pneumonia in civilian adults, *Am. J. Epidemiol.,* 86, 526, 1967.

18. **Fransén, H.,** Clinical and laboratory studies on the role of viruses, bacteria, *Mycoplasma pneumoniae* and Bedsonia in acute respiratory illness, *Scand. J. Infect. Dis. Suppl.,* 1, 1, 1970.

19. **Fekety, F. R., Jr., Caldwell, J., Gump, D., Johnson, J. E., Maxson, W., Mulholland, J., and Thoburn, R.,** Bacteria, viruses and mycoplasmas in acute pneumonia in adults, *Am. Rev. Respir. Dis.,* 104, 499, 1971.

20. **Sullivan, R. J., Jr., Dowdle, W. R., Marine, W. M., and Hierholzer, J. C.,** Adult pneumonia in a general hospital: etiology and host risk factors, *Arch. Intern. Med.,* 129, 935, 1972.

21. **Bisno, A. L., Giffen, J. P., Van Epps, K. A., Niell, H. B., and Rytel, M. W.,** Pneumonia and Hong Kong influenza: a prospective study of the 1968—1969 epidemic, *Am. J. Med. Sci.,* 261, 251, 1971.

22. **Dorff, G. J., Rytel, M. W., Farmer, S. G., and Scanlon, G.,** Etiologies and characteristic features of pneumonias in a municipal hospital, *Am. J. Med. Sci.,* 266, 349, 1973.

23. **Garb, J. L., Brown, R. B., Garb, J. R., and Tuthill, R. W.,** Differences in etiology of pneumonias in nursing home and community patients, *JAMA,* 240, 2169, 1978.

24. **White, R. J., Blainey, A. D., Harrison, K. J., and Clarke, S. K. R.,** Causes of pneumonia presenting to a district general hospital, *Thorax,* 36, 566, 1981.

25. **Fick, R. B., Jr. and Reynolds, H. Y.,** Changing spectrum of pneumonia — news media creation or clinical reality?, *Am. J. Med.,* 74, 1, 1983.

26. **McNabb, W. R., Williams, T., Shanson, D. C., and Land, A. F.,** Adult community-acquired pneumonia in central London, *J. R. Soc. Med.,* 77, 550, 1984.

27. **MacFarlane, J. T., Finch, R. G., Ward, M. J., and MacRae, A. D.,** Hospital study of adult community-acquired pneumonia, *Lancet,* 2, 255, 1982.

28. **Kerttula, Y., Leinonen, M., Koskela, M., and Mäkelä, P. H.,** The aetiology of pneumonia. Application of bacterial serology and basic laboratory methods, *J. Infect.,* 14, 21, 1987.

29. **Larsen, R. A. and Jacobson, J. A.,** Diagnosis of commmunity-acquired pneumonia: experience at a community hospital, *Compr. Ther.,* 10, 20, 1984.

30. Research Committee of the British Thoracic Society and the Public Health Laboratory Service, Community-acquired pneumonia in adults in British hospitals 1982—1983: a survey of aetiology, mortality, prognostic factors and outcome, *Q. J. Med.,* 62, 195, 1987.

31. **Mohamed, A. R. E. and Price Evans, D.,** The spectrum of pneumonia in 1983 at the Riyadh Armed Forces Hospital, *J. Infect.,* 14, 3, 31, 1987.

32. **Cunha, B. A. and Quintiliani, R.,** The atypical pneumonias, *Postgrad. Med.,* 66, 95, 1979.

33. **Marrie, T. J., Haldane, E. V., Noble, M. A., Faulkner, R. S., Martin, R. S., and Lee, S. H. S.,** Causes of atypical pneumonia: results of a 1-year prospective study, *Can. Med. Assoc. J.,* 125, 1118, 1981.

34. **Huebner, R. J., Jellison, W. L., and Beck, M. D.,** Q fever — a review of current knowledge, *Ann. Intern. Med.,* 30, 495, 1949.

35. **Meyer, R. D. and Edelstein, P. H.,** *Legionella* pneumonias, in *Respiratory Infections: Diagnosis and Management,* Pennington, J. E., Ed., Raven Press, New York, 1983, 283.

36. **Fraser, D. W., Tsai, T. R., Orenstein, W., Parkin, W. E., Beecham, H. J., Sharrar, R. G., Harris, J., Mallison, G F., Martin, S. M., McDade, J. E., Shepard, C. C., and Brachman, P. S.,** Legionnaires' disease: description of an epidemic of pneumonia, *N. Engl. J. Med.,* 297, 1189, 1977.

37. **Zuravleff, J. J., Yu, V. L., Shonnard, J. W., Davis, B. K., and Rihs, J. D.,** Diagnosis of Legionnaires' disease. An update of laboratory methods with new emphasis on isolation by culture, *JAMA,* 250, 1981, 1983.

38. **D'Angelo, L. J. and Hetherington, R.,** Q fever treated with erythromycin, *Br. Med. J.,* 2, 305, 1979.

39. **Ellis, M. E. and Dunbar, E. M.,** In vivo response of acute Q fever to erythromycin, *Thorax,* 37, 867, 1982.

40. **Yeaman, M. R., Mitscher, L. A., and Baca, O. G.,** *In vitro* susceptibility of *Coxiella burnetii* to antibiotics, including several quinolones, *Antimicrob. Agents Chemother.,* 31, 1079, 1987.

41. **Powell, O. W., Kennedy, K. P., McIver, M., and Silverstone, H.,** Tetracycline in the treatment of "Q" fever, *Australas. Ann. Med.,* 11, 184, 1962.

42. **Varma, M. P. S., Adgey, A. A. J., and Connolly, J. H.,** Chronic Q fever endocarditis, *Br. Heart J.,* 43, 695, 1980.

43. **Tobin, M. J., Cahill, N., Gearty, G., Maurer, B., Blake, S., Daly, K., and Hone, R.,** Q fever endocarditis, *Am. J. Med.,* 72, 396, 1982.

44. **Freeman, R. and Hodson, M. E.,** Q fever endocarditis treated with trimethoprim and sulphamethoxazole, *Br. Med. J.,* 1, 419, 1972.

45. **Gordon, J. D., MacKeen, A. D., Marrie, T. J., and Fraser, D. B.,** The radiographic features of epidemic and sporadic Q fever pneumonia, *J. Can. Assoc. Radiol.,* 35, 293, 1984.

46. **Kosatsky, T.,** Household outbreak of Q fever pneumonia related to a parturient cat, *Lancet,* 2, 1447, 1984.

47. **Millar, J. K.,** The chest film findings in 'Q' fever — a series of 35 cases, *Clin. Radiol.,* 29, 371, 1978.

48. **Jacobson, G., Denlinger, R. B., and Carter, R. A.,** Roentgen manifestations of Q fever, *Radiology,* 53, 739, 1949.

49. **Seggev, J. S., Levin, S., and Schey, G.,** Unusual radiological manifestations of Q fever, *Eur. J. Respir. Dis.,* 69, 120, 1986.

50. **Feizi, T.,** Cold agglutinins, the direct Coombs' test and serum immunoglobulins in *Mycoplasma pneumoniae* infection, *Ann. N.Y. Acad. Sci.,* 143, 801, 1967.

51. **Murray, H. W., Masur, H., Senterfit, L. B., and Roberts, R. B.,** The protean manifestations of *Mycoplasma pneumoniae* infection in adults, *Am. J. Med.,* 58, 229, 1975.

52. **Fischman, R. A., Marschall, K. E., Kislak, J. W., and Greenbaum, D. M.,** Adult respiratory distress syndrome caused by *Mycoplasma pneumoniae, Chest,* 74, 471, 1978.

53. **Jastremski, M. S.,** Adult respiratory distress syndrome due to *Mycoplasma pneumoniae, Chest,* 75, 529, 1979.

54. **Torres, A., de Celis, M. R., Roisin, R. R., Vidal, J., and Aqusti Vidal, A.,** Adult respiratory distress syndrome in Q fever, *Eur. J. Respir. Dis.,* 70, 332, 1987.

55. **Kirby, B. D., Snyder, K. M., Meyer, R. D., and Finegold, S. M.,** Legionnaires' disease: report of sixty-five nosocomially acquired cases and review of the literature, *Medicine (Baltimore),* 59, 188, 1980.

56. **Woodhead, M. A. and MacFarlane, J. T.,** Comparative clinical and laboratory features of *Legionella* with pneumococcal and *Mycoplasma pneumonias, Br. J. Dis. Chest,* 82, 133, 1987.

57. **Domingo, P., Orobitg, J., Colomina, J., Alvarez, E., and Cadafalch, J.,** Liver involvement in acute Q fever, *Chest,* 94, 895, 1988.

58. **Cardellach, F., Font, J., Agusti, A. G. N., Ingelmo, M., and Balcells, A.,** Q fever and hemolytic anemia, *J. Infect. Dis.,* 148, 769, 1983.

59. **Strikas, R., Seifert, M. R., and Lentino, J. R.,** Autoimmune hemolytic anemia and *Legionella pneumophila* pneumonia, *Ann. Intern. Med.,* 99, 345, 1983.

60. **King, J. W. and May, J. S.,** Cold agglutinin disease in a patient with Legionnaires' disease, *Arch. Intern. Med.,* 140, 1537, 1980.

61. **Powell, O.,** "Q" fever: clinical features in 72 cases, *Australas. Ann. Med.,* 9, 214, 1960.

62. **Williams, J. C.,** personal communications, 1988.

63. **Turck, W. P. G.,** Q fever, *Medical Microbiology and Infectious Diseases,* Braude, A. I., Davis, C. E., and Fierer, J., Eds., W. B., Saunders, Toronto, 1981, 932.

64. **Marrie, T. J.,** Pneumonia and meningo-encephalitis due to *Coxiella burnetii, J. Infect.,* 11, 59, 1985.

65. **Laing-Brown, G.,** Q fever, *Br. Med. J.,* 2, 43, 1973.

66. **Ladurner, G., Stünzner, D., Lechner, H., and Sixl, W.,** Q-fieber-meningoencephalitis, *Nervenarzt,* 46, 274, 1975.

67. **Brooks, R. G., Licitra, C., and Peacock, M. G.,** Encephalitis caused by *Coxiella burnetii, Ann. Neurol.,* 20, 91, 1986.

68. **Peacock, M. G., Philip, R. N., Williams, J. C., and Faulkner, R. S.,** Serological evaluation of Q fever in humans: enhanced phase I titers of immunoglobulins G and A are diagnostic for Q fever endocarditis, *Infect. Immun.,* 41, 1089, 1983.

69. **Gomez-Aranda, F., Diaz, J. P., Acebal, M. R., Cortes, L. L., Rodriguez, A., and Moreno, J. M.,** Computed tomographic brain scan findings in Q fever encephalitis, *Neuroradiology,* 26, 329, 1984.

70. **Wallace, S. J. and Zealley, H.,** Neurological, electroencephalographic and virological findings in febrile children, *Arch. Dis. Child.,* 45, 611, 1970.

71. **Schwartz, R. B.,** Manic psychosis in connection with Q fever, *Br. J. Psychiatry,* 124, 140, 1974.

72. **Gallaher, W. H.,** Q fever, *JAMA,* 177, 187, 1961.

73. **Whittick, J. W.,** Necropsy findings in a case of Q fever in Britain, *Br. Med. J.,* 1, 979, 1950.

74. **Masbernard, A.,** Les localisations neurologiques des rickettsioses, *Bull. Soc. Pathol. Exot.,* 56, 714, 1963.

75. **Siegert, R., Simrock, W., and Stroeder, R.,** Über einen epidemische Ausbruch von Q-fieber in einem Krankenhaus, *Z. Tropenmed. Parasitol.,* 2, 1, 1950.

76. **Deller, M. and Streiff, E. B.,** Névrite rétrobulbaire et fièvre Q, *Bull. Soc. Ophtalmol. Fr.,* 69, 309, 1956.
77. **Catros, A. and Hoëll, J.,** Névrite optique axiale bilatérale à fièvre Q, *Bull. Soc. Ophtalmol. Fr.,* 60, 325, 1960.
78. **Fontan, P. and Desbordes, P.,** Névrite optique et fièvre Q, *Ann. Ocul.,* 194, 971, 1961.
79. **Schuil, J., Richardus, J. H., Baarsma, G. S., and Schaap, G. J. P.,** Q fever as a possible cause of bilateral optic neuritis, *Br. J. Ophthalmol.,* 69, 580, 1985.
80. **Somlo, F. and Kovalik, M.,** Acute thyroiditis in a patient with Q fever, *Can. Med. Assoc. J.,* 95, 1091, 1966.
81. **Sheridan, P., MacCaig, J. N., and Hart, R. J. C.,** Myocarditis complicating Q fever, *Br. Med. J.,* 2, 155, 1974.
82. **Caughey, J. E.,** Pleuropericardial lesion in Q fever, *Br. Med. J.,* 1, 1447, 1977.
83. **Brada, M. and Bellingham, A. J.,** Bone-marrow necrosis and Q fever, *Br. Med. J.,* 281, 1108, 1980.
84. **Delsol, G., Pellegrin, M., Familiades, J., and Auvergnat, J. C.,** Bone marrow lesions in Q fever, *Blood,* 52, 637, 1978.
85. **Okun, D. B., Sun, N. C. J., and Tanaka, K. R.,** Bone marrow granulomas in Q fever, *Am. J. Clin. Pathol.,* 71, 117, 1979.
86. **Hitchins, R., Cobcroft, R. G., and Hocker, G.,** Transient severe hypoplastic anemia in Q fever, *Pathology,* 18, 254, 1986.
87. **Ende, N. and Gelpi, A. P.,** Pathological changes noted in bone marrow in a case of Q fever, *Arch. Intern. Med.,* 100, 793, 1957.
88. **Lillie, R. D., Perrin, T. L., and Armstrong, C.,** Institutional outbreak of pneumonitis; histopathology in man and rhesus monkeys in pneumonitis due to virus of Q fever, *Public Health Rep.,* 56, 149, 1941.
89. **Bernstein, M., Edmondson, H. A., and Barbour, B. H.,** The liver lesion in Q fever, *Arch. Intern. Med.,* 116, 491, 1965.
90. **Voigt, J. J., Delsol, G., and Fabre, J.,** Liver and bone marrow granulomas in Q fever, *Gastroenterology,* 84, 887, 1983.
91. **Schleicher, E. M.,** *Bone Marrow Morphology and Mechanics of Biopsy,* Charles C Thomas, Springfield, IL, 1973, 149.
92. **Estrov, Z., Bruck, R., Schtalrid, M., Berrebi, A., and Resnitzky, P.,** Histiocytic hemophagocytosis in Q fever, *Arch. Pathol.,* 108, 7, 1984.
93. **Ramos, H. S., Hodges, R. E., and Meroney, W. H.,** Q fever: report of a case simulating lymphoma, *Ann. Intern. Med.,* 47, 1030, 1957.
94. **Uriarte, S. A., Peña, G. M., Gordo, J. M. A., and Serrano Rios, C. S. R. T.,** Paniculitis mesenterica, *Rev. Clin. Esp.,* 171, 347, 1983.
95. **Biggs, B. A., Douglas, J. G., Grant, I. W. B., and Crompton, G. K.,** Prolonged Q fever associated with inappropriate secretion of antidiuretic hormone, *J. Infect.,* 8, 61, 1984.
96. **Weir, W. R. C., Bouchet, V. A., Mitford, E., Taylor, R. F. H., and Smith, H.,** Kawasaki disease in European adult associated with serological response to *Coxiella burnetii, Lancet,* 2, 504, 1985.
97. **Swaby, E. D., Fisher-Hoch, S., Lambert, H. P., and Stern, H.,** Is Kawasaki disease a variant of Q fever?, *Lancet,* 2, 146, 1980.
98. **Lambert, H. P., Fisher-Hoch, S. P., and Grover, S. A.,** Kawasaki disease and *Coxiella burnetii, Lancet,* 2, 844, 1985.
99. **Conget, I., Mallolas, J., Mensa, J., and Rovira, M.,** Erythema nodosum and Q fever, *Arch. Dermatol.,* 123, 867, 1987.
100. **Heard, S. R., Ronalds, C. J., and Heath, R. B.,** *Coxiella burnetii* infection in immunocompromised patients, *J. Infect.,* 11, 15, 1985.
101. **Sidwell, R. W., Thorpe, B. D., and Gebhardt, L. P.,** Studies of latent Q fever infections. I. Effects of whole body x-irradiation upon latently infected guinea pigs, white mice and deer mice, *Am. J. Hyg.,* 79, 113, 1964.
102. **Sidwell, R. W., Thorpe, B. D., and Gebhardt, L. P.,** Studies of latent Q fever infections. II. Effects of multiple cortisone injections, *Am. J. Hyg.,* 79, 320, 1964.
103. **Lev, V. I., Shachar, A., Segev, S., Weiss, P., and Rubinstein, E.,** Quiescent Q fever endocarditis exacerbated by cardiac surgery and corticosteroid therapy, *Arch. Intern. Med.,* 148, 1531, 1988.
104. **Kanfer, E., Farrag, N., Price, C., MacDonald, D., Coleman, J., and Barrett, A. J.,** Q fever following bone marrow transplantation, *Bone Marrow Transplant.,* 3, 165, 1988.
105. **Loudon, M. M. and Thompson, E. N.,** Severe combined immunodeficiency syndrome, tissue transplant, leukaemia, and Q fever, *Arch. Dis. Child.,* 63, 207, 1988.
106. **Murphy, P. P. and Richardson, S. G.,** Q fever presently as an eosinophilic pleural effusion, *Thorax,* 44, 228, 1989.

Chapter 7

PATHOLOGY OF Q FEVER PNEUMONIA

David T. Janigan and Thomas J. Marrie

TABLE OF CONTENTS

I. INTRODUCTION

Reported fatalities from Q fever pneumonia are rare, and since not all patients were necropsied, information about its pathology is limited. In four patients a mixed interstitial and alveolar pneumonia, with a predominance of mononuclear inflammatory cells, diffusely involved one lower,[1] one upper, or all lobes.[3,4] The mixed pneumonic consolidation in these patients correlates well with observations that, radiologically, Q fever pneumonia exhibits an alveolar pattern much more often than the diffuse interstitial markings characteristic of atypical pneumonias.[5] In a fifth case diagnosed (but not proved) as fatal Q fever pneumonia, a necrotizing alveolar pneumonia and bronchiolitis, with microabscesses, were observed in all lobes.[6] The lobar consolidation in these fatal cases may reflect secondary bacterial infection since it contrasts with the characteristic segmental or rounded opacities observed radiologically in many nonfatal cases of Q fever pneumonia. However, we know of only three reported cases in which lung biopsies were taken during life. One patient had a persistent pulmonary infiltrate of 5-week duration, and the pathology was described only as "chronic non-specific inflammatory disease".[7] The histopathology of a transbronchial biopsy from another patient was not described except for the staining of "coccobacillary bodies within alveolar macrophages".[8] In a third patient, who presented clinically with adult respiratory distress syndrome, the biopsy revealed "non-specific diffuse alveolar damage" with hyaline membranes.[9]

A lung biopsy and a whole lung lobe were obtained from two of our patients with Q fever pneumonia relatively early in the course of their disease when evidence of superimposed bacterial infection was absent. These tissues provided an opportunity to characterize and localize the histological changes. In addition to examining random histology sections from multiple blocks of these materials, serial sections 5 μm thick were also cut from a single block. Every fifth serial section was collected and stained until a total of 35 sections were collected, that is, one at every 25-μm level. Thus, these sections represented interval samples through a 900-mm-thick sample of lung tissue. In brief, the pathology consisted of (1) an interstitial bronchioloalveolitis associated with necrosis and regeneration of lining epithelium, (2) filling of bronchioloalveolar air spaces by mononuclear inflammatory cells, and (3) variable fibroblastic narrowing of bronchioles causing distal obstructive changes. The autopsy lung findings in a third patient dying with Q fever pneumonia demonstrated the changes of secondary bacterial infection which obscured the primary pathology.

II. PATIENT 1

A. FINDINGS
This 27-year-old insurance salesman became ill with fever, myalgia, chest pain, sweats, and fatigue. A chest radiograph (Figure 1 A and B) showed a left lower lobe rounded opacity. Despite treatment with erythromycin he became progressively short of breath, and a chest radiograph on day 5 of his illness showed multiple rounded opacities (Figure 1C). An open lung biopsy was carried out on day 6. Cultures of the pulmonary tissue were negative for aerobic and anaerobic bacteria, viruses, and fungi. Serological studies revealed a >4-fold rise in antibody to *C. burnetii* phase II antigen from <1:8 in the acute phase sample to 1:1024 in the convalescent sample. Examination of both multiple random and serial sections of the biopsy revealed two major features.

1. Interstitial Bronchioloalveolitis
There were multifocal necrosis and associated regenerative and/or reactive proliferation of bronchiolar epithelium, most marked in terminal and respiratory bronchioles (Figures 2 to 4). While necrosis of alveolar epithelium was not as readily evident, some of these cells were enlarged and/or hyperplastic, suggesting regenerative and/or reactive changes (Figure 2B). The

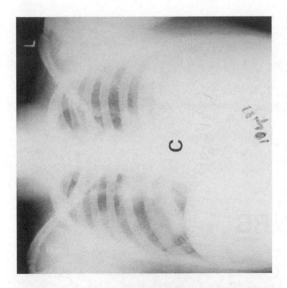

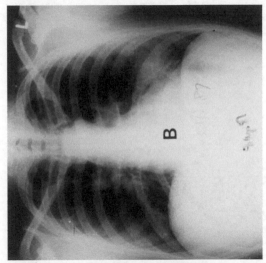

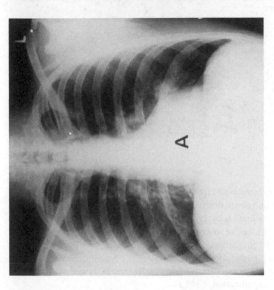

FIGURE 1. (A) Chest radiograph, Case 1. Obtained on the first day of his illness. Note the small round alveolar opacity involving the left lingula. (B) Chest radiograph, Case 1. Day 3 of his illness. There was a slight increase in the size of the opacity involving the left lingula. (C) Chest radiograph, Case 1. Day 5 of his illness. Note the marked increase in the size of the opacity in the left lower lobe and the appearance of multiple rounded opacities throughout the right lung field. This is a portable chest film. It is underpenetrated compared to those in 1A and 2B. It is also an anteroposterior projection while the others are posteroanterior.

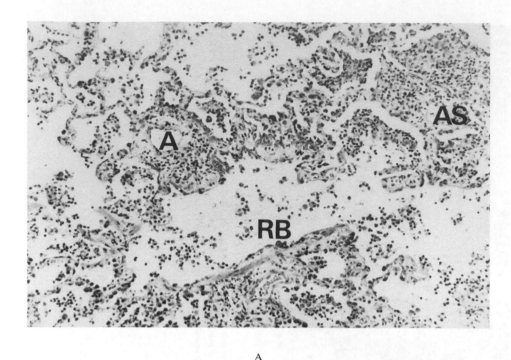

A

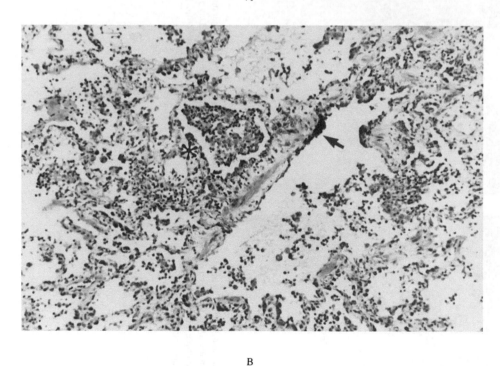

B

FIGURE 2.   (A) (Serial section level 150). Respiratory bronchioles (rb), arteriole (a), and surrounding alveolated tissue. The bronchiolar epithelial lining is denuded, in part due to necrosis. The interstitium around the bronchiole and arteriole shows inflammation, i.e., edema and infiltration by lymphocytes and macrophages. Interstitial inflammation in alveoli is less marked in this level. The alveolar space (as) to the right is filled with lymphocytes and macrophages, the other spaces much less so. (H.E.; magnification × 445.) (B) (Serial section level 160). In comparison with A, there is a greater degree of alveolar interstitial inflammation (asterisk). Note the encroachment of alveolar spaces by the interstitial reaction. Reactive and/or regenerative bronchiolar (arrow) as well as alveolar epithelia are more evident at this serial level. (H.E.; magnification × 445.)

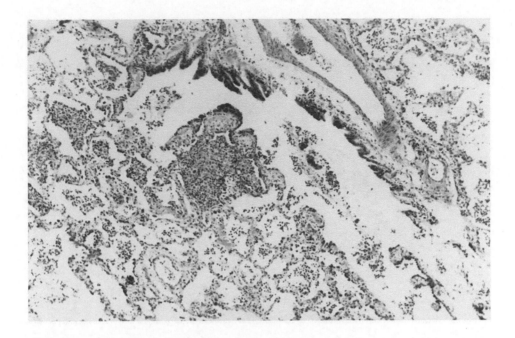

FIGURE 3. (Serial section level 140). The section is at or near the origin of a respiratory bronchiole from a terminal bronchiole. The focus of intense inflammation represents a mix of interstitial and alveolar inflammation, the interstitial changes producing a narrowing of bronchiolar and alveolar lumens. Note the gradient of decreasing severity of both the interstitial and alveolar inflammation away from the bronchiole, i.e., toward the lower right. (H.E.; magnification × 280.)

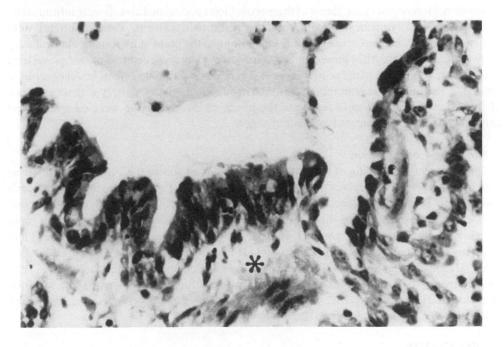

FIGURE 4. Respiratory bronchiolar epithelium is relatively normal on the left. There is focal early necrosis and proliferation in the middle segments, and there are regenerative changes on the right side. Note the subepithelial edema, (asterisk) and the exudate in the lumen. (H.E.; magnification × 1780.)

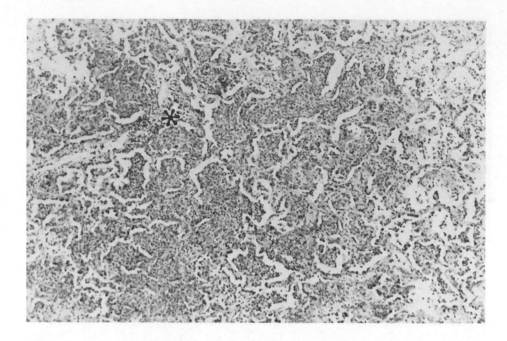

FIGURE 5. Alveolar spaces are filled with macrophages, some with swollen cytoplasm, lymphocytes, and other cells resembling detached epithelium. The asterisk locates approximate centriacinar area where the changes are more marked than those in the peripheral acinar locations toward the right. (H.E.; magnification × 280.)

subepithelial interstitial tissue planes of these bronchioles and related alveoli were infiltrated by macrophages and lymphocytes, with only rare neutrophils evident. These infiltrates were sometimes focally marked, expanding the interstitium to the point of narrowing of respiratory bronchioles (Figure 3). All of these changes were most evident in the centriacinar zones and least prominent in distal parts of the acini (Figure 3). There was no arterial necrosis or venoocclusive disease. A recently formed thrombus was found in one small muscular artery without evidence of infarction. The intralobular septa and the pleura connective tissue were edematous and exhibited dilated or prominent lymphatic channels.

2. Air Space Consolidation

There was variable filling of bronchiolar and alveolar air spaces (i.e., alveolar pneumonia) with macrophages, lymphocytes, and other mononuclear cells (Figures 3 and 5), some of which appeared to be detached, reactive lining cells. Only few neutrophilic polymorphonuclear leukocytes were present, and there was no suppuration. Again, these changes were more prominent in the centriacinar zones (Figure 5). The lumens of a few inflamed terminal and respiratory bronchioles were narrowed or occluded due to expansion of the inflamed interstitium. Alveolar spaces distal to these narrowed airways contained some macrophages with swollen pale cytoplasm, features suggestive of a mild obstructive pneumonia. There was little or no edemal fluid or material in air spaces. However, a few proteinaceous hyaline membranes lining alveolar ducts were found in some congested lobules.

B. DISCUSSION

The biopsy showed a mixture of interstitial and alveolar mononuclear cell pneumonia, apparently in response to the necrosis of bronchiolar and alveolar epithelium. These changes differ from those of acute bacterial pneumonia in which interstitial pneumonia is usually absent or mild and alveolar pneumonia with neutrophils is characteristic. While the necrosis of bronchiolar and alveolar lining epithelium and the interstitial changes in this biopsy cannot be distinguished histologically from those seen in some acute viral pneumonias, the degree of air space filling by lymphocytes and macrophages was, by comparison, more marked here. This filling accounts in large part, if not entirely, for the radiographic features of alveolar consolidation. While that air space filling reflects an inflammatory reaction to the parenchymal injury inflicted by the *Coxiella*, it appears that the narrowing of some bronchioles may contribute to the local persistence or accumulation of these cells, by interfering with the normal air space clearing mechanisms. Hyaline membranes generally reflect injury to alveolar walls sufficient to cause increased microvascular permeability. These are characteristically and extensively present in the adult respiratory distress syndrome.[9] However, this histological change was limited here, and the syndrome had not developed clinically.

III. PATIENT 2

A. FINDINGS

The clinical and some pathological features were previously reported.[10] This patient presented with an acute febrile respiratory illness caused by *Coxiella burnetii*. Pneumonia had developed in the posterior segment of the left upper lobe and, at the time of presentation, appeared radiologically as a mass lesion resembling a carcinoma. The lobe was resected, and a relatively circumscribed, rounded inflammatory segmental mass was found (Figure 6).

Within the mass many terminal and respiratory bronchioles were partially or totally occluded by plugs of fibroblasts and collagen intermixed with variable numbers of lymphocytes, plasma cells, and macrophages (Figure 7). Bronchiolar epithelium was focally necrotic and regenerative. There was also an interstitial mononuclear cell infiltration of these bronchioles and alveolated tissues. Within the mass, alveolar spaces distal to occluded bronchioles contained variable numbers of "foam" cells (i.e., macrophages with phagocytosed lipid residues from cell debris) and/or organizing fibrocellular exudates. The remainder of the apical-posterior segment exhibited an interstitial bronchioloalveolitis similar to that found in Patient 1.

B. DISCUSSION

The degree of bronchiolar occlusion and narrowing was much greater than that found on biopsy from Patient 1. Additionally, while the mild bronchiolar narrowing in Patient 1 was mainly due to interstitial changes, here it was largely due to fibroblastic granulation tissue within the lumens resulting in "bronchiolitis obliterans", a characteristic but nonspecific reaction to injury.[11] This complication resulted in obstructive pneumonia superimposed on the primary changes. These mixed changes contributed to the development of the localized mass lesion, a so-called "inflammatory pseudotumor". This complication may account for the prolonged persistence of radiographic lung changes in some patients with Q fever pneumonia[12,13] and for the appearance of these changes as mass lesions.[7,12-14]

IV. PATIENT 3

C. FINDINGS

This 60-year-old male was found in an unconscious state. On hospitalization he was in shock, uremic, and febrile. His chest radiograph showed diffuse pulmonary edema compatible with acute alveolar damage. He died within 4 h of admission. Blood cultures were positive for *Escherichia coli*. Shortly after his death, it was learned that he and several of the members of his poker-playing group had developed Q fever pneumonia following recent exposure to an infected parturient cat.[15] The lungs at autopsy were very heavy and edematous and showed widespread hemorrhagic confluent bronchopneumonia, more marked in the lower lobes. Histologically, there was a filling of alveoli by neutrophils with focal suppuration. Numerous

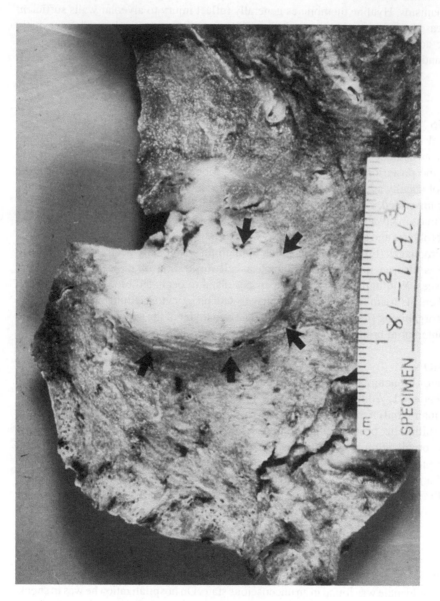

FIGURE 6. Left upper lobe, Case 2. Note the mass ("pseudotumor") indicated by the arrows.

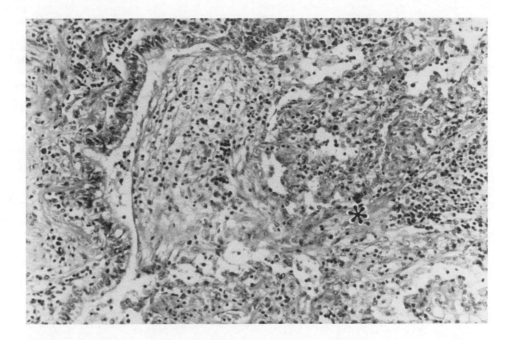

FIGURE 7. Patient 2. Bronchiolitis obliterans. Part of a terminal bronchiole with relatively intact lining epithelium is to the left. Its lumen is filled with a plug of loose granulation and inflammatory tissue which extends into a respiratory bronchiole (asterisk). The remainder of the lumen contains mononuclear inflammatory cells. (H.E.; magnification × 715.)

Gram-positive and Gram-negative mixed bacteria were present. Thrombi were present in two small pulmonary muscular arteries, but there were no infarcts. Material scraped from the cut surfaces of formaldehyde-fixed lung was stained by the direct immunofluorescence method for *C. burnetii* using a direct immunofluoresence technique, and positive-stained organisms were found.

B. DISCUSSION

The pathology findings in this case were dominated by bacterial pneumonia superimposed upon the Q fever pneumonia, an event which appears to have occurred in a previously reported fatal case[6] and possibly in others.[2] Thus, both the histological and radiological features in this case were quite different from those found in Patients 1 and 2 and radiologically were different from most patients with nonfatal Q fever pneumonia. The complication of secondary bacterial infection may obfuscate the histological recognition of the primary histopathology and, therefore, the etiology of the lung injury. This emphasizes the value of immunofluorescence staining with specific antibody to *C. burnetti*, a step which was also diagnostically valuable in Patient 1.

REFERENCES

1. **Whittick, J. W.,** Necropsy findings in a case of Q fever in Britain, *Br. Med. J.,* 1, 979, 1950.
2. **Lillie, R. D., Perrin, T. L., and Armstrong, C.,** An institutional outbreak of pneumonitis. III. Histopathology in man and rhesus monkeys in the pneumonitis due to virus of Q fever, *Public Health Rep.,* 56, 149, 1941.
3. **Perrin, T. L.,** Histopathologic observations in a fatal case of Q fever, *Arch. Path.,* 47, 361, 1949.

4. **Tonge, J. L. and Derreck, E. H.,** A fatal case of Q fever associated with hepatic necrosis, *Med. J. Australas.,* 1, 594, 1959.

5. **Warren, J. W. and Hornick, R. B.,** *Coxiella burnetii* (Q fever), in *Principles and Practice of Infectious Diseases,* Mandell, G. L., Douglas, R. G., Jr., and Bennett, J. E., Eds., John Wiley & Sons, New York, 1979, 1516.

6. **Urso, F. P.,** The pathologic findings in Rickettsial pneumonia, *Am. J. Clin. Pathol.,* 64, 335, 1975.

7. **Musher, D. M.,** Q fever: a common treatable cause of endemic non-bacterial pneumonia, *JAMA,* 204, 111, 1968.

8. **Pierce, T. H., Yucht, S. C., Gorin, A. B., Jordan, G. W., Tesluk, H., and Lillington, G. A.,** Q fever pneumonitis. Diagnosis by transbronchoscopic lung biopsy, *West. J. Med.,* 130, 453, 1979.

9. **Torres, A., de Celis, M. R., Rodriques Roisin, R., Vidal, J., and Vidal, A. A.,** Adult respiratory distress syndrome in Q fever, *Eur. J. Respir. Dis.,* 70, 322, 1987.

10. **Janigan, D. T. and Marrie, T. J.,** An inflammatory pseudotumor of lung in Q fever pneumonia, *N. Engl. J. Med.,* 30, 86, 1983.

11. **Epler, E. R. and Colby, T. V.,** The spectrum of bronchiolitis obliterans, *Chest,* 83, 161, 1983.

12. **Millar, J. K.,** The chest film findings in Q fever - a series of 35 cases, *Clin. Radiol.,* 29, 371, 1978.

13. **Ramos, H. S. and Hodges, R. E.,** Q fever: report of a case simulating lymphoma, *Ann. Intern. Med.,* 47, 1030, 1957.

14. **Johnson, J. E. and Kadull, P. J.,** Laboratory-acquired Q fever, *Am. J. Med.,* 41, 391, 1966.

Chapter 8

Q FEVER HEPATITIS

Thomas J. Marrie

TABLE OF CONTENTS

I. INTRODUCTION

A separate chapter is devoted to Q fever hepatitis because the liver may be involved in both the acute and chronic forms of the disease.

II. EXPERIMENTAL STUDIES

Experimental studies in both animals and man indicate that the liver is probably involved in all patients with acute Q fever. Lillie and Perrin[1] inoculated four strains of *Coxiella burnetii* into the right lung of eight rhesus monkeys. The liver was histologically normal in two monkeys, and a slight to moderate focal interstitial or periportal lymphocyte infiltration or both was present in five monkeys.

In another study Gonder and colleagues[2] assigned 18 cynomolgus monkeys *(Macaco fascicularis)* to four experimental groups. One group of four monkeys served as a control group and received only heart infusion broth; the second group was challenged with a small dose of *C. burnetii,* and the third group with a large dose. Two monkeys given the large dose were examined and killed for pathological examination. All monkeys in the experimental groups received *C. burnetii* phase I Henzerling strain via an aerosol. All the monkeys exposed to the large dose became ill, while only one of the four monkeys exposed to the small dose developed fever. Increases in serum alkaline phosphatase were noted from day 3 to day 13. The SGOT (serum glutamic oxaloacetic transaminase also known as asparate transaminase [AST]) and total bilirubin were also significantly higher than normal. The two monkeys who died on days 7 and 14, respectively, had scattered areas of subacute hepatitis. *C. burnetii* was demonstrated in frozen sections of lung, liver, spleen, and kidney by the indirect fluorescent antibody technique. There is no mention that it was cultured from the liver.

Heggers et al.[3] infected guinea pigs with the Nine Mile strain of phase I *C. burnetii.* The route of inoculation was not stated. Severe microscopic changes were seen in the liver. These changes were related to the duration of infection and progressed from fatty metamorphosis to reticuloendothelial hyperplasia, necrosis, and mineralization. Microorganisms were found in the cytoplasms of Kupffer cells.

Mice inoculated via the intraperitoneal or pulmonary route developed nodular and granulomatous lesions in the liver.[4]

Finally, hepatic lesions were demonstrated in two male volunteers who received an aerosol challenge of 3000 50% guinea pig intraperitoneal infectious dose of *C. burnetii.* Liver biopsies were obtained on days 8, 12, and 36 following infection. Both volunteers were treated with doxycycline starting on day 12 after inoculation. Changes were present in the initial biopsy and were still present in the biopsy at 36 d. Small inflammatory foci were evident early on. These consisted of lymphocytes, histiocytes, and polymorphonuclear leukocytes. Some liver cells had undergone necrosis. Rare hepatic parenchymal cells had pyknotic nuclei (Councilman-like bodies). The portal triads were involved. The degree of inflammatory infiltrate varied. In one volunteer it was mild, and in the other the triads were heavily infiltrated with mononuclear inflammatory cells.[5]

III. CLINICAL SYNDROMES OF Q FEVER HEPATITIS

A. ACUTE Q FEVER
These are

1.   An infectious hepatitis-like picture
2.   Pyrexia of unknown origin with typical "doughnut granulomas" evident on biopsy
3.   Incidental hepatitis, that is, biochemical evidence of hepatitis in a patient whose major manifestation of Q fever involves another organ system such as pneumonia.

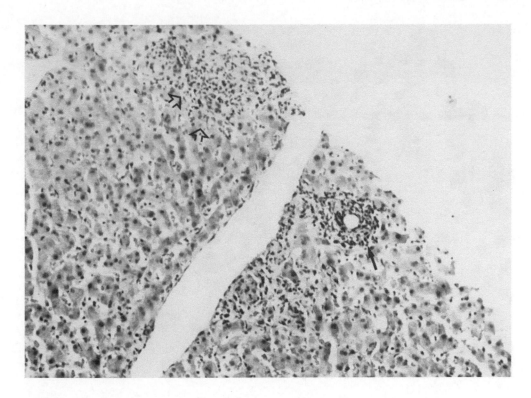

FIGURE 1. Low-power view of liver biopsy from patient with Q fever hepatitis. A granuloma (arrow) is evident as is an inflammatory infiltrate (arrowhead).

B. CHRONIC Q FEVER
1. Q Fever Hepatitis in Association with Endocarditis

From the very beginning it was evident that one of the manifestations of acute Q fever could be hepatitis. Indeed, one of Derrick's original nine patients was jaundiced.[6]

It was not until 1956, however, that Gerstl and co-workers[7] described the distinctive histopathological changes. Subsequently, the "doughnut granuloma" has become synonymous with acute Q fever hepatitis.[8-20] Unfortunately, this granuloma is not pathognomonic for Q fever as it is also seen on occasion in Hodgkin's disease,[21] in allopurinol hypersensitivity,[22] in visceral leishmaniasis,[23] in Epstein-Barr virus,[24] and in cytomegalovirus hepatitis.[25] The granulomas in Q fever are not confined to the liver but are also found in the bone marrow.[26-28]

The doughnut granuloma consists of a granuloma with a central vacuole (presumed to be lipid) and fibrin deposits (Figures 1 and 2). The fibrin deposits may be within or at the periphery of the granuloma. Fibrin is also deposited within the sinusoids. Neutrophils are common in early granulomas, and giant cells are present in later lesions.[29] A variety of other nondiagnostic changes are also found including fatty change, Kupffer cell hyperplasia, mononuclear cell infiltration of the portal tracts, and focal parenchymal inflammation. The importance of Q fever as a cause of granulomatous hepatitis is evident from the study of Voigt et al.[30] These workers reported on 112 cases of granulomatous hepatitis: tuberculosis was the most common cause, 24 cases; Hodgkin's disease, 21; sarcoidosis, 14; and Q fever, 14.

IV. CLINICAL PRESENTATION(S) OF ACUTE Q FEVER HEPATITIS

The most common picture in our experience is that of biochemical changes of hepatitis in patients whose major manifestation of Q fever is pneumonia.[31] The alkaline phosphatase is

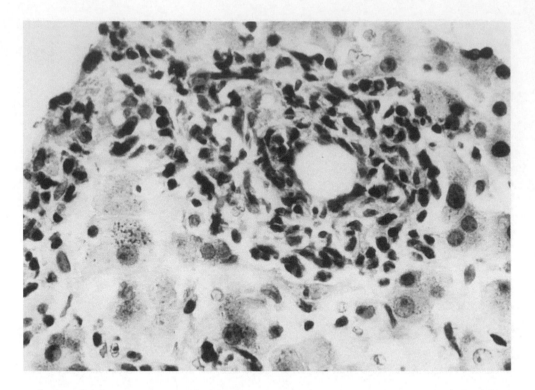

FIGURE 2. Photomicrograph of liver biopsy showing the characteristic doughnut granuloma. The central clear space is dissolved lipid.

elevated in 57% of patients, the AST in 53%, and the ALT (alanine transaminase) in 43%. The elevations of these enzymes are usually mild, two to three times normal.

Alkan et al.[32] in a study carried out from 1960 to 1962 in Israel found that 13% (17 of 128) of patients who presented with "infectious hepatitis" had Q fever, while 2 of 13 with infectious mononucleosis, 4 of 119 with pyrexia of unknown origin, 7 of 108 (6.5%) with pneumonia, and 2 of 207 with a variety of other illnesses had Q fever. Severe headache was the only differentiating symptom between Q fever and viral hepatitis. In 1985, Sawyer et al.[33] tested for Q fever 959 serum samples that had been submitted to laboratories in six states for hepatitis testing. Six cases (0.6%) of Q fever were identified.

Rarely, Q fever hepatitis is fatal.[34,35] The patient reported by Derrick[34] was a 52-year-old meatworker who presented with a mass in the right hypochondrium. At laparotomy the liver was swollen, and there was a necrotic area on the surface. Postoperatively, he developed jaundice and at autopsy extensive liver necrosis was present. A 2-year-old Arab boy was admitted to the hospital with a 14-d history of abdominal pain, fever, anorexia, diarrhea, and progressive jaundice. The liver enzymes were elevated to ten times normal and the total bilirubin reached a high of 411 mmol/l. On hospital day 18, bleeding from the groins occurred following minor trauma, and the next day hepatic coma developed.[35]

The major manifestation of Q fever during an outbreak in Ontario, Canada, was granulomatous hepatitis.[36] These patients all had fever of unknown origin. Pneumonia was not present in these patients. Epidemiologically, the Q fever was associated with exposure to goats. This outbreak suggests that strain differences may be important in the manifestation of Q fever hepatitis. We have never seen pyrexia of unknown origin due to Q fever granulomatous hepatitis in Nova Scotia, despite more than 179 cases of Q fever. Furthermore, patients who develop Q fever pneumonia following exposure to contaminated products of feline conception do not have hepatomegaly (see Chapter 6). In other countries, however, hepatomegaly is a common manifestation of acute Q fever ranging from 11% for patients from California, to 65% of Australian patients with Q fever.[37]

V. HEPATIC INVOLVEMENT IN CHRONIC Q FEVER

Westlake et al.[17] were able to find in their review of 220 published cases of Q fever endocarditis information on the liver histology in 30 cases. "Doughnut granulomas" of acute Q fever hepatitis were not described. Only 7 of the 30 cases had granulomas. Mononuclear cell infiltration of the portal tracts was most common, occurring in 22 cases. Five cases had spotty necrosis of parenchymal cells.

One of 16 patients with Q fever endocarditis studied by Turck et al.[37] showed progression to micronodular cirrhosis over 10 months. Atienza et al.[39] described a 73-year-old woman who had serological evidence of chronic Q fever, hepatomegaly, splenomegaly, fever, and a history of heterograft replacement for aortic stenosis. Extensive portal fibrosis with many granulomas and steatosis were present on liver biopsy. The fibrosis increased initially but then remained static on follow-up liver biopsy to 18 months.

From time to time there is confusion in the literature, and cases of Q fever hepatitis are labeled as "chronic Q fever". Two such cases were reported by Weir and colleagues.[40] However, the serological profile of both patients was that of acute Q fever. Some workers maintain that the serological profile of patients with granulomatous Q fever hepatitis is different from that of acute or chronic Q fever. Peacock et al.,[41] based on a study of five patients with granulomatous hepatitis, concluded that phase I titers by the indirect immunofluorescence assay were significantly higher than in patients with acute Q fever. "Significantly" is not used in this paper to indicate statistical significance. Indeed, only one of their five patients had very high phase I IgG antibody titers — this patient's titer peaked at 1:131, 072. Otherwise, the highest titer was 1:2048. Further study is needed to determine if, indeed, patients with granulomatous Q fever hepatitis have an antibody response that is different from acute Q fever.

The suggestion has been made that the liver may act as a reservoir for *C. burnetii*, explaining the long delay that frequently occurs between the initial episode of acute Q fever and the diagnosis of endocarditis.[42]

There may be a form of chronic Q fever hepatitis, in the absence of endocarditis. Yebra and co-workers[43] reported the case of a 31-year-old man who had acute Q fever hepatitis with fibrin ring granulomas on liver biopsy. Two years later, liver function tests were still abnormal, and ring granulomas were still present on biopsy. The phase I antibody titer by complement fixation was 1:8, and by IFA it was 1:2048. This case is chronic in terms of the duration of the liver disease, but the serological profile is not that of chronic Q fever.

VI. TREATMENT

The treatment for acute fever hepatitis is tetracycline, doxycycline, or rifampin for 2 weeks. In chronic Q fever the hepatitis is usually an incidental part of the disease and the treatment is really that of chronic Q fever (see Chapter 9).

VII. SUMMARY

Hepatic involvement is common in Q fever. In acute Q fever this may manifest itself as an infectious hepatitis-like illness, as mild to moderate biochemical abnormalities suggestive of hepatitis in a patient with Q fever pneumonia, or as granulomatous hepatitis in a patient with fever of unknown origin. Hepatic granulomas with a central vacuole and a fibrin ring (doughnut granulomas) are seen in acute Q fever hepatitis but probably not in the liver involvement that occurs with chronic Q fever. These doughnut granulomas are not pathognomonic for Q fever as they are also found in Hodgkin's disease, Epstein-Barr virus and cytomegalovirus hepatitis, in visceral leishmaniasis, and allopurinol hypersensitivity. Fulminant hepatic necrosis occurs as part of acute Q fever hepatitis but is extremely rare.

REFERENCES

1. **Lillie, R. D. and Perrin, T. L.,** An institutional outbreak of pneumonitis. III. Histopathology in man and rhesus monkeys in the pneumonitis due to the virus of "Q" fever, *Public Health Rep.,* 56, 149, 1941.
2. **Gonder, J. C., Kishimoto, R. A., Kastello, M. D., Pedersen, C. E., Jr., and Larson, E. W.,** Cynomolgus monkey model for experimental Q fever infection, *J. Infect. Dis.,* 139, 191, 1979.
3. **Heggers, J. P., Billups, Z. H., Hinrichs, D. I., and Mallavia, L. P.,** Pathophysiologic features of Q fever-infected guinea pigs, *Am. J. Vet. Res.,* 36, 1047, 1975.
4. **Perrin, T. L. and Bengtson, I. A.,** The histopathology of experimental "Q" fever in mice, *Public Health Rep.,* 57, 790, 1942.
5. **Dupont, H. L., Hornick, R. B., Levin, H. S., and Woodward, T. E.,** Q fever hepatitis, *Ann. Intern. Med.,* 74, 198, 1971.
6. **Derrick, E. H.,** Q fever, a new fever entity. Clinical features, and laboratory investigation, *Med. J. Aust.,* 2, 281, 1937.
7. **Gerstl, B., Movitt, E. R., and Skahen, J. R.,** Liver function and morphology in "Q" fever, *Gastroenterology,* 30, 813, 1956.
8. **Qizilbash, A. H.,** The pathology of Q fever as seen on liver biopsy, *Arch. Pathol. Lab. Med.,* 107, 354, 1983.
9. **Pellegrin, M., Delsol, G., Auvergnat, J. C., Familiades, J., Faure, H., Guiu, M., and Voigt, J. J.,** Granulomatous hepatitis in Q fever, *Hum. Pathol.,* 11, 51, 1980.
10. **Voigt, J. J., Delsol, G., and Fabre, J.,** Liver and bone marrow granulomas in Q fever, *Gastroenterology,* 84, 887, 1983.
11. **Hofmann, C. E. and Heaton, J. W., Jr.,** Q fever hepatitis. Clinical manifestation and pathological findings, *Gastroenterology,* 83, 474, 1982.
12. **Picchi, J., Nelson, A. R., Waller, E. E., Razavi, M., and Clizer E. E.,** Q fever associated with granulomatous hepatitis, *Ann. Intern. Med.,* 53, 1965, 1960.
13. **Bernstein, M., Edmondson, H. A., and Barbour, B. H.,** The liver lesion in Q fever, *Arch. Intern. Med.,* 116, 491, 1965.
14. **Powell, O. W.,** Liver involvement in "Q" fever, *Aust. Ann. Med.,* 10, 52, 1961.
15. **Baranda, M. M., Carranceja, J. C., and Errasti, C. A.,** Q fever in the Basque country; 1981—1984, *Rev. Infect. Dis.,* 7, 700, 1985.
16. **Silver, S. S. and McLeish, W. A.,** "Doughnut" granulomas in Q fever, *Can. Med. Assoc. J.,* 130, 102, 1984.
17. **Westlake, P., Price, L. M., Russell, M., and Kelly, J. K.,** The pathology of Q fever hepatitis, *J. Clin. Gastroenterol.,* 9, 357, 1987.
18. **Travis, L. B., Travis, W. D., Li, C.-Y., and Pierce, R. V.,** Q fever. A clinicopathologic study of five cases, *Arch. Pathol. Lab. Med.,* 110, 1017, 1986.
19. **Pierrugues, R., Buttigieg, R., Barnon, G., Parelon, G., Payes, A., and Michel, H.,** Hepatite de la fievre Q: diagnostic differential de l' hepatite alcoolique aigue sur cirrhose, *Gastroenterol. Clin. Biol.,* 12, 583, 1988.
20. **Raoult, D., Drancourt, M., DeMicco, C., Durand, J. M., Nesri, M., Charrel, C., Bernard, J. P., Gallais, H., and Casanova, P.,** Les hepatites de la fievre Q a propos de quatorze cas revue de la litterature, *Sem. Hop. Paris.,* 62, 997, 1986.
21. **Delsol, G., Pellegrin, M., Voigt, J. J., and Fabre, J.,** Diagnostic valve of granuloma with fibrinoid ring, *Am. J. Clin. Pathol.,* 73, 289, 1981.
22. **Vanderstigel, M., Zafrani, E. S., Lejonc, J. L., Schaeffer, A., and Portos, J. L.,** Allopurinol hypersensitivity syndrome as a cause of hepatic fibrin ring granulomas, *Gastroenterology,* 90, 188, 1986.
23. **Moreno, A., Marazuela, M., Yebra, M., Hernandez, M. J., Hellin, T., Montalban, C., and Vargas, J. A.,** Hepatic fibrin — ring granulomas in visceral leishmaniasis, *Gastroenterology,* 95, 1122, 1988.
24. **Nenert, M., Mariner, P., Dubuc, N., Deforges, L., and Zafroni, E. S.,** Epstein-Barr virus infection and hepatic fibrin — ring granulomas, *Hum. Pathol.,* 19, 608, 1988.
25. **Lobdell, D. H.,** 'Ring' granulomas in cytomegalovirus hepatitis, Arch. Pathol. Lab. Med., 111, 881, 1987.
26. **Delsol, G., Pellegrin, M., Familiades, J., and Auvergnat, J. C.,** Bone marrow lesions in Q fever, *Blood,* 52, 637, 1978.

27. **Okun, D. B., Sun, N. C. J., and Tanaka, K. R.,** Bone marrow granulomas in Q fever, *Am. J. Clin. Pathol.,* 71, 117, 1979.
28. **Ende, J. and Gelpi, A. P.,** Pathological changes noted in bone marrow in a case of Q fever, *Arch. Intern. Med.,* 100, 793, 1957.
29. **Srigley, J. R., Vellend, H., Palmer, N., Phillips, M. J., Geddie, W. R., Van Nostrand, A. W., and Edwards, U. D.,** Q fever. The liver and bone marrow pathology, *Am. J. Surg. Pathol.,* 9, 752, 1985.
30. **Voigt, J. J., Cassigneul, J., Delsol, G., Vinet, J. P., Paci, H., and Fabre, J.,** Granulomatous hepatitis. Apropos of 112 cases in adults, *Ann. Pathol.,* 4, 78, 1984.
31. **Marrie, T. J.,** Liver involvement in acute Q fever, *Chest,* 94, 896, 1988.
32. **Alkan, W. J., Evenchik, Z., and Eshchar, J.,** Q fever and infectious hepatitis, *Am. J. Med.,* 38, 54, 1965.
33. **Sawyer, L. A., Fishbein, D. B., and McDade, J. E.,** Q fever in patients with hepatitis and pneumonia: results of laboratory-based surveillance in the United States, *J. Infect. Dis.,* 158, 497, 1988.
34. **Derrick, E. H.,** The course of infection with *Coxiella burnetii, Med. J. Aust.,* 1, 1051, 1973.
35. **Berkovitch, M., Aladjem, M., Beer, S., and Cohar, K.,** A fatal case of Q fever hepatitis in a child, *Helv. Paediatr. Acta,* 40, 87, 1985.
36. **Vellend, H.,** personal communication.
37. **Sawyer, L. A., Fishbein, D. B., and McDade, J. E.,** Q fever: current concepts, *Rev. Infect. Dis.,* 9, 935, 1987.
38. **Turck, W. P. G., Howitt, G., Turnberg, L. A., Fox, H., Longson, M., Matthews, M. B., and Das Gupta, R.,** Chronic Q fever, *Q. J. Med.,* 178, 193, 1976.
39. **Atienza, P., Ramond, M.-J., Degott, C., Lebrec, D., Rueff, B., and Benhamou, J.-P.,** Chronic Q fever hepatitis complicated by extensive fibrosis, *Gastroenterology,* 95, 478, 1988.
40. **Weir, W. R. C., Bannister, B., Chambers, S., Cock, K., and De Mistry, H.,** Chronic Q fever associated with granulomatous hepatitis, *J. Infect.,* 8, 56, 1980.
41. **Peacock, M. G., Philip, R. N., Williams, J. C., and Faulkner, R. S.,** Serological evaluation of Q fever in humans: enhanced phase I titers of immunoglobulin A and G are diagnostic for Q fever endocarditis, *Infect. Immun.,* 41, 1989, 1983.
42. **Geddes, A. M.,** Q fever, *Br. Med. J.,* 287, 927, 1983.
43. **Yebra, M., Marazuela, M., Albarran, F., and Moreno, A.,** Chronic Q fever hepatitis, *Rev. Infect. Dis.,* 10, 1229, 1988.

Chapter 9

Q FEVER ENDOCARDITIS AND OTHER FORMS OF CHRONIC Q FEVER

Didier Raoult, Asma Raza, and Thomas J. Marrie

TABLE OF CONTENTS

I. INTRODUCTION

Q fever endocarditis is the most serious complication of *Coxiella burnetii* infection. There has been an increase in the incidence of Q fever endocarditis, but this may be due to increased reporting.[1] Some aspects, such as the reasons for the occurrence of the chronic form in some patients, the specific increase in phase I antibodies, and the inability of antibiotic treatment to eradicate the organism in less than 1 year remain obscure. The clinical picture of the disease does not resemble classic endocarditis, and the diagnosis is generally made because of previous valvular damage. We have studied 25 cases of Q fever endocarditis diagnosed in France. We now feel that some aspects of this disease, such as the potential role of immunosuppression, may have been underestimated. An unresolved point is the course of infection. Some authors suggest that a primary infection, sometimes asymptomatic could be followed years later by endocarditis, whereas other investigators think that there are specific strains which can cause an acute or chronic form of the disease. Historically, Evans et al.[2] reported the first case of Q fever endocarditis.

II. PATHOPHYSIOLOGY

C. burnetii is a strict intracellular pathogen living in the phagolysosome[3] of the host cell. It is capable of remaining viable in host cells and can even persist in daughter cells.[4-6] *C. burnetii* does not affect the viability of the persistently infected cells.

Recently, it has been suggested that the strains responsible for chronic and acute cases are different.[7,8] The chronic type strain, Priscilla, was isolated from a goat, and the acute type strain, Nine Mile, was isolated from a tick. Strains isolated from animals and acute and chronic human cases could be divided into three different genetic types. One pattern was identical to that of Q_p H_1 — the plasmid of the Nine Mile strain. The second was the same as several endocarditis isolates and the goat abortion strain (Priscilla) plasmid $Q_p R_s$, and in the third group of endocarditis isolates no plasmid was found, and it was assumed that it was integrated.[9,10]

The protein pattern and the lipopolysaccharide patterns differ in the prototype Nine Mile and Priscilla strains.[11,12] These findings correlate with the genomic analysis.[12,13] From these data it is likely that the strains from acute cases could not cause chronic infection and that strains from chronic cases may not cause acute infections. Clinical observations led us to believe that this may not be true. Cases of endocarditis have been reported to appear years after symptomatic acute Q fever.[14,73] In our population, a woman with an aortic graft suffered from acute Q fever pneumonia. She was treated for 3 weeks with doxycycline and was apparently cured. Five months later she was asymptomatic, but serological examination showed the typical profile of chronic infection. She might have developed endocarditis if she had not been treated.

There are two host factors that are important in the development of Q fever endocarditis. The first of these is underlying heart disease. Most patients who develop Q fever endocarditis have abnormal valves or blood vessels. This seems to be a necessary condition for endocarditis and for vascular infection which appears in patients with grafts or aneurysms.[15,16] Second, the role of the immune system in preventing the occurrence of endocarditis is not completely known. Patients suffering from Q fever endocarditis have profound lymphocyte unresponsiveness to *Coxiella*,[17] but this may be secondary to the disease. We have recently had the opportunity to study[74] three patients who had cancer (a tonsil cancer in one and breast cancer in the other two patients) and Q fever endocarditis. Two of these patients had previous valvular damage, and in two cases the endocarditis was diagnosed just after the cancer was recognized. Q fever endocarditis had been reactivated by cortisone treatment.[18] In guinea pigs immunosuppression by cortisone or irradiation has resulted in generalized infection in cases of latent infection.[19,20]

Latent infection has been demonstrated in the human placenta. Thus, in immunocompromised hosts,[19,20,22] latent infection may relapse or cause chronic infection depending on the strain

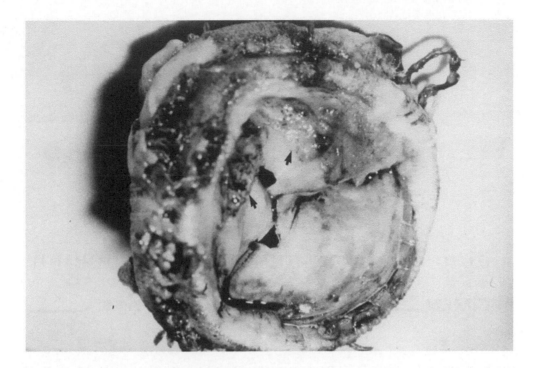

FIGURE 1.   Bioprosthetic aortic valve from Case 2. Vegetative material is evident on the surface of the valve (arrows).

involved and/or the cardiovascular situation. The humoral immune response in Q fever endocarditis is manifested by very high levels of specific antibodies. These antibodies are directed against both phase I and phase II antigens in the IgG and IgA fractions. Antibodies to phase I antigen predominate. These antibodies are not protective.[23] The reason why phase I antibodies are increased to such a high level in contrast to acute cases is not known, nor is the reason for the augmentation of IgA antibodies understood. These antibodies, in association with *C. burnetii* antigens result in immune complexes which are responsible for many aspects of the disease.

The general pathological findings of the valve in Q fever endocarditis differ greatly from those of other causes of endocarditis. In Q fever endocarditis, the vegetations often have a smooth surface (Figures 1 to 3) and may have a nodular appearance (Figures 2 and 3). In one case we were able to observe *C. burnetii,* using immunofluorescence, in a valve which was considered normal on gross examination (Figure 4). Valvular aneurysms are more frequent in Q fever endocarditis.[24,25] Wilson et al.[14] reported that the appearance of the affected valve in Q fever endocarditis was typical. The valves were calcified with small areas of necrotic, thrombotic vegetations. There was often an aneurysm of the aortic wall, and the authors suggest the use of angiography in the diagnosis of Q fever endocarditis. *C. burnetii* can be found in the valves using immunofluorescence or electron microscopy.

In our experience microscopic examination of the vegetations in Q fever endocarditis reveals focal calcification which may be due to degenerative valvular disease (Figure 5). The inflammatory infiltrate in the vegetation may be composed of aggregates of neutrophils, as in Figure 6, or subacute and chronic inflammatory exudate together with granulation tissue including many plasma cells, as in Figure 7. Large "foamy" macrophages, some of which contain vacuolated refractile and granular material (Figures 7 and 8) are characteristic of Q fever endocarditis. These macrophages contain *C. burnetii,* as shown by electronmicroscopy (Figure 9). In some areas these distended macrophages appeared to have ruptured with extrusion of aggregates of finely punctate basophilic material.

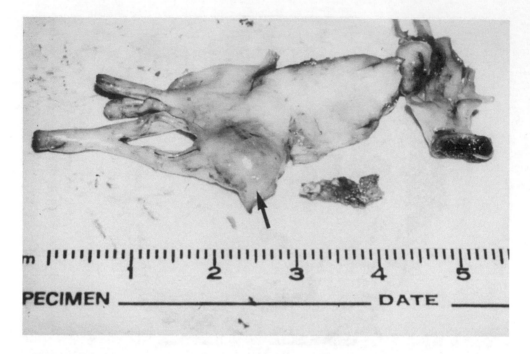

FIGURE 2. Mitral valve of a patient with Q fever endocarditis. The nodule on the valve (arrow) was "packed" with *C. burnetii*. See Figure 9. (From Marrie, T. J., Q fever pneumonia, *Medical Grand Rounds*, 3, 354, 1985. With permission.)

FIGURE 3. *C. burnetii* vegetation (arrows) on a prosthetic valve. Note the ridge of material that comprises the vegetation.

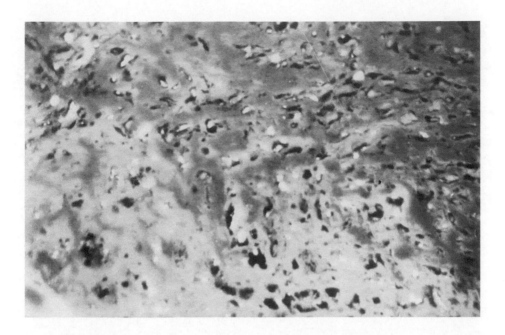

FIGURE 4. Positive direct immunofluorescence showing *C. burnetii* within an infected valve.

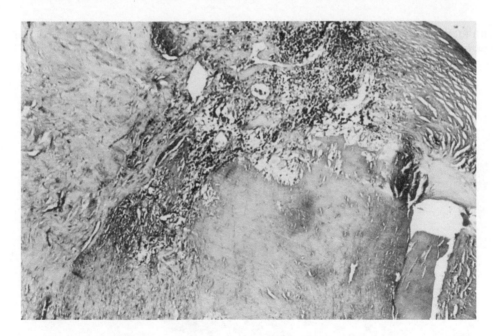

FIGURE 5. Photomicrograph of a *C. burnetii*-infected aortic valve. Notice the inflammatory infiltrate and the calcification within the valve. (Magnification × 100.)

III. EPIDEMIOLOGY

Q fever endocarditis is found in countries where acute fever is well known.[26] Cases have been described in Australia,[27] U.K.,[28] Ireland,[29] France,[30] Spain,[31] Canada,[25] and Switzerland.[32] A few cases have been reported from the U.S., Germany, Greece, Nigeria,[36] Libya,[37] and Jordan.[38] We have observed it in patients from Senegal, Algeria, and Tunisia.

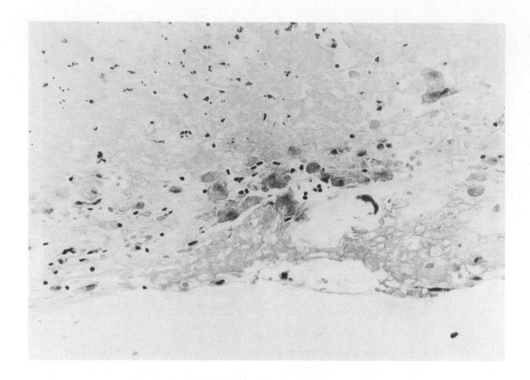

FIGURE 6.   Photomicrograph of a *C. burnetii*-infected native aortic valve. The inflammatory infiltrate consists of neutrophils. Foamy macrophages are also evident. (Magnification × 510.)

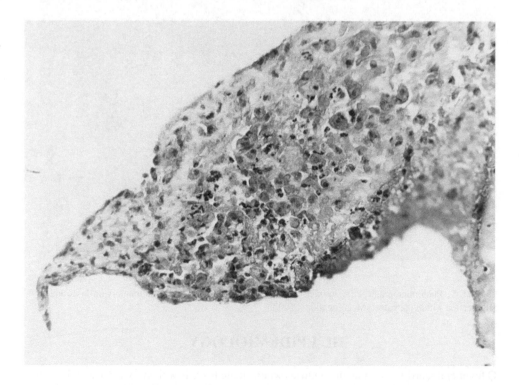

FIGURE 7.   Subacute and chronic inflammatory exudate in a *C. burnetii*-infected native bicuspid aortic valve. Note also the large foamy macrophages. (Magnification × 510.)

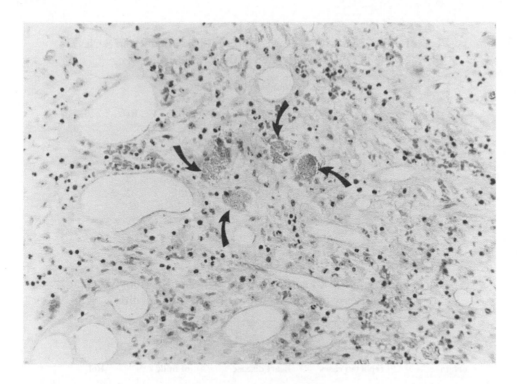

FIGURE 8. *C. burnetii*-infected native aortic valve showing more detail of the foamy macrophages (arrows). (Magnification × 510.)

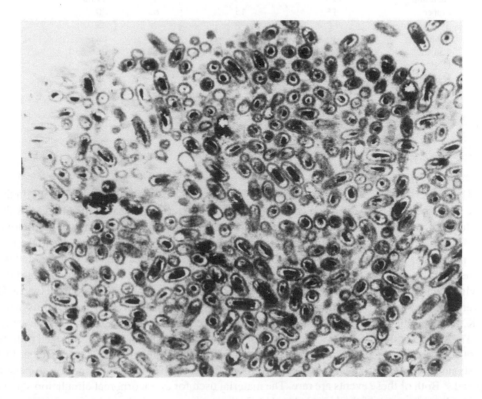

FIGURE 9. Electronmicrograph of vegetation in Figure 2. Note the many microorganisms. (Magnification × 5500.)

TABLE 1
Percent of Patients with Q Fever Who Had Endocarditis in Series of Q Fever Reported in the Literature

Country	Year of publication	Number of cases of Q fever	Number (%) of cases of endocarditis		Ref.
U.S.	1951	180	0	(0)	64
Germany	1987	5300	2	(0.04)	34
U.K.	1987	1351	118	(8.7)	29
Spain	1985	145	10	(6.9)	68
Switzerland	1985	415	0	(0)	32
Hungary	1987	87	0	(0)	69
Netherlands	1987	51	3	(5.9)	70
Australia	1982	111	1	(0.9)	27

TABLE 2
Geographic Origin, Risk Factors, and Incidence of Previous Valvular Heart Disease in Patients with Q Fever Endocarditis

Geographic origin	Number of reported cases	Previous valvular heart disease	Exposures to cattle or consumption of milk	Ref.
U.K. & Ireland	48	42(4–ns)[a]	26	15,35,40—42, 52,65,71
Australia	17	16	16	14,63
France	15	15	6	30
Switzerland	8	8	?	45
Spain	10	9	6	68
Canada	5	5	1	25
U.S.	1	1	1	33
Nigeria	1	1	—	36
Libya	2	2	—	37,47
Tunisia	1	1	—	47
Total cases	108	100/108(93%)	56/96(58%)	

[a]   ns = not stated.

The proportion of endocarditis in series of Q fever varies from 0% in the U.S. to 9% in the U.K.[29] (Table 1). This percentage could be biased by the lack of a search for Q fever as the etiology of endocarditis and by the overestimation of severe disease. In our laboratory, endocarditis represents 20% of the patients diagnosed as having Q fever. This is due to the fact that physicians test patients who have blood-culture negative endocarditis for Q fever, and we systematically test sera from cardiology wards. From our experience in France it is clear that, in cities where this disease was unknown, increased interest makes it possible to diagnose Q fever endocarditis. Q fever endocarditis may represent 1 to 2% of all cases of endocarditis in the U.K. In the South of France where it is systematically searched for, the percentage is the same.

The source of the Q fever was contact with cattle or consumption of unpasteurized milk in 56 of 96 cases (58%) (Table 2). The possibility of nosocomial acquisition of Q fever endocarditis exists since acute Q fever has been acquired through blood transfusion[22] and by person-to-person spread.[39] Both of these events are rare. The material used for extracorporeal circulation should be sterilized because of the knowledge that *C. burnetii* may partially resist disinfection with formaldehyde.

TABLE 3
Clinical Features of Q Fever Endocarditis as Summarized
from References 14, 15, 24, 25, 30, 31, 38, 40—45

Feature	Total	(%)
No. of pts	84	
No. males	64	(76)
No. with known		
exposure factors	51	(61)
Aortic valve	28	(33)
Mitral valve	42	(50)
Mitral and aortic	14	(17)
No. (%) who presented with		
Fever	57	(68)
Cardiac failure	37/55	(67)
Hepatomegaly	47	(56)
Splenomegaly	46	(55)
Digital clubbing	31	(37)
Purpuric rash	16	(19)
Arterial embolism	18	(21)
No. (%) who died	31	(37)
Diagnostic delay		
(mean in months)	12	—

IV. CLINICAL FEATURES[14,15,24,25,30,31,38,40-45]

The symptoms and signs of Q fever are protean, and the onset of the disease is frequently difficult to determine. Q fever endocarditis is generally subacute or chronic, and the diagnosis is frequently delayed. Our first case diagnosed in Marseille[46] was a patient with a prosthetic mitral valve who had a 3-year history of intermittent fever. He received several courses of empiric therapy with beta-lactam antibiotics and aminoglycosides, some of which were coincidentally associated with remission of fever. Finally, he was diagnosed as having Q fever endocarditis when he had a relapse 3 years after the first episode, and the previous sera were examined and confirmed the diagnosis. When the diagnosis of blood-culture negative endocarditis is considered, Q fever endocarditis is readily diagnosed. Frequently, the diagnosis of endocarditis is not considered, especially since the patient may be afebrile (33% of the cases) and because the echocardiographic examination usually shows no vegetation. Even when febrile, the patient may have remissions. In such cases, the diagnosis is chronic hepatitis, encephalitis, or inflammatory disease.

One hundred of 108 (93%) patients with Q fever endocarditis had preexisting abnormal valves (see References in Table 2). The underlying heart disease may be congenital, rheumatic, degenerative, or syphilitic. More and more cases are occurring on prosthetic valves. The patients are male in 75% of cases and generally older than 40 years. Children may also develop Q fever endocarditis.[37,47] An exposure factor such as professional contact with cattle is frequently found, but the fact that the incubation time is unknown may cause underestimation of the exposure factors.

Fever is present at the beginning of the disease in 68% of cases (Table 3). It is usually a low grade fever, 38 to 38.5°C, which is remittent and well tolerated. The most frequently observed symptoms are malaise, weakness, fatigue, weight loss, chills, anorexia, and night sweats. Cardiac failure in patients with previous valvulopathies is reported in 67% of documented cases. Valvular dysfunction and modification of a previously known murmur may be observed as well as the appearance of a new murmur. Cardiac failure is the most common symptom with the general manifestations of dyspnea, acute pulmonary edema, angina, and palpitations. The clinical picture may be suggestive of myocardial infarction.[48,50] The chest X-ray may show

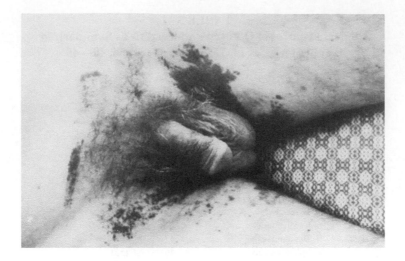

FIGURE 10. Purpuric exanthema involving the thighs of a patient with Q fever endocarditis. Biopsy revealed leukocytoclastic vasculitis.

cardiomegaly. The electrocardiographic exam may confirm arrhythmia and ventricular hypertrophy. Echocardiography is frequently inconclusive. In fact, it may confirm the existence of a valvulopathy but frequently fails to demonstrate a vegetation. In our experience the most commonly observed echocardiographic change was the dysfunction of a prosthetic valve. The role of angiography in detecting the existence of an aneurysm suggestive of Q fever endocarditis[14] remains to be evaluated.

Peripheral manifestations of endocarditis are frequently found. Digital clubbing is found in 37% of cases, which is a higher rate than is generally observed in other types of endocarditis.[51] In one case we described unilateral digital clubbing associated with unilateral purpura of the left arm in chronic *C. burnetii* infection of a vascular graft.[16]

A purpuric rash is noticed in 19% of cases. It generally occurs on the extremities and the mucosa. In one case, diagnosed immediately after a tonsil cancer was found, we observed a purpuric rash with the same distribution as the patient's underpants (Figure 10). When biopsied these purpuric lesions revealed immune complex vasculitis. We have never seen an Osler's node in a patient with Q fever endocarditis, and it is rarely reported, but some patients mentioned such a lesion retrospectively.

Splenomegaly occurs in 55% of cases and may be very prominent.[47,52] Hepatomegaly is also frequently seen. The liver is generally hard and often considerably enlarged. It is frequently misdiagnosed as cirrhotic. The pathological findings of the liver are nonspecific: lymphocytic infiltration and foci of spotty necrosis.[34,54] These liver abnormalities were extensively studied by Turck et al.[24] in eight patients, and granulomas were observed in three. He described a 10-month evolution of cirrhosis in a patient in whom sequential biopsies were carried out. This case remains unique. No author has reported the observation of the typical fibrinoid ring,[55,56] surrounding the granuloma as observed in acute Q fever hepatitis. Most cases of chronic hepatitis in Q fever are associated with valvular damage, and one should suspect endocarditis when diagnosing chronic Q fever hepatitis.

Renal involvement as indicated by microscopic hematuria is frequent, being present in eight of ten patients studied by Tobin et al.[41] In a case reported by Ramos-Dias et al.,[67] glomerulonephritis with deposits of IgM and C3 lead to progressive anuria. Musculoskeletal manifestations are quite common. These include arthritis, which may be misdiagnosed due to the presence of rheumatoid factor.

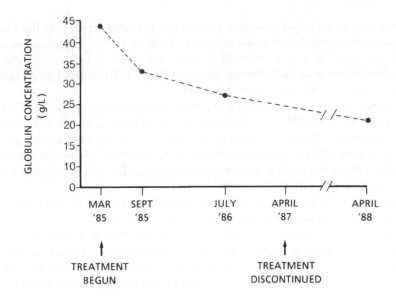

FIGURE 11. Decline in serum globulin concentration with time, Case 1. Treatment began in March 1985. Note the rapid decline over the first 6 months and the slower decline thereafter.

Embolic manifestations occurred in 21% of the patients. Emboli may involve cerebral, arm, or leg vessels. Embolectomy or amputation may be required.

Pulmonary and pleural manifestations may be observed as a complication of endocarditis.[57] Neurological manifestations of Q fever endocarditis are rarely reported except in cases of stroke. Brooks et al.[58] reported a case of encephalitis in a patient with valvular damage that may have been a subclinical clinical form of endocarditis.

The following cases illustrate the protean manifestations of Q fever endocarditis.

A. CASE 1

A 51-year-old white female was admitted on March 23, 1985. Two years prior to admission she had a valve replacement for mitral stenosis. She was well until August 1984, at which time she had a cholecystectomy for indigestion. Following this she complained of fatigue, anorexia, and a general sense of malaise. She was found to have microscopic hematuria after presenting to her family doctor with left flank pain. She denied fever, chills, and rigors but was found to be intermittently febrile. She had a temperature of 38°C following cystoscopy on March 12, 1985. Prior to the cystoscopy she was treated prophylactically with ampicillin and gentamicin. In Antigonish, where she had the cystoscopy, she remained intermittently febrile. This was a low-grade fever of 37.6°C. Because of this she was admitted to the Victoria General Hospital for investigation.

On examination she had a grade 3/6 systolic ejection murmur heard loudest at the apex which radiated to the left sternal border and the axilla. The spleen was felt at the left costal margin. The consultants who examined her (a cardiologist and an infectious disease physician) felt there was no evidence of endocarditis. On April 1, 1985, a second infectious disease physician noted that her work was delivering the mail to local farms. Because of this and the hyperglobulinemia (Figure 11), he thought that she might have Q fever endocarditis. Investigations revealed the following: blood cultures were negative; an echocardiogram showed no vegetations; the spleen was mildly enlarged on ultrasound; bone marrow exam was normal; the WBC on admission was 9×10^9/l, Hgb 115 g/l, platelet count 204×10^9/l. Immunoglobulins were IgG 25 g/l (normal 6.9 to 13.5), IgA 4.17 g/l (normal 1 to 3.5), IgM 3.66 g/l (normal 1.65 to 3). Electrophoresis revealed a polyclonal increase in IgG. Cryglobulins were positive, brucella titers were negative. The *C*.

burnetii complement titer (phase II antigen) was positive to titer of 1:1046. The third and fourth complements were normal, latex fixation positive to a titer of 1:5000. The urinalysis showed 3 to 4 RBCs per hpf. Titers to *C. burnetii* phase I and phase II antigens by the indirect immuofluorescence test were the following: phase I — IgG 1:262,144; IgA 1:4096; IgM 1:2048 and phase II — IgG 1:8192; IgA 1:4096; IgM 1:512.

Treatment was begun with tetracycline and trimethoprimsulfamethoxazole (TMP-SMX). Within 1 month the TMP-SMX was discontinued because of a skin rash and rifampin 600 mg/d was substituted. Therapy was continued for 2 years. Throughout this period and subsequently (to June 1989) she has remained well. Forty months following diagnosis and onset of therapy the antibody titers by IFA were the following: phase I — IgG 1:8192; IgA 1:256; IgM <1:8 and phase II — IgG 1:2048; IgA 1:256; IgM <1:8.

The concentration of the serum globulins is shown in Figure 11. Note the sharp decline during the first 6 months of treatment. Normal globulin level is 21 to 30 g/l.

Comment—The clues to the diagnosis of Q fever endocarditis in this patient were the malaise, fatigue, low-grade fever, the high globulin concentration, her occupation, and the presence of a prosthetic valve. If one of us had not had considerable experience with Q fever endocarditis, the diagnosis would have been delayed much longer as is illustrated by the next case.

B. CASE 2

A 59-year-old male was admitted on September 9, 1977, with severe aortic stenosis, for which he underwent a porcine valve replacement. His preoperative left ventricular ejection fraction was 24%. In August of 1978, he developed microscopic hematuria. In November 1979, he was admitted to the urology service for evaluation of the microscopic hematuria. At that time his hemoglobin was 98 g/l, and the LDH 1380 IU/l. He had a normal IVP and was discharged. On December 11, 1979, he was admitted to the department of medicine at the Victoria General Hospital with a 3-month history of fatigue, shortness of breath, fever, chills, and sweats. The only relevant historical information that was obtained was that in August of 1978 he developed a febrile illness while visiting his daughter. His daughter worked on a farm and he had helped her milk the cows.

Physical examination revealed that he was chronically ill. He had marked clubbing of his fingers and toes and had tender splenomegaly. During this admission he had an extensive diagnostic workup which revealed that he was anemic, had an increased LDH, and had positive cryoglobulins. Thirty-five sets of blood cultures were negative. A temporal artery biopsy was negative. A renal biopsy showed mesangioproliferative glomerulonephritis. Because of this he was started on corticosteroid therapy. His hemoglobin increased, but he developed steroid psychosis. The temporal course of several laboratory parameters from the time of replacement of his first valve is shown in Figure 12.

In May 1981, he developed amaurosis-fugax, and at that time Q fever endocarditis was diagnosed. While in the hospital he had an intercurrent febrile illness with atypical lymphocytes and left lower lobe infiltrate. He also had an episode of pulmonary edema. Cardiac catheterization showed a 35 to 40-mmHg gradient across his aortic valve with a decreased ejection fraction of 26%. A petechial rash was noted on his legs. He was begun on treatment with tetracycline and TMP-SMX. Three weeks later his aortic valve was replaced.

His aortic valve is shown in Figure 1. *C. burnetii* was isolated from the valve. The postoperative course was complicated by a right hemisphere stroke. Antibiotic treatment was discontinued after 2 years.

In May 1988, he again had replacement of his aortic valve because of the development of aortic insufficiency due to degeneration of the valve. There was no evidence of infection.

Comment—Diagnosis was delayed in this patient because Q fever endocarditis had not been previously recognized in Nova Scotia. His clinical picture was that of culture-negative endocarditis.

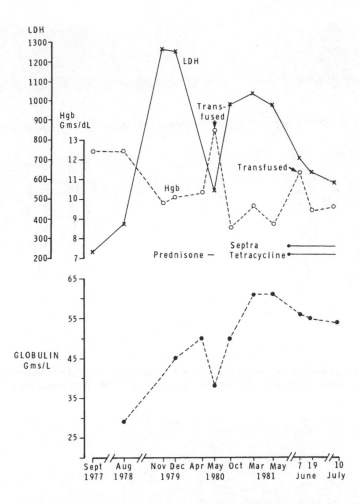

FIGURE 12. Graph of time course of hemoglobin (Hgb), lactic dehydrogenase (LDH), and serum globulin concentration, Case 2. Q fever endocarditis
was diagnosed in May 1981. Note that the Hgb and LDH were normal when his
native valve was replaced in September 1977. The serum globulin was 29 g/l
in August 1978, at which time he had a febrile illness. Note the rapid decline in
LDH and globulin concentration when therapy with prednisone was begun. (A
portion (globulin concentration) of this figure is reproduced from Marrie, T. J.,
Q fever — clinical signs, symptoms, and pathophysiology, in *Biology of
Rickettsial Diseases*, Walker, D. H., Ed., CRC Press, Boca Raton, FL. With
permission.)

The marked clubbing of the fingers that is seen in Q fever endocarditis is shown in Figure 13,
while in Figure 14 a maculopapular purpuric rash (palpable purpura) on the legs of a patient with
Q fever endocarditis is evident. A biopsy showed leukocytoclastic vasculitis. Contrast this rash
with the one shown in Figure 10.

The temperature curves of two patients with Q fever endocarditis are shown in Figures 15 and
16. The only patient in our series with this disease that had such hectic fever is shown in Figure
15.

V. LABORATORY FINDINGS

Hematologic parameters are often abnormal. The white blood cell count is elevated in one
third, decreased in one third, and normal in one third of patients in our experience. Anemia is

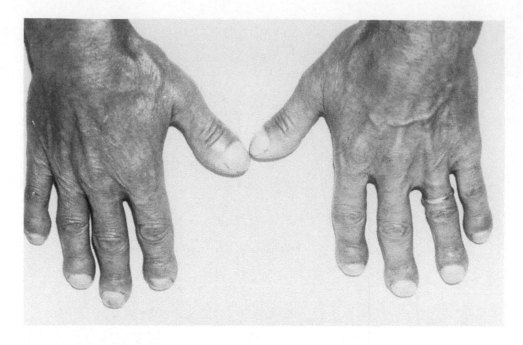

FIGURE 13. Marked clubbing of the fingers in a patient with Q fever endocarditis. (From Marrie, T. J., *Biology of Rickettsial Diseases,* Walker, D. H., Ed., CRC Press Inc., Boca Raton, FL, 1988. With permission.)

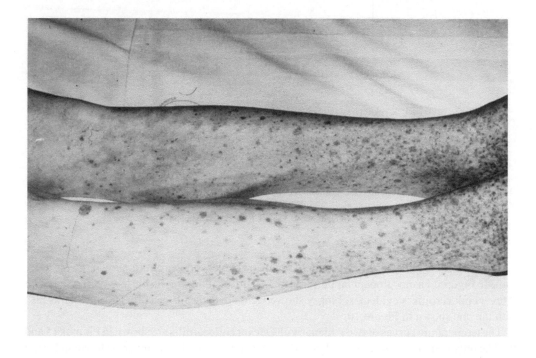

FIGURE 14. Palpable purpura on the legs of a patient with Q fever endocarditis. (From Marrie, T. J., Q fever — clinical signs, symptoms, and pathophysiology, in *Biology of Rickettsial Diseases,* Walker, D. H., Ed., CRC Press, Boca Raton, FL. With permission.)

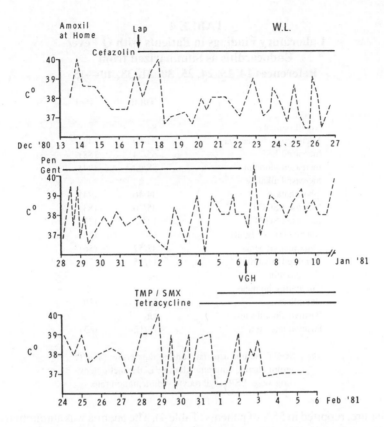

FIGURE 15. Temperature readings during a 2-month period in a patient with Q fever endocarditis, showing no response to treatment with amoxicillin (at home), cefazolin, penicillin, and gentamicin in another hospital. Following a diagnosis of Q fever endocarditis at our hospital, January 31, 1981, trimethoprimsulfamethoxazole and tetracycline therapy was initiated. Within 4 d he was afebrile and remained so. Unfortunately, he was killed in a car accident 1 year later. At that time his prosthetic mitral valve showed no evidence of infection. Lap = laparotomy, VGH = admission to Victoria Hospital. (From Marrie, T. J., et al. Q fever in Maritime Canada, *Can. Med. Assoc. J.*, 126, 1295, 1982. With permission.)

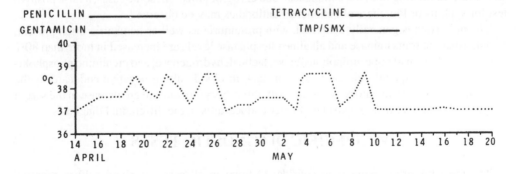

FIGURE 16. Temperature curve of another patient with Q fever endocarditis. Contrast this with Figure 19. This patient had a bicuspid aortic valve and presented with a clinical picture of culture-negative endocarditis.

TABLE 4
Laboratory Findings in Patients with Q Fever
Endocarditis as Summarized from
References 14, 15, 24, 25, 30, 31, 38, 40—45

	Total	(%)
No. of patients	84	
Microscopic hematuria	34/69	(49)
Increased ESR	65/74	(88)
Increased globulin	58/62	(94)
Increased alkaline phosphatase	34/46	(74)
Increased SGOT	25/30	(83)
Increased SGPT	14/38	(37)
Anemia (Hb <11g/dl)	46	(55)
Thrombocytopenia	32/57	(56)
Increased creatinine	11/15	(73)
Cryoglobulins	4	
Circulating immune complexes	8/9	(89)
Positive Coombs test	1/6	
Positive latex test	10/12	(83)

Note: SGOT: serum aspartate aminotransferase; SGPT: serum alanine aminotransferase; LDH: lactic dehydrogenase; ESR: erythrocyte sedimentation rate.

a common feature, reported in 55% of patients (Table 4). The anemia was autoimmune in a case reported by Turck and Matthews[59] and in two of our cases. Initially the anemia may be worsened by therapy. Thrombocytopenia is a frequent event occurring in 56% of cases and may help in the diagnosis.

The erythrocyte sedimentation rate is elevated in nearly all cases. Hyperglobulinemia is noticed in 94% of cases. It is helpful diagnostically, especially when the globulins represent more than 50% of the total protein concentration. In these cases a polyclonal increase of IgG, IgM, and IgA is noticed. The longer the diagnostic delay, the higher the globulin concentration (Figure 12). The presence of rheumatoid factor, as detected by the latex test, is very frequent as well as are circulating immune complexes[30] and cryoglobulins.[25] In some cases, a false positive test for syphilis or low titers of antinuclear antibodies may be observed.

The urinalysis is generally abnormal, with proteinuria as well as hematuria often present. Serum aspartate transaminase and alkaline phosphatase levels are increased in more than 80% of the cases. Serum alanine aminotransferase, lactic dehydrogenase, and creatinine phosphokinase levels are frequently increased also; in fact, in blood culture negative endocarditis the association of an inflammatory syndrome with thrombocytopenia and an increase in the serum enzymes should lead one to consider Q fever endocarditis in the differential diagnosis.

VI. SPECIFIC DIAGNOSTIC TESTS

The most important point is to consider Q fever in all cases of blood-culture negative endocarditis. The diagnosis can be made by serology. Antibodies to phase I and phase II antigens may be determined using a variety of tests.[71] Complement fixation has been used, and a titer of ≥1:200 to phase I antigen is considered diagnostic. The most commonly used technique is the microimmunofluorescence (IFA) test. With this technique antibodies to both phase I and phase II antigen can be determined. In addition, the titers of IgG, IgM, and IgA antibodies are determined. It is important to remove the IgG prior to determining IgA and IgM levels. Omitting

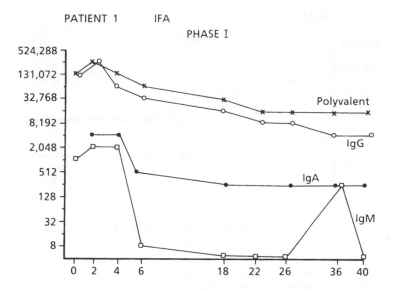

FIGURE 17.   The time course of the level of IgM, IgG, IgA antibodies to phase I *C. burnetii* in a patient with Q fever endocarditis. Polyvalent refers to a conjugate containing anti IgM, IgG, IgA. The antibody titers are on the Y axis and the time in months following the onset of treatment on the X axis.

this adsorption results in false positive IgM antibodies (due to the rheumatoid factor) and false negative IgA antibodies (related to competition on antigen sites by the IgG). A single serum sample is diagnostic for Q fever endocarditis. Phase I antibodies are elevated in 95% or more of cases of endocarditis. IgG phase I and phase II titers are ≥1:800. The titer of IgG antibodies may be as high as 1×10^6 and is frequently >1×10^5 (see Case 1). IgA anti-phase I antibodies are ≥1:50 and generally are higher than IgA anti-phase II antibodies. These high levels of IgG anti-phase I antibodies are observed only in chronic cases. The titers of IgM are variable, in some cases very high and sometimes very low. These observations are unexplained, and the prevalence of IgM antibodies varies from one series to another. The antibody titer may be used to monitor the course of treatment. In general, the antibody titers diminish slowly with treatment. In our experience, after 2 years of antibiotic thereapy, IgG levels are still high in almost all cases. Essentially the same results are observed using ELISA. Figures 17 and 18 illustrate the antibody titers to *C. burnetii* in a patient with Q fever endocarditis.

Other methods of diagnosis of *C. burnetii* endocarditis include the isolation of the microorganism or the demonstration of the microorganism in the tissues. The isolation of *C. burnetii* should be restricted to high security laboratories because it represents an important biohazard. The easiest way to demonstrate the presence of *C. burnetii* is to inoculate mice intraperitoneally with blood or a valve extract. A mouse is sampled on day 10, and its spleen is used to make subsequent isolation (in embryonated eggs, or cells); another mouse is bled on day 21 to be tested serologically. The seroconversion of the mouse provides evidence of the presence of *C. burnetii*. Histologically, *C. burnetii* may be observed in valves using immunofluorescence[62] or electron microscopy (Figures 4 and 9).

VII. TREATMENT

The antimicrobial susceptibility of *C. burnetii* and a review of the antibiotics used in treatment of chronic Q fever are dealt with in another chapter. In this chapter we give our approach to the treatment of this illness. Two groups of antibiotics have been shown to be more effective for intracellular *C. burnetii*: rifampin and the fluoroquinolones.

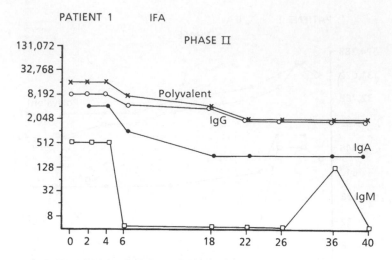

FIGURE 18. Time course of the level of IgM, IgG, IgA antibodies to phase II *C. burnetii* in a patient with Q fever endocarditis.

In the medical treatment of Q fever endocarditis one should remember that relapses and positive valve cultures have been observed after more than 1 year of tetracycline therapy.[63] As yet, there is no specific test that makes it possible to conclude that a patient is definitely cured.

Medical treatment is still based on the prescription of doxycycline which temporarily cures the patient as long as it is prescribed. In order to obtain better results and shorten the duration of therapy, combination therapy has been suggested. The combination of doxycyline (or tetracycline) and rifampin has been most useful in our experience, but the rifampin may render anticoagulation difficult in patients who are receiving this form of treatment. In the last 2 years, we have been using quinolones such as ciprofloxacin (1.5 g/d) or ofloxacin (0.6 g/d) to treat Q fever endocarditis. This work is too preliminary to draw conclusions. In one patient the valve was removed after 4 months of this therapy for hemodynamic reasons, and it still contained *C. burnetii* cells. The place of cotrimoxazole in the therapy of *C. burnetii* endocarditis remains controversial. Some authors report good results, others report failure. Based on *in vitro* data, we feel this agent should not be used alone. The duration of medical therapy is difficult to determine. Therapy can be discontinued when IgA antibodies disappear and when a level of IgG anti-phase I antibody <1:400 is reached. We have never observed these values in less than 3 years of antibiotic therapy. The absolute minimum duration of therapy is 2 years. Some investigators recommend treating Q fever endocarditis for life.

Surgical therapy, i.e., valve replacement, is generally indicated for hemodynamic reasons. If possible, at least 3 weeks of antimicrobial therapy should be given prior to valve replacement. Following initiation of antibiotic therapy the patient becomes afebrile within 4 to 5 d, rarely up to 4 weeks (Figures 15 and 16). The hepatomegaly and the splenomegaly disappear slowly in 2 to 12 weeks. Other parameters such as elevated serum enzymes, thrombocytopenia, and inflammatory proteins slowly return to normal. During treatment serological testing should be carried out once monthly for 6 months and every 3 months thereafter. The levels of antibodies decrease very slowly (Figures 17 and 18). The IgM, when present, disappears first, then the IgA, but the IgG antibodies remain positive for years.[60-62,66,71]

A prospective multicenter study is required to evaluate new therapies and to define the criteria for stopping antibiotic therapy.

VIII. OTHER FORMS OF CHRONIC Q FEVER

Ellis et al.[15] described three patients with osteomyelitis due to *C. burnetii* (vertebral osteomyelitis in two of the three). Two patients with bone infection were recently reported from France.[72] *C. burnetii* was isolated from the right hip of a 7-year-old child. The second patient was a 51-year- old man with infection of the right hip.

IX. SUMMARY

In areas where Q fever is endemic, clinicians must consider *C. burnetii* in the differential diagnosis of all patients with culture-negative endocarditis. Malaise and anorexia with or without low grade fever in a patient with an abnormal value or a vascular prosthesis are clinical clues suggestive of this illness. Hyperglobulinemia, increased erythrocyte sedimentation rate, and elevated lactic dehydrogenase are additional clues. The diagnosis is readily confirmed serologically when a phase I complement fixation titer of $\geq 1:200$ is found or when the phase I titer in the IgG or IgA fraction is equal to or greater than the phase II titer by the IFA test.

Treatment with combination antibiotics (e.g., doxycycline and rifampin) should be continued for at least 2 years, and preferably a decision to discontinue therapy should be based on the disappearance of IgA antibodies and the decline of IgG antibodies to <1:400.

REFERENCES

1. Editorial, Chronic Q fever, *J. Infect.*, 82, 1, 1984.
2. **Evans, A. D., Powell, D. E., and Burrel, C. D.,** Fatal endocarditis asociated with Q fever, *Lancet*, 1, 864, 1959.
3. **Baca, O. G. and Parestsky D.,** Q fever and *Coxiella burnetii:* a model for host-parasite interactions, *Microbiol. Rev.*, 47, 127, 1983.
4. **Roman, M. J., Coritz, P. O., and Baca, O. G.,** A proposed model to explain persistent infection of host cells with *Coxiella burnetii, J. Gen. Microbiol.*, 132, 1415, 1986.
5. **Baca, O. G. and Crissman, H. A.,** Correlation of DNA, RNA and protein content by flow cytometry in normal and *Coxiella burnetii*-infected L 929 cells, *Infect. Immun.*, 55, 1731, 1987.
6. **Baca, O. G., Scott, T. O., Akporiaye, E. T., De Blassie, R., and Crissman, H. A.,** Cell cycle distribution patterns and generation times of L929 fibroblast cells persistently infected with *Coxiella burnetii, Infect. Immun.*, 47, 366, 1985.
7. **Samuel, J. E., Frazier, M. E., and Mallavia, L. P.,** Correlation of plasmid type and disease caused by *Coxiella burnetii, Infect. Immun.*, 49, 775, 1985.
8. **Vodkin, M. H., Williams, J. C., and Stephenson, E. H.,** Genetic heterogeneity among isolates of *Coxiella burnetii, J. Gen. Microbiol.*, 132, 455, 1986.
9. **Samuel, J. E., Frazier, M. E., Kahn, M. L., Thomashow, L. S., and Mallavia, L. P.,** Isolation and characterisation of plasmid from phase I *Coxiella burnetii, Infect Immun.*, 41, 488, 1983.
10. **Mallavia, L. P., Samuel, J. E., Kahn, M. L., Thomashow, L. S., and Frazier, M. E.,** *Coxiella burnetii* plasmid DNA, in *Microbiology,* Lieve, L. and Schlessinger, D., Eds., American Society for Microbiology, Washington, D.C., 1984, 293.
11. **Moos, A. and Hackstadt, T.,** Comparative virulence of intra- and interstrain lipopolysaccharide variants of *Coxiella burnetii* in guinea pig model, *Infect. Immun.*, 55, 1144, 1987.
12. **Hackstadt, T., Peacock, M. G., Hitchcock, P. J., and Cole, R. L.,** Antigenic variation in Phase I lipopolysaccharide of *Coxiella burnetii* isolates, *Infect. Immun.*, 52, 337, 1986.
13. **Hackstadt, T.,** Lipopolysaccharide variation in *Coxiella burnetii:* intrastrain heterogeneity in structure and antigenicity, *Infect. Immun.*, 48, 359, 1985.
14. **Wilson, H. G., Neilson, G. H., Galea, E. G., Stafford, G., and O'Brien, M. F.,** Q fever endocarditis in Queensland, *Circulation,* 53, 680, 1976.
15. **Ellis, M. E., Smith, C. C., and Moffatt, M. A. J.,** Chronic or fatal Q fever infection: a review of 16 patients in North East Scotland (1967—1980), *Q. J. Med.,* LII, 54, 1983.
16. **Raoult, D., Piquet, P., Gallais, H., De Micco, C., Drancourt, M., and Casanova, P.,** *Coxiella burnetii* infection of vascular prosthesis, *N. Engl. J. Med.*, 315, 1358, 1986.
17. **Koster, F. T., Williams, J. C., and Goodwin, J. S.,** Cellular immunity in Q fever: specific lymphocyte unresponsiveness in Q fever endocarditis, *J. Infect. Dis.*, 152, 1283, 1985.
18. **Lev, B., Shachar, A., Segev, S., Weiss, P., and Rubinsten, E.,** Quiescent Q fever endocarditis exacerbated by cardiac surgery and corticosteroid therapy, *Arch. Intern. Med.*, 148, 1531, 1988.

19. **Sidwell R. W., Thorpe B. D., and Gebhardt, L. P.,** Studies of latent Q fever infections. I. Effects of whole body X-irradiation upon latently infected guinea pigs, white mice and deer mice, *Am. J. Hyg.,* 79, 113, 1964.

20. **Sidwell, R. W., Thorpe, B. D., and Gebhardt, L. P.,** Studies of Q fever infections. II. Effects of multiple cortisone injections, *Am. J. Hyg.,* 79, 320, 1964.

21. **Syucek, L., Sobelavsky, O., and Gurvith, I.,** Isolation of *Coxiella burnetii* from human placentas, *J. Hyg. Epidemiol. Microbiol. Immunol.,* 2, 693, 1958.

22. **Heard, S. R., Ronalds, C. S., and Heath, R. B.,** *Coxiella burnetii* infection in immunocompromised patients, *J. Infect.,* 11, 15, 1985.

23. **Humphres, R. C. and Hinrichs, D. J.,** Role of antibody in *Coxiella burnetii* infection, *Infect. Immun.,* 31, 641, 1981.

24. **Turck, W. P. G., Howitt, G., Turnberg, L. A., Fox, M., Longson, M., Matthews, M. G., and Das Gupta, R.,** Chronic Q fever, *Q. J. Med.,* 45, 193, 1987.

25. **Haldane, E. V., Marrie, T. J., Faulkner, R. S., Lee, S. H. S., Cooper, J. H., Mac Pherson, D. D, and Montague, T. J.,** Endocarditis due to Q fever in Nova Scotia: experience with five patients in 1981—1982, *J. Infect. Dis.,* 148, 978, 1983.

26. **Babudieri, B.,** Q fever: a zoonosis, *Adv. Vet. Sci.,* 5, 81, 1959.

27. **Spelman, D. W.,** Q fever a study of 111 consecutive cases, *Med. J. Australas.,* 26, 547, 1982.

28. **Palmer, S. R. and Young, S. E.,** Q fever endocarditis in England and Wales 1975—1982, *Lancet,* 2, 1448, 1981.

29. **Aitken, I. D.,** Q fever in United Kingdom and Ireland, *Zentralbl. Bakteriol. Parasitenkd. Infektionskr. Hyg.,* 267, 37, 1987.

30. **Raoult, D., Etienne, J., Massip, P., Iaocono, S., Prince, M. A., Beaurain, P., Benichou, S., Auvergnat, J. C., Mathieu, P., Bachet, P., and Serramigni, A.,** Q fever endocarditis in South of France, *J. Infect. Dis.,* 155, 570, 1987.

31. **Fernández Guerrero, M. L., Muelas, J. M., Aguado, J. M., Renedo, G., Fraile, J., Soriano, F., and De Villalobos, E.,** Q fever endocarditis on porcine bioprosthetic valves, *Ann. Intern. Med.,* 108, 209, 1988.

32. **Dupuis, G., Peter, O., Pedroni, D., and Petite, J.,** Aspects cliniques observes lors d'une epidemie de 415 cas de fievre Q, *Schweiz. Med. Wochenschr.,* 115, 814, 1985.

33. **Kimbrough, R. L., Ormsbee, R. A., and Peacock, M. G.,** Q fever endocarditis: a three and one-half year follow up, in *Rickettsiae and Rickettsial Diseases,* Burgdorfer, W. and Anacker, R. L., Eds., Academic Press, New York, 1981, 125.

34. **Krauss, M., Schmeer, N., and Schieffer, H. G.,** Epidemiology and significance of Q fever in the Federal Republic of Germany, *Zentralbl. Bakteriol. Parasitenkd. Infektionskr. Hyg. Abt. I Orig. Reihe A,* 267, 42, 1987.

35. **Kristinsson, A. and Bentall, M. M.,** Medical and surgical treatment of Q fever endocarditis, *Lancet,* 2, 693, 1967.

36. **Walters, J.,** Rickettsial endocarditis, *Br. Med. J.,* 4, 770, 1968.

37. **Jones, R. W. A. and Pitcher, D. W.,** Q fever endocarditis in a 6 year old child, *Arch. Dis. Child.,* 55, 312, 1980.

38. **Ross, P. J., Jacobson, J., and Muir, J. R.,** Q fever endocarditis of porcine xenograph valves, *Am. Heart J.,* 105, 151, 1983.

39. **Mann, J. S., Douglas, J. G., Inglis, J. M., and Leitch, A. G.,** Q fever person to person transmission within a family, *Thorax,* 41, 974, 1986.

40. **Varma, M. P., Adgey, A. A. J., and Connolly, J. H.,** Chronic Q fever endocarditis, *Br. Heart J.,* 43, 695, 1980.

41. **Tobin, M. J., Cahill, N., Gearty, G., Maurer, B., Blake, S., Daly, K., and Hone, R.,** Q fever endocarditis, *Am. J. Med.,* 72, 396, 1982.

42. **Grist, N. R. and Ross, C. A. C.,** Four cases of Rickettsial endocarditis, *Br. Med. J.,* 2, 119, 1968.

43. **Saginur, R., Silver, S. S., Bonin, R., Carlier, M., and Orizaga, M.,** Q fever endocarditis, *Can. Med. Assoc. J.,* 133, 1228, 1985.

44. **Sawyer, L. A., Fishbein, D. B, and Mc Dade, J. E.,** Q fever: current concepts, *Rev. Infect. Dis.,* 9, 935, 1987.

45. **Dupuis, G., Peter, O., Lüthy, R., Nicolet, J., Peacock, M., and Burgdorfer, W.,** Serological diagnosis of Q fever endocarditis, *Eur. Heart J.,* 7, 1062, 1986.

46. **Raoult, D., Prince, M. A., Baragan, P., Bernard, J. P., Durand, J. M., Bory, M., Djiane, P., Edlinger, E., and Serradimigni, A.,** Endocardites a Coxiella burnetii: deux observations, *Med. Mal. Infect.,* 15, 436, 1985.

47. **Laufer, D., Lew, P. D., Oberhansli, I., Cox, J. N., and Longson, M.,** Chronic Q fever endocarditis with masssive splenomegaly in childhood, *J. Pediatr.,* 108, 535, 1986.

48. **Watt, A. H., Fraser, A. G., and Stephens, M. R.,** Q fever endocarditis presenting as myocardial infarction, *Am. Heart J.,* 112, 1333, 1986.

49. **Etienne, J., Delahaye, F., Raoult, D., Frieh, J. P., Loure, R., and Delaye, J.,** Acute heart failure due to Q fever endocarditis, *Eur. Heart J.,* 9, 923, 1988.

50. **Willey, R. F., Matthews, M. B., Peutherer, J. F., and Marmion, B. P.,** Chronic cryptic Q fever infection of the heart, *Lancet,* 2, 270, 1979.

51. **Scheld, W. M. and Sande, M. A.,** Endocarditis and intravascular infections, in *Principles and Practice of Infectious Diseases,* 2nd ed., Mandell, G. L., Douglas, R. G., Jr., and Bennett, J. E., Eds., John Wiley & Sons, New York, 1985, 504.

52. **Spring, W. J. C. and Hampson, J.,** Chronic Q fever endocarditis causing massive splenomegaly and hypersplenism, *Br. Med. J.,* 285, 1244, 1982.
53. **Delaney, J. C. and Roberts, H. L.,** Q fever endocarditis and chronic liver involvement, *Practitioner,* 214, 243, 1975.
54. **Qizilbash, A. H.,** The pathology of Q fever as seen in liver biopsy, *Arch. Pathol. Lab. Med.,* 107, 364, 1983.
55. **Dupont, H. L., Hornick, R. B., Levin, H. S., Rapoport, M. I., and Woodward, T. E.,** Q fever hepatitis, *Ann. Intern. Med.,* 74, 198, 1971.
56. **Hoffmann, C. E. and Heaton, S. W., Jr.,** Q fever hepatitis. Clinical manifestations and pathological findings, *Gastroenterology,* 83, 474, 1982.
57. **Marmion, B. P., Higgins, F. E., and Edwards, A. T.,** A case of subacute rickettsial endocarditis; with a survey of cardiac patients for this infection, *Br. Med. J.,* 2, 1264, 1960.
58. **Brooks, R. G., Licitra, L. M., and Peacock, M. G.,** Encephalitis caused by *Coxiella burnetii, Ann. Neurol.,* 20, 91, 1986.
59. **Turck, W. P G. and Matthews, M. B.,** Rickettsial endocarditis, *Br. Med. J.,* 1, 185, 1969.
60. **Coyle, P. V., Thompson, J., Adgey, A. A. J., Rutter, D. A., Fay, A., McNeill, T. A., and Connolly, J. H.,** Change in circulating immune complex concentrations and antibody titers during treatment of Q fever endocarditis, *J. Clin. Pathol.,* 38, 743, 1985.
61. **Peacock, M. G., Philip, R. N., Williams, J. C., and Faulkner, R. S.,** Serological evaluation of Q fever in humans: enhanced Phase I titers of Immunoglobulins G and A are diagnostic for Q fever endocarditis, *Infect. Immun.,* 41, 108, 1983.
62. **Raoult, D., Urvolgyi, J., Etienne, J., Roturier, M., Puel, J., and Chaudet, H.,** Diagnosis of endocarditis and acute Q fever by immunofluorescence serology, *Acta Virol.,* 32, 70, 1988.
63. **Tunstall Pedoe, H. D.,** Apparent recurrence of Q fever endocarditis following homograph replacement of aortic valve, *Br. Heart J.,* 32, 568, 1970.
64. **Clark, W. H., Lenette, E. H., Railback, O. C., and Romer, M. S.,** Q fever in California — clinical features in one hundred eighty cases, *Arch. Intern. Med.,* 88, 155, 1951.
65. **Subramanya, N. I., Wright, J. S., and Khan, M. A. R.,** Failure of rifampicin and cotrimoxazole in Q fever endocarditis, *Br. Med. J.,* 285, 343, 1982.
66. **Worswick, D. and Marmion, B. P.,** Antibody responses in acute and chronic Q fever and in subjects vaccinated against Q fever, *J. Med. Microbiol.,* 19, 281, 1985.
67. **Ramos-Díaz, M., Creagh Cerquera, R., Martin Herrera, C., Castillo Jiménez, J., Fernández Alonso, J., and Gómez Camacho, E.,** Glomerulofritis mesangiocapilar asociada a endocarditis por fiebre Q, *Rev. Clin. Esp.,* 180, 314, 1987.
68. **Tellez, A., Sainz, L., Echevaria, C., De Carlos, S., and Fernandes, M. V.,** Q fever in Spain. Evaluations of acute and chronic cases seen during 1982-1983, in *Rickettsiae and Rickettsial Diseases,* Kazar, J., Ed., Slovak Academy of Sciences, Bratislava, 1985, 398.
69. **Rady, M., Glavits, R., and Nagy, G.,** Epidemiology and significance of Q fever in Hungary, *Zentralbl. Bakteriol. Parasitenkd. Infectionskr. Hyg.,* 267, 10, 1987.
70. **Houvers, D. J. and Richardus, J. H.,** Infections with *Coxiella burnetii* in man and animals in Netherlands, *Zentralbl. Bakteriol. Parasitenkd. Infectionskr. Hyg.,* 267, 30, 1987.
71. **Péter, O., Dupuis G., Bee, D., Lüthy, R., Nicolet, J., and Burgdorfer, W.,** Enzyme-linked immunosorbent assay for diagnosis of chronic Q fever, *J. Clin. Microbiol.,* 26, 1978, 1988.
72. **Raoult, D., Bollini, G., and Gallais, H.,** Osteoarticular infection due to *Coxiella burnetii, J. Infect. Dis.,* 159, 1159, 1989.
73. **Ascher, M.,** personal communication.
74. **Raoult, D., Raza, A., and Marrie, T.,** unpublished data.

Chapter 10

IS *COXIELLA BURNETII* A HUMAN PERINATAL PATHOGEN?

Joanne M. Langley

TABLE OF CONTENTS

I. INTRODUCTION

Q fever, a zoonosis caused by the rickettsial organism *Coxiella burnetii* is endemic throughout the world.[1] The most common animal reservoirs are cattle, including sheep and goats, but many other mammalian, avian, and arthropod hosts have been described.[2] Rickettsiae are excreted in cattle urine, feces, and milk, but are most concentrated in birth products, especially placenta, in which up to 10 hamster infective doses (ID_{50}) per gram of tissue may be present in late gestation.[3] Small particles of *C. burnetii* are aerosolized at the time of parturition from the placenta, amniotic fluid, and fetal membranes.[3] Human infection, which presents on a spectrum from subclinical or nonspecific febrile illness to atypical pneumonia, hepatitis, or more rarely endocarditis,[1] occurs when small particle aerosols of *C. burnetii* are inhaled.

Although it is during parturition that this highly infectious organism is most concentrated in mammals, there are few reports on the manifestations of perinatal Q fever in the human host. This is perhaps not surprising given that the disease is, in general, believed to be significantly underreported.[4] In some endemic areas heightened surveillance has led to increased diagnosis,[4,5] with up to 20% of rural community-acquired pneumonia attributed to *C. burnetii* infection.[5] Since failure to look for this organism may explain lack of diagnosis,[6] it is theoretically possible that *C. burnetii* is an unrecognized pathogen in the perinatal period. The purpose of this review is to address the following questions: is perinatal Q fever infection a clinical problem for the human mother or infant that is underrecognized, or is it not a problem?

II. METHODS

The strategy for obtaining primary research from the world literature, English and non-English, on perinatal Q fever infection, included use of three bibliographic data bases, scrutinizing all references cited in past reviews and in primary research obtained, and personal communication with an expert in the area of Q fever infection. The first bibliographic data base used was the U.S. National Library of Medicine Medline, a computerized data base, through BRS Afterdark, back to 1966 using the medical subject headings (MeSH) term "Q fever". More specific searches with the addition of terms such as "pregnancy", "newborn", "abortion", or "teratogenesis" were nonproductive. The Science Citation Index was manually searched forward in time using Reference 3, the first primary research paper establishing the concentration of *C. burnetii* in birth products and the role of parturition in the generation of infective aerosols. *Current Contents* was manually searched using the term "Q fever". This search strategy yielded over 800 citations.

The criteria for selecting titles from the list of 800 were intentionally general because of the seeming paucity of primary literature on human perinatal Q fever and the concern that the indexing strategy might not reference perinatal phenomena if it were not a major focus of the article. Titles were chosen for evaluation if they mentioned perinatal infection in man or in mammals regardless of research design, if they were reviews or case series of human Q fever infection, or if the term "epidemiology" was in the title. The perinatal period was defined as during pregnancy, parturition, and the month following or the period including breastfeeding. Methodological criteria were initially considered but discarded as the literature was found to be mostly descriptive in nature. Articles in English or non–English articles, for which translation was available, were eligible. This strategy generated 60 articles.

Articles selected from this search were included in the current review if they reported natural or experimental Q fever infection in man or animals in the perinatal period, as defined above. Second, to ensure the validity of the diagnosis of Q fever, the diagnostic criteria had to fulfill at least one of the following features: isolation of the organism by culture or by seroconversion of test animals injected with the specimen in question; a fourfold rise in antibody titer or a phase I titer of at least 1:200; or presence of *C. burnetii* antigen on direct staining of tissue. This yielded 17 articles.

Results were not combined statistically because most papers were without statistical analysis and because research design varied widely. Reproducibility of the criteria for inclusion was not tested for logistical reasons.

The validity of the primary research was assessed by classifying each study according to its "level of evidence"[7] and by applying those of the "nine diagnostic tests of causation"[8] that were relevant.

The classification of levels of evidence ranging from I to V was developed for a consensus conference attempting to summarize what was known about the etiology, clinical course, and management of a particular clinical problem.[7] The levels of evidence are

Level I	Randomized trials with low false-positive and false-negative errors
Level II	Randomized trials with high false-positive and/or high false-negative errors
Level III	Nonrandomized concurrent cohort comparisons between contemporaneous patients who did and did not receive (the exposure)
Level IV	Nonrandomized historical cohort comparisons between current patients who did receive (the exposure) and former patients who did not
Level V	Case series without controls

The nine diagnostic tests for evaluating claims of causation,[8] listed in order of importance, are

1. Is there evidence from true experiments in humans?
2. Is the association strong?
3. Is the association consistent from study to study?
4. Is the temporal relationship correct?
5. Is there a dose-response gradient?
6. Does the association make epidemiological sense?
7. Does the association make biological sense?
8. Is the association specific?
9. Is the association analogous to a previously proven causal association?

Questions 6, 7, and 9 were excluded, as all the research would deal with a known pathogen, and thus the association of the pathogen with perinatal infection would make epidemiological and biological sense, and be analogous to previously proved disease. Question 5 was excluded unless a report mentioned quantitative cultures. A subjective assessment of each study after applying these tests led to their evaluation as "supportive"(S) or "inconclusive" of (I) or "contrary" (C) to evidence for causation.

III. RESULTS

The primary research literature on human and mammalian perinatal Q fever infection consists of reports of epidemics, case reports, case series, a few cohort studies, and experimental studies in animals. The evidence to evaluate the primary question (is perinatal Q fever a clinical problem for the human mother or infant) is evaluated in the following categories: placental infection and its role in abortion, teratogenesis, maternal and neonatal illness, and transmissibility of *C. burnetii* through milk.

TABLE 1
Q Fever as a Perinatal Pathogen: Evidence for Causation

Evidence	Research design	Level of evidence	Evidence for causation	Ref.
Abortion				
Cattle epidemics	Descriptive	V	I	9, 18, 19
Goat, sheep epidemic	Descriptive	V	I	20
Goat epidemic	Descriptive	V	I	21
Goat, sheep epidemic	Case-control	IV	S	22
Human				
Present in placenta	Descriptive	V	C	9, 25, 26
Present in placenta	Descriptive	V	I	23, 24
Serological survey	Cross-sectional, controlled	IV	I	26
Teratogenicity				
Rat teratogen	Controlled experiment	III	S	29
Normal human	Descriptive	V	C	9, 30, 31
Minor anomaly	Descriptive	V	I	30
Maternal and neonatal illness				
Subclinical illness	Descriptive	V	S	9, 25, 30, 31
Maternal illness	Descriptive	V	S	45
Maternal illness	Descriptive	V	I	47,49
Neonatal illness	Descriptive	V	I	49
Breast milk transmission				
Animal seroconversion	Descriptive	V	S	40
Presence in milk	Descriptive	V	I	9, 41, 42
Infected infant	Descriptive	V	S	43

Note: I = *inconclusive*; S = supportive; and C = contrary. Level I = randomized trials with low false-positive and false-negative errors; Level II = randomized trials with high false-positive and/or high false-negative errors; Level III = nonrandomized concurrent cohort comparisons between contemporaneous patients who did and did not receive (the exposure); Level IV = nonrandomized historical cohort comparisons between current patients who did receive (the exposure) and former patients who did not; and Level V = case series without controls.

Each study or group of similar studies is classified according to its level of evidence and evidence for causation, as previously described in Table 1.

A. PLACENTAL INFECTION AND ITS ROLE IN ABORTION

The epidemiological observation that many humans who acquired Q fever had a history of exposure to livestock led to the search for *C. burnetii* in the excreta and secretion of cows, sheep, and goats.[9] Rickettsia were first shown to be present in the placentas of cows artificially infected by intranasal atomizer,[10] and then from the placentas of naturally infected dairy cows.[11] *C. burnetii* was then isolated from the placentas of naturally[12-14] and experimentally infected[15] sheep. The organism was subsequently documented in bovine amniotic fluid[16] and shown to be transplacentally transmitted in the guinea pig.[17]

Epidemics of abortion were first attributed to *C. burnetii* in goats[9] and later in cattle.[18,19] These uncontrolled studies reported a higher abortion rate than expected and concomitant isolation of the organism; causation was then alleged by association of the two outcomes. This evidence is judged inconclusive since coexistence of two attributes may suggest but clearly does not prove a causal relationship.

More recently, placentitis and abortion in goats and sheep in Ontario has been attributed to *C. burnetii* based on serological confirmation by complement fixation (CF) and presence of the organism in placental smears.[20] Abortion rates between herds varied from 5 to 91%.

Concomitant *chlamydia* infection, based on serological and microscopic evidence, occurred in an unspecified number of these animals. No information on seropositivity rates in a nonaborting or control population is given. Without knowledge of the rates of isolation in placentas and seropositivity in a control population, however, the association between the presence of *C. burnetii* and abortion is inconclusive.

An epidemic of abortion in dairy goats, which occurred in Idaho in 1978 in which 20 to 39 pregnant does in one flock delivered 2 to 4 weeks early, was attributed to *C. burnetii* infection.[21] All of the fetal livers, spleens, and stomach contents were culture negative for *Brucella* and *Campylobacter*. Two of the 20 placentas and all of the fetal livers stained positive for flourescein-labeled *C. burnetii*, and organisms were seen in these preparations. A placental tissue suspension when inoculated into yolk sac revealed rickettsia-like organisms in the dead embryos, and guinea pigs inoculated with placental tissue after three egg passages seroconverted to *C. burnetii*. Sera were obtained from only 7 and 15 of the 20 aborters 7 months apart. In May of 1976, three animals had a CF titer of 1:16 to a mix of phase I and II *C. burnetii* antigen, one had a titer of 1:8, and three were negative. In November, two animals had a titer of 1:8, 3 of 1:16, and 10 were negative. None of the sera had antibodies to *Toxoplasma gondii*. The magnitude of the abortion rate (20/39 or 51%) and the specific isolation of *C. burnetii* alone in microbiological investigation are suggestive that this agent was the etiology of this epidemic. One cannot, however, discount another noninfectious cause confounding this conclusion. The rate and degree of seropositivity is also less than expected. Again, the comparison of infection rates in a nonaborting population is essential to delineate whether infection is subclinical or a cause of symptomatic disease. This evidence is judged inconclusive.

More convincing evidence was available following a case-control study of an epidemic of abortions in sheep and goats, consisting of 21 outbreaks in Cyprus in 1974 to 1975.[22] In a serological survey of flocks from whom rickettsial placentas had been received by the Central Veterinary Laboratory 3 months after these outbreaks, sera was taken from ten aborters and ten nonaborters in each of ten affected flocks. CF testing showed 72% (70/90) of the aborting animals were serologically positive for antibodies to phase II antigen of *C. burnetii*, whereas 44% (45/102) of the nonaborters were positive, a statistically significant difference at $p<0.01$. The strength of association is perhaps better illustrated by the odds ratio of 4.43. The inclusion of a control group, greatly strengthens the argument that abortion may be attributable to exposure to *C. burnetii*. This evidence is therefore judged supportive.

Evidence for *C. burnetii* as a cause of abortion in humans is less clear. While the organism has been isolated from placenta and birth products of pregnancies that spontaneously aborted,[23,24] it does not follow, for reasons previously stated, that the organism was the etiologic agent for the abortion. As well, the agent has been isolated from normal[9,25,26] placentas of normal pregnancies, which would be judged as "contrary" evidence.

One retrospective cross-sectional study compared placentas and sera of women with pathological pregnancy to that of women with normal pregnancy.[26] Four percent of the women with spontaneous abortions, premature deliveries, and stillbirths had positive CF to *C. burnetii*. Rickettsiae were not seen in the placentas or post-mortem specimens of dead infants of seropositive mothers although other organisms were. The authors concluded that *C. burnetii* infection is not an etiologic factor in adverse pregnancy outcome.

The demonstration of CF antibodies in maternal sera at a single point in time, however, is not an accurate diagnostic test for acute or recrudescent perinatal infection, since antibodies fall slowly and may persist for years.[27] We cannot be certain that the "exposed" group really had active Q fever in the perinatal period. This may explain why placental and postmortem specimens were negative. As well, there exists the possibility of a falsely negative conclusion, or type II error.[28] That is, even if the pathological pregnancy group had perinatal Q fever, the total sample size of 687 was likely too small to detect a difference in seropositivity if the proportion in one group was 4.05% and absent in the other. This study's conclusion is not supported by the evidence presented since we cannot be sure the exposed group had disease, and the power is small. This study then provides "inconclusive" evidence that *C. burnetii* is a perinatal pathogen.

In summary, the evidence that *C. burnetii* causes abortion in animals is suggestive in one study (level IV evidence) and inconclusive in four studies (level V evidence). In humans the data are contrary to the hypothesis of causation in three reports (level V evidence), and inconclusive in three (level IV and V evidence). Thus, there is supportive evidence for causation in animals; none exists for humans. If we accept the animal model as direct evidence of causation and extrapolate to humans, it is theoretically possible that this organism causes abortion. This theory finds some support in the finding that women with a history of recurrent pregnancy loss may have a higher incidence of serum anticomplementary activity, one cause of which may be antigen-antibody complexes due to prior Q fever infection.[48]

To determine if the organism is a pathogen in or present in a subclinical carrier state one would ideally prospectively follow pregnant women and compare the rate of isolation of *C. burnetii* in placentas from normal pregnancies to that in spontaneous abortions.

It is also not clear from these studies whether Q fever infection can become latent with recrudescence in pregnancy, as is known to occur in animals. This issue is addressed later.

B. TERATOGENICITY

Teratogenicity has not been attributed to human perinatal infection with *C. burnetii*. However, there are some animal experimental data indicating that this may occur.[29] Although there is at least one case report of an abnormal baby born to a woman with Q fever,[30] the association of this infection with the anomaly could be purely coincidental. In the case cited, the anomaly, hypospadias, is among the more common to occur in any newborn population[31] and thus not at all specific for this agent. This is judged "inconclusive" evidence.

Many normal infants have been delivered of women with perinatal Q fever,[30,31,51] suggesting its existence as a commensal organism, contrary to the hypothesis of pathogenicity. Babudieri, for example, in a review of Q fever in 1959, reported a woman who acquired Q fever in a laboratory setting in Italy and gave birth to a normal child 6 months later.[51] The manifestations of maternal illness are not mentioned.

The experimental evidence that *C. burnetii* is a teratogen was demonstrated in a rat model of perinatal Q fever.[29] Twenty-two female rats were inoculated with strain C9 *C. burnetii* with yolk passage on the ninth day of pregnancy and sacrificed on day 21, while ten control rats received no injection. It is not stated if animals were randomized or not. The mothers remained clinically well after inoculation but seroconverted with antibodies present up to a 1:160 dilution. The fetuses had no visible external malformation, and their internal organs were said to look normal on gross examination. Of the offspring there were 133 living fetuses and 12 dead in the experimental group and 112 living with no dead fetuses in the controls. On histological examination cataracts were seen in 46 of the experimental group; none were seen in the controls. Abnormalities of the retina and optic nerve were seen in 105 of the experimental group; none were seen in the controls. Although no statistical analysis was presented in this paper, a one-sided significance test for independent proportions[26] on the three outcomes of fetal death, cataracts, and abnormal retinas, with a Bonferoni adjustment[34] for multiple outcomes (three) shows that the probability of these occurring by chance is less than 0.05 ($Zc = 3.1, 6.34$ and 11.6, respectively). We cannot, then, reasonably attribute these outcomes to chance sampling variation, so this is judged "supportive" evidence.

As with any new experimental finding this result would need to be confirmed in other laboratories. In particular, it would be necessary to confirm that the control and experimental groups were from comparable populations such that confounding by a concomitant noninfectious teratogen was avoided. As well it is conceivable that the experimental dose is larger and perhaps more teratogenic than that which occurs in natural infection. It would be premature to extrapolate from direct evidence of teratogenicity in a single animal experiment to human teratogenicity without consistent results from animal models.

In summary, several descriptive studies document human perinatal Q fever infection without teratogenicity evident in the newborn. Whether adverse events occur in humans is speculative at present.

C. MATERNAL AND NEONATAL ILLNESS

Research supporting the existence of maternal or congenitally acquired infection in humans consists of case reports[25,30,45,47,49] and one cross-sectional serological survey.[31]

In the course of a survey of humans and dairy cattle for evidence of *C. burnetii* infection in Maryland, investigators discovered a pregnant woman, who lived in an area of concentrated dairying, with a CF titer of 1:256.[25] A repeat sample 50 d later at the time of delivery was positive at 1:512, and *C. burnetii* was isolated from the placental material. The woman denied any symptoms of Q fever in the preceeding 18 months. Mention of the outcome of the infant is not made.

In Czechoslovakia investigators attempted to demonstrate *C. burnetii* in five women who became pregnant within 1 to 2 years of infection.[30] It is not stated if these patients received antimicrobial therapy. The interval between Q fever infection and delivery ranged from 2 years and 8 months to 3 years and 2 months. Four of these women worked in cowsheds, and one was a veterinary surgeon.

Three women had normal children, and in two of these, *C. burnetii* was isolated from placenta. One had her pregnancy interrupted because of rubella, and placental *C. burnetii* was isolated. The fifth woman had a male with slight hypospadias. The placenta was positive for *C. burnetii*. Breast milk was examined in two of the women and found negative. The method for isolation of the agent consisted of injecting an emulsion of placenta into the peritoneum of guinea pigs and measuring CF antibody response, considered significant over a titer of 1:16. Antibody response was also measured after reinoculation with a strain from guinea pig spleens.

This paper documents chronic placental infection in one patient and demonstrates that subclinical infection in parturient women can occur with normal neonatal outcome. The authors conclude that *C. burnetii* can persist for long periods in the human organism; although the possibility of reinfection in some of these women could not be excluded, this conclusion seems sound. It is possible the "chronic placental infection" may, in fact, be recrudescent infection occurring during pregnancy, as is known to occur in cattle. Since excretion of rickettsiae was not monitored outside of the pregnancy, this cannot be determined. It may be that antimicrobial therapy does not eradicate the "carrier" or "latent" state of Q fever infection. Although shedding of rickettsiae may not pose a risk to mother or fetus, it is a potential risk to health care workers exposed to aerosolized particles at delivery.

Maternal illness was clearly demonstrated in a case report of Q fever associated with thrombocytopenia from Israel.[45] A 29-year-old previously healthy woman was admitted at 21 weeks gestation with a 1-month history of fever, headache, weakness, and diaphoresis. She lived on a kibbutz and had been exposed to breeding sheep 1 month before the onset of symptoms. The main physical findings consisted of fever, mild hepatomegaly, and an appropriately sized uterus. A chest radiograph and liver function tests were normal. A fourfold rise in titers to *C. burnetii* phase 1 and 2 antigens was demonstrated, the organism was isolated from the necrotic placenta, and guinea pigs inoculated with placental suspension seroconverted to phase 1 and 2 antigens.

The patient developed a severe thrombocytopenia thought to be of autoimmune etiology as the bone marrow was normal and transfused platelets were rapidly consumed. Tetracycline therapy resulted in improvement of symptoms; the thrombocytopenia, however, was refractory to antimicrobial and steroid therapy, platelet transfusions, and only transiently responsive to intravenous immunoglobulin. Serological tests indicated ongoing infection with Q fever, and, thus, labor was induced at 28 weeks because of the potential risk to the fetus of placental insufficiency and to the mother of complications of chronic infection. The baby was not infected and is well at 1 year except for a yellow tooth. The mother's thrombocytopenia resolved after 9 months of tetracycline and danazol.

This case report provides supportive evidence of clinical illness associated with maternal Q fever infection. The authors speculate that antibodies produced as a sequela of chronic Q fever infection may have reacted immunologically with the platelets. This hypothesis is compatible with other research indicating that immune complex disease should be considered a feature of Q fever illness.[46]

Concomitant Q fever and *Chlamydia* infection in pregnancy were diagnosed in a 30-year-old wife of a sheep farmer in England at 14 weeks gestation who presented with fever, dyspnea, and jaundice. She demonstrated a fourfold rise in titers to *C. burnetii* and *Chlamydia psittaci*, and *C. psittaci* antigen was detected in smears from the placenta. The fetus died *in utero*, and *C. psittaci* was isolated in cell culture from post-mortem specimens of fetal organs. It is not possible to attribute adverse fetal outcome in this case to maternal infection with Q fever, nor is it possible to delineate to what extent it contributed to overt maternal illness compared with *Chlamydia*. One can only conclude that infection occurred.

A case report of three premature infants with respiratory disease attributed to Q fever was reported in the Italian literature in 1956. The infants, all of whom were initially well, developed cough, tachypnea, and fever, and pulmonary consolidation was seen on chest films in the first month of life. Acute and convalescent sera showed titers between 1:8 and 1:16 on these children. One infant died, and *Staphylococcus aureus* was grown from lung specimens. No other bacterial or viral etiology was found in these three cases. The parents had titers between 1:32 and 1:64 to *C. burnetti*, but a fourfold rise was not documented. The epidemiologic risk factors included habitation in a rural area through which sheep were transported. It is not possible to attribute definitively the neonatal infection to this agent. The measured antibody could have been transplacentally acquired and should have changed quantitatively over time if this were the major etiology of the infants respiratory distress. This is supportive evidence of maternal clinical infection but not of congenital infection.

More convincing evidence of congenital infection was demonstrated in a serological survey of maternal and cord blood in Egypt which identified *C. burnetii* specific IgM.[31] Maternal and cord blood pairs were screened for antibody by phase II agglutination titers. Eight pairs were found to have titers ranging from 1:16 to 1:256. Treatment of these specimens with ethanediol reduced all titers to <1:5 indicating the agglutinins were IgM. Four of the eight samples were of sufficient quantity to be further characterized by absorption with phase II antigen and IgM quantitation by radioimmunodiffusion, before and after absorption. Absorption eliminated agglutination and reduced IgM levels to normal. At birth all of the infants were normal, but follow-up was not further pursued. Maternal manifestations of Q fever are not discussed.

This survey demonstrates that human fetal response to maternal Q fever infection can occur. IgM is not a specific measure of this response, as cross-reactivity with other organisms does not occur.[32] While these newborns were asymptomatic, the possibility of neonatal Q fever or later effects in the developing child cannot be dismissed. The true incidence of asymptomatic and symptomatic infection in those newborns exposed would only be shown by a larger prospective study of infected parturient women.

In summary, there is evidence that maternal infection occurs in a subclinical and overt form. *In utero* infection may occur but seems to be subclinical in nature in the neonate. The spectrum of clinical manifestations in mother or child is not known.

D. TRANSMISSION BY BREAST MILK

Coxiellae were first shown to be present in the milk of cows infected experimentally,[36] and numerous serological surveys demonstrating *C. burnetii* antibody in bovine milk followed.[37-39]

The prevalence of antibody-positive milk varied from 0.8% in Scotland in 1952[39] to 72.4% of samples from Maryland in 1965.[25] Shedding of the organism was found to be continuous or intermittent for one or more periods of lactation, with individual samples containing up to 100,000 organisms per milliliter. It has been identified in the milk of goats, sheep, and other mammals.[9]

In a guinea pig model of oral transmission of Q fever the animals were fed naturally infected milk, following which 8 of 25 animals had CF antibodies to *C. burnetii* with titers between 1:5 and 1:280.[40] In 5 of 8 who developed antibodies rickettsiae were demonstrated in internal organs.

In humans there is evidence that antibody to *C. burnetii* and the organism itself may be present in breast milk;[9,41,42] however, it is not clear if this may cause disease in the infant.

The first report of *C. burnetii* in human breast milk was by Babudieri in 1954.[51] He isolated the organism in the milk of a woman 1 month postpartum who had had Q fever 6 months before delivery. No mention is made of adverse outcome in the newborn.

Breastfeeding was speculated as the source of infection in the case of a 5-month-old Dutch boy who presented with a 4-month history of fever.[43] He had a fourfold rise in antibody titer to phase II antigen by CF, immunoflourescent testing for IgM rose from <16 to 128 to phase II, and his IgG from <16 to 256. The family had not been outside Holland, nor had they had any animal contact. They had, however, traveled to Austria, an area of high endemicity, when the mother was 6 months pregnant. She had IgG but not IgM antibodies by immunoflourescent testing. Lactogenic transmission may have occurred in this case. It is also conceivable that transplacental transmission occurred.

Two serological surveys of human milk samples in India have demonstrated antibody and the agent in breast milk. In 1981, Kumar et al.[41] found antibody by capillary agglutination testing, or indirect evidence of the organism by seroconversion in mice, and intracytoplasmic rickettsiae in 5.2% of 97 samples tested.

In a study of 153 milk samples from lactating women in India, 22 (14.4%) were found to have antibodies by capillary agglutination testing (CAT) using *C. burnetii* Nine Mile strain phase I antigen ranging in titer from 1:8 to 1:256.[42] The CAT positive samples were then inoculated into embryonated eggs by yolk sac route and into two guinea pigs per sample intraperitoneally. Presence of at least a 1:8 CAT titer together with intracytoplasmic rickettsial organisms in impression smears of spleen or liver of both guinea pigs were considered to be specific seroconversion indicating the presence of *C. burnetii* in the original sample. By this process 4 of the 22 positive samples were found to contain the organism.

In neither of these two studies are the results correlated with clinical manifestations in infants fed infected milk. Thus, the possibility of lactogenic transmission of Q fever, and the clinical significance of these findings, are not explored. We can conclude only that *C. burnetii* can be excreted in human milk and that infants are theoretically at risk of acquiring Q fever through the breast milk of infected mothers based on one case report.[43]

The actual prevalence of *C. burnetii* or antibody to it in human milk cannot be estimated from these studies since the populations are hospital based and not generalizable to the general population. The proportion of infants exposed to infected milk who acquire disease or the proportion of infants who receive passive immunity through maternal breast milk could only be determined in a prospective study.

IV. SUMMARY

The study of etiologic agents of disease, as has been noted by eminent epidemiologists,[44] is made difficult because suspected exposures or interventions cannot be tested experimentally in human populations. The evidence is, as a result, observational in nature. This problem is exemplified by the current attempt to answer the question posed in this review; the evidence in humans consists of observational data, whereas experimental data were found only in animal models. Nonetheless, we are able to make several conclusions.

Human perinatal Q fever infection occurs. *C burnetii* has been demonstrated in placenta and breast milk; maternal antibody has been shown in breast milk and fetal antibody in cord blood. Maternal infection can occur in a clinical and subclinical form, but teratogenicity or abortion risk has not been demonstrated in humans. Lactogenic transmission of disease is theoretically possible but not yet convincingly demonstrated. Whether this organism is associated with early pregnancy loss is speculative at present. It is theoretically possible that this organism is teratogenic, based on animal data. Subclinical fetal infection can occur. Whether *C. burnetii* is a cause of clinical neonatal disease cannot be determined from the evidence presented. Perinatal Q fever may be a true perinatal pathogen, but overall it seems to pose a small burden of illness.

Treatment of Q fever in the perinatal period would be problematic, since tetracycline, the antimicrobial of choice, is generally contraindicated in children under 7 years of age and in pregnancy because of transplacental passage. Tetracycline forms tetracycline-calcium orthophosphate complexes in bones and teeth resulting in permanent discoloration.[50] A reversible depression of bone marrow in prematures has also been attributed to this agent. The preferred therapy in children under 7 is chloramphenicol, an antimicrobial also associated with untoward effects including blood dyscrasia.

Future research in endemic areas for Q fever, however, could assess prevalence of seropositivity in women of childbearing age. If these women are susceptible to infection during pregnancy, they could be followed as a cohort to determine if outcomes such as abortion, teratogenicity, or neonatal illness occur and if so in what frequency. Only then would the true role of *C. burnetii* as a perinatal pathogen be known.

REFERENCES

1. **Sawyer, L. S., Fishbein, D. B., and McDade, J. E.,** Q fever: current concepts, *Rev. Infect. Dis.*, 9, 935, 1987.
2. **Baca, O. G. and Paretsky, D.,** Q fever and *Coxiella burnetii*: a model for host-parasite reactions, *Microbiol. Rev.*, 47, 127, 1983.
3. **Welsh, H. H., Lennette, E. H., Abinati, F. R., and Winn, J. F.,** Air-borne transmission of Q fever: the role of parturition in the generation of infective aerosols, *Ann. N.Y. Acad. Sci.*, 70, 528, 1958.
4. **Sienko, D. G., Barlett, P. C., McGee, H. B., Wentworth, B. B., Herndon, J. L., and Hall, W. N.,** Q fever: a call to heighten our index of suspicion, *Arch. Intern. Med.*, 148, 609, 1988.
5. **Marrie, T. J., Haldane, E. V., Faulkner, R. S., Kwan, C., Grant, B., and Cook, F.,** The importance of *Coxiella burnetii* as a cause of pneumonia in Nova Scotia, *Can. J. Public Health*, 76, 233, 1985.
6. **Angelo, L. J., Baker, E. F., and Schlosser, W.,** Q fever in the United States, 1948—1977, *J. Infect. Dis.*, 139, 613, 1979.
7. **Sackett, D. L.,** Rules of evidence and clinical recommendations on the use of antithrombotic agents, *Chest*, 89, 2S, 1986.
8. **Hill, A. B.,** *Principles of Medical Statistics*, 9th ed., Lancet, London, England, 1971, 312.
9. **Babudieri, B.,** Q fever: a zoonosis, *Adv. Vet. Sci.*, 5, 81, 1959.
10. **Parker, R. R., Bell, E. J., and Lackman, D. B.,** Experimental studies of Q fever in cattle. I. Observations on four heifers and two milk cows, *Am. J. Hyg.*, 48, 191, 1948.
11. **Luoto, L. and Heubner, R. J.,** Q fever studies in Southern California. IX. Isolation of Q fever organisms from parturient placentas of naturally infected dairy cows, *Public Health Rep.*, 65, 541, 1950.
12. **Welsh, H. H., Lennette, E. H., Abinati, F. R. and Winn, J. F.,** Q fever in California. IV. Occurrence of "*Coxiella burnetii*" in the birth fluids of naturally infected sheep, *Am. J. Hyg.*, 58, 385, 1953.
13. **Rosati, T.,** Semmaria.sull epidemiologia della febbre Q in Italia, *Ann. Sanita Pubblica.*, 13, 88, 1952.
14. **Rosati, T. and Morozzi, A.,** Febbre Q: Indigini sperimentali sul contagio fragli animali di specie ovina, *Ig. Sanita Pubblica.*, 8, 269, 1952.
15. **Abinati, F. R., Welsh, H. H., Lennette, E. H., and Brunetti O.,** Q fever studies. XVI. Some aspects of experimental infection induced in sheep by the intratracheal route of inoculation, *Am. J. Hyg.*, 57, 170, 1953.
16. **Abinati, F. R., Welsh, H. H., Lenette, E. H., and Brunetti, O.,** Q fever studies. XVII. Presence of "Coxiella burnetti" in birth fluids of naturally infected sheep, *Am. J. Hyg.*, 58, 385, 1953.
17. **Combiesco, D.,** Recherches experimentales sur le mecanisme de transmission de l'infection dans le typhus pulmonaire ("fievre Q"), *Sem. Hop.*, 32, 267, 1956.

18. **Stoenner, H. G.**, Experimental Q fever in cattle. Epizootiologic aspects, *J. Am. Vet. Med. Assoc.*, 118, 170, 1951.
19. **Stoker, M. G. P. and Thompson, J. F.**, An explosive outbreak of Q fever, *Lancet*, 1, 137, 1953.
20. **Palmer, N. C., Kierstead, M. M., Key, D. W., Williams, J. C., Peacock, M. G., and Vellend, H.**, Placentitis and abortion in sheep and goats in Ontario caused by *Coxiella burnetii*, *Can. Vet. J.*, 24, 60, 1983.
21. **Waldham, D. G., Stoenner, H. G., Simmons, R. E., and Thomas, L. A.**, Abortion associated with *Coxiella burnetii* infection in dairy goats, *J. Am. Vet. Med. Assoc.*, 173, 1580 , 1978.
22. **Crowther, R. W. and Spicer, A. J.**, Abortion in sheep and goats in Cyprus caused by Coxiella burneti (sic), *Vet. Rec.*, 99, 29, 1976.
23. **Prasad, B. N., Chandiramani, N. K., and Wagle, A.**, Isolation of *Coxiella burnetii* from human sources, *Int. J. Zoon.*, 13, 112, 1986.
24. **Ellis, M. E., Smith, C. C., and Moffat, M. A.**, Chronic or fatal Q-fever infection: a review of 16 patients seen in Northeast Scotland (1967-80), *Q. J. Med.*, 205, 54, 1983.
25. **Wagstaff, D. J., Janney, J. H., Crawford, K. L., Dimijian, G. G., and Joseph, M. J.**, Q fever studies in Maryland, *Public Health Rep.*, 80, 1095, 1965.
26. **Mincev, A., Nikolov, Z., Gancev, S., Mateeva, E., Atanasov, D., and Stavreva, D.**, Q-fever in pregnant women and its influence on the fetus and newborn, *Akush. Ginekog*, 22, 21, 1983.
27. **Ormsbee, R. A.**, Rockettsiae, in *Manual of Clinical Microbiology*, 3rd ed., Lennett, E. H., Balows, A., Hausler, W. J., Jr., Truant, J. B., Eds, American Society for Microbiology, Washington, D. C., 1980, 922.
28. **Colton, T.**, *Statistics in Medicine*, Little, Brown, Boston, 1974, 120.
29. **Giroud, A., Giroud, P., Martinet, M., and Deluchat, C.**, Inapparent maternal infection by *Coxiella burnetii* and fetal repercussions, *Teratology*, 1, 257, 1968.
30. **Syrucek, L., Sobeslavsky, O., and Gutvirth, I.**, Isolation of *Coxiella burnetii* from human placentas, *J. Hyg. Epidemiol. Microbiol. Immunol.*, 2, 29, 1958.
31. **Fiset, P., Wisseman, C. L., Jr., and Batawi, Y. E.**, Immunologic evidence of human fetal infection with *Coxiella burnetii*, *Am. J. Epidemiol.*, 101, 65, 1975.
32. **Ormsbee, R. A.**, Rickettsiae, in *Manual of Clinical Microbiology*, 3rd ed., Lennett, E. H., Balows, A., Hausler, W. J., Jr., Truant, J. B., Eds., American Society for Microbiology, Washington, D. C., 1980, 851.
33. **Behrman, R. E., Vaughan, V. C., III, Eds.**, *Nelson Textbook of Pediatrics*, 13th ed., W. B. Saunders, Philadelphia, 1987, 1163.
34. **Colton, T.**, *Statistics in Medicine*, Little, Brown, Boston, 1974, 165.
35. **Godfey, K.**, Comparing the means of several groups, *N. Engl. J. Med.*, 313, 1450, 1985.
36. **Bell, E. J., Parker, R. R., and Stoenner, H. G.**, Q fever-experimental Q fever in cattle, *Am. J. Public Health*, 39, 478, 1949.
37. **Heubner, R. J., Jellison, W. L., Beck, M. D., Parker, R. R., and Shepard, C. C.**, Q fever studies in Southern California. I. Recovery of *Rickettsia burnetii* from raw milk, *Public Health Rep.*, 63, 214, 1948.
38. **Evans, A. D.**, Q fever in South Wales, *Mon. Bull. Minist. Health Lab. Serv.*, 15, 215, 1956.
39. **Slavin, G.**, "Q" fever: the domestic animal as a source of infection for man, *Vet. Rec.*, 64, 743, 1952.
40. **Schaal, V. E. and Kleinsorgen, A.**, Zur oralen ubertragen des Q-fieber-erregers durch infizierte milch, *Dtsch. Tierarztl. Wochenschr.*, 80, 393, 1973.
41. **Kumar, A., Yadav, M. P., and Kakkar, S.**, Human milk as a source of Q-fever infection in breast-fed babies, *Indian J. Med. Res.*, 73, 510, 1981.
42. **Prasad, B. N., Chandiramani, N. K., and Wagle, A.**, Isolation of *Coxiellae burnetii* from human sources, *Int. J. Zoon.*, 13, 1986.
43. **Richardus, J. H., Dumas, A. A. M., Huisman, J., and Schaap, G. J. P.**, Q fever in infancy: a review of 18 cases, *Pediatr. Infect. Dis.*, 4, 369, 1985.
44. **Feinstein, A. R. and Horwitz, R. I.**, Double standards, scientific methods, and epidemiologic research, *N. Engl. J. Med.*, 307, 1611, 1982.
45. **Reichmann, N., Raz, R., Keysary, A., Goldwasser, R., and Faltau, E.**, Chronic Q fever and severe thrombocytopenia in a pregnant woman, *Am. J. Med.*, 85, 253, 1988.
46. **Peacock, M. G., Philip, R. N., Williams, J. C., and Faulkner, R. S.**, Serologic evaluation of Q fever in humans: enhanced phase I titers of immunoglobins G and A are diagnostic for Q fever endocarditis, *Infect. Immun.*, 41, 1089, 1983.
47. **McGivern, D., White, R., Paul, I. D., Caul, E. O., Roome, A. P. C. H., and Westmoreland, D.**, Concomitant zoonotic infections with ovine *Chlamydia* and "Q" fever in pregnancy: clinical features, diagnosis, management and public health implications. Case report, *Br. J. Obstet. Gynecol.*, 95, 294, 1988.
48. **Quinn, P. A. and Petric, M.**, Anticomplementary activity in serum of women with a history of recurrent pregnancy loss, *Am. J. Obstet. Gynecol.*, 158, 368, 1988.
49. **Gaburro, D. and Del Campo, A.**, Considerazioni epidemiologiche e cliniche su un'infezione da Coxiella Burneti in tre gemelle immature, *G. Mal. Infett. Parassit.*, 8, 384, 1956.
50. **Goodman Gilman, A., Goodman, L. S., Gilman, A.**, *Goodman and Gilman's The Pharmacological Basis of Therapeutics*, 6th ed, MacMillan, New York, 1980, 1187.
51. Cited in **Babudieri, B.**, *Adv. Vet. Sci.*, 5, 81, 1959.

Chapter 11

ANTIBIOTIC SUSCEPTIBILITY OF *COXIELLA BURNETII*

Michael R. Yeaman and Oswald G. Baca

TABLE OF CONTENTS

I. INTRODUCTION

While antibiotic management of acute Q fever is generally successful in nonimmunocompromised patients, treatment of chronic Q fever endocarditis is poor at best. Antibiotic treatment of chronic Q fever has not been consistently effective and has been directed toward elimination of the microorganisms from cardiac valve leaflets. Despite prolonged antibiotic therapy, relapses occur. Valve replacement may become necessary but is no guarantee of cure since they, too, may become infected. The source of the recolonizing rickettsiae is unknown but may include extracardiac sites such as the liver. The present state of knowledge of antibiotic management of acute and chronic Q fever will be discussed. The relatively recent discovery of genetically distinct *Coxiella burnetii* isolates associated with different disease manifestations, including chronic and acute disease, implies that they may also be differentially susceptible to antibiotics; recent results obtained with the use of persistently infected cell lines substantiate this hypothesis.

II. HISTORICAL PERSPECTIVE

In 1937, Derrick reported an outbreak of a fever of unknown etiology that had occurred 2 years earlier among a group of slaughterhouse workers in Queensland, Australia.[1] Later termed "Q" fever, nine clinical cases were described in terms of incubation period, onset, course of illness, fever, pulse rate, headache, shivers and sweats, rash, jaundice, conjunctival congestion, splenic involvement, blood involvement, other signs and symptoms, and convalescence. Derrick described the use of Congo red, calcium gluconate, and sodium thiosulfate as means of treatment of the mysterious disease. In retrospect, the patients were clearly suffering from acute infection caused by the organism now called *C. burnetii*, and the treatments employed were used to alleviate symptoms rather than to eliminate the etiologic agent. None of the cases reported ended in fatality. In the half century since that time, great strides have been made in the treatment of bacterial infections — acute Q fever itself is generally well managed in noncompromised patients with a variety of antibiotics. However, despite these strides, antibiotic management of chronic Q fever or Q fever endocarditis remains elusive.

III. ACUTE Q FEVER: ANTIBIOTIC TREATMENT

The majority of reports indicate that in most cases Q fever endocarditis usually results in death within 3 years of onset despite both chemotherapeutic and surgical efforts.[2,3] In contrast, the majority of cases involving acute Q fever — a respiratory infection accompanied by severe headache, fever, and malaise[4] — are successfully treated with a variety of antibiotics including lincomycin,[2] erythromycin,[5] cotrimoxazole (sulfamethoxazole and trimethoprim)[6,7] chloramphenicol,[8] and several tetracyclines.[2,3,8-10] It seems apparent that in acute cases diagnosed promptly, *C. burnetii* is well managed through antibiotic therapy, relying principally upon the tetracycline family of drugs.

IV. CHRONIC Q FEVER: ANTIBIOTIC TREATMENT

The poor management of Q fever endocarditis seems to be, at least in part, due to a basic failure of antibiotic regimens employed in the treatment of this disease. This does not appear to be physician error, but rather to a much more basic problem: the absence of antibiotics which effectively control or lead to the permanent inhibition of *C. burnetii*. The basic question concerning the treatment of acute vs. chronic Q fever has yet to be answered: Why is *C. burnetii*, the etiologic agent of both acute and chronic Q fever syndromes, sensitive to antibiotics in the acute disease but highly resistant in the chronic disease? One possibility, as suggested by Samuel et al.,[11] is that different isolates of *C. burnetii* are responsible for different Q fever syndromes:

acute or chronic. Recent evidence indicates this to be the case. Chronic isolates of *C. burnetii* exhibit a dramatic antibiotic resistance when compared to acute isolates *in vitro* (see below).[54] Prior to the discovery of such *C. burnetii* isolates associated with specific disease, explanations for the lack of effective antibiotic management of chronic Q fever included strict patient predisposition in the form of defective cell-mediated immunity[12-16] and the possibility of persisting reservoir sites (such as liver, spleen, or kidney) within chronically infected patients.[2] It is likely that chronic Q fever results from a more complex combination of patient predisposition along with infection by specific *C. burnetii* isolates.

Controlled studies of the antibiotic management of Q fever endocarditis have not been performed, and, therefore, the most effective means of its chemotherapeutic management (antibiotic[s] of choice, duration, combination, etc.) are not clear. However, there are two themes which seem to emerge in cases which result in apparent remission of Q fever endocarditis:

1. Abnormal valves are usually infected. The infection results in accelerated valvular deterioration necessitating replacement when hemodynamic impairment occurs, and antibiotics do not reverse this damage.
2. Antibiotics used in the treatment of Q fever endocarditis have thus far proved to be rickettsiostatic and must be administered indefinitely to help prevent relapse. Prolonged use of some antibiotics is contraindicated, however (inhibition of hematopoiesis by chloramphenicol and rifampin's anticoagulation interference properties), thereby reducing the potential array of drugs available for treatment of chronic Q fever. The fact remains that even using these regimens, Q fever endocarditis is fatal in the majority of cases due to primary endocarditis or secondary relapse.

Table 1 is a summary of antibiotics and their outcomes used in the treatment of chronic Q fever.

V. LIMITED SUCCESS IN THE TREATMENT OF CHRONIC Q FEVER

In spite of the poor prognosis for a person with confirmed Q fever endocarditis, successful management has been achieved in a limited number of cases. Walters described a case in which tetracycline and a combination of novobiocin and sulphamethizole (albamycin) were effective in controlling a verified case of Q fever endocarditis.[17] Other patients were afforded remission of fever, fatigue, and high antibody titers to *C. burnetii* antigens through the use of a combination of sulfamethoxazole and trimethoprim (this combination is known as cotrimoxazole).[18,19] Tetracycline and doxycycline were successful in managing two cases of childhood Q fever endocarditis in conjunction with corrective surgery to repair damaged heart valve tissue.[20] Tobin et al.[21] found that a combination of tetracycline and cotrimoxazole achieved remission of Q fever endocarditis while other antibiotics (including both tetracycline and cotrimoxazole alone) failed to prevent death from valvular incompetence and congestive heart failure. These clinicians found an equal prevalence of both mitral and tricuspid valve involvement.

Tetracyclines in combination with other antibiotics have been reported to be successful in controlling Q fever endocarditis in some other cases as well.[6,19,22,23] In clinical experience from the south of France, Raoult and others observed that the combination of rifampin and either tetracycline or doxycycline was more effective than other antibiotic regimens used in the treatment of 15 Q fever endocarditis patients.[24] Clindamycin, gentamicin, and lincomycin have apparently been successful in combination with tetracycline, doxycycline, or oxytetracycline in a number of other cases.[2,25]

TABLE 1
Antibiotics Used in the Treatment of Chronic Q Fever[a]

Antibiotic(s) Outcome	Ref.
Remission	
Albamycin[b]	17
Cotrimoxazole[c]	18,19
Doxycycline	20
Minocycline	24
Tetracycline	20
Tetracycline/Chloramphenicol	37
Tetracycline/Clindamycin	2, 25
Tetracycline/Cotrimoxazole	21
Tetracycline/Gentamicin	25
Tetracycline/Lincomycin	2
Tetracycline/Rifampin	23, 24
Relapse	
Chloramphenicol	2
Chloramphenicol/Methicillin	23
Doxycycline/Cotrimoxazole	23
Tetracycline	32—34
Tetracycline/Chloramphenicol	37
Tetracycline/Cotrimoxazole	23
Tetracycline/Doxycycline	25
Tetracycline/Lincomycin	25
Failure	
Beta lactam	24
Beta lactam/Aminoglycoside (7)[d]	24
Chloramphenicol	37
Cotrimoxazole	30
Cotrimoxazole/Erythromycin	30
Cotrimoxazole/Rifampin	30
Erythromycin	24
Penicillin	29
Penicillin/Gentamicin	40
Penicillin/Streptomycin	41, 43
Sulfonamides (23)	39
Tetracycline (3)	25
	29
	37
	43
(3)	21
	9
Tetracycline/Erythromycin	24
Tetracycline/Lincomycin	21

[a] Incomplete, but comprehensive summary of cases.
[b] A combination of novobiocin and sulphamethizole.
[c] A combination of sulfamethoxazole and trimethoprim.
[d] Number of cases using this treatment (others 1 case).

VI. RELAPSE OF CHRONIC Q FEVER FOLLOWING ANTIBIOTIC CESSATION

Relapse is not an uncommon occurrence following cessation of antibiotic treatment of chronic Q fever endocarditis patients. Relapse is probably due to reinfection with organisms from an extracardiac site such as the liver or reactivation of suppressed organisms residing in the heart tissue itself.[2] Reinfection from external sources, however, cannot be excluded.

Relapse may be the result of the bacteriostatic nature of the antibiotics used and the *C. burnetii* isolate involved. The accepted drugs of choice against *C. burnetii*, the tetracyclines (especially doxycycline, chlortetracycline, and oxytetracycline),[10] are bacteriostatic. Such drug selection has been based on early experimental and clinical evidence provided by Wong and Cox,[26] Ormsbee and Pickens,[27] and others.[2,8,9] Cotrimoxazole, a combination of the antifolate drugs sulfamethoxazole and trimethoprim, although frequently used in the treatment of Q fever endocarditis, remains a controversial combination for use in the treatment of this disease. *C. burnetii* was discovered to have folate-dependent metabolism by Mattheis et al. in 1963.[28] The antimetabolites sulfamethoxazole and trimethoprim selectively inhibit folate metabolism in prokaryotes and, therefore, would theoretically be effective in the treatment of *C. burnetii*. With this in mind, physicians began experimenting with these antimetabolite drugs in the treatment of Q fever endocarditis. Some workers claimed that cotrimoxazole was bactericidal against *C. burnetii*,[21,22] while others speculated that this drug combination stimulated growth of the rickettsia and, therefore, should be contraindicated in the treatment of chronic Q fever.[24,29] Generally, cotrimoxazole is accepted as a bacteriostatic drug combination against sensitive organisms. Both remission[18] and complete failure[30] have resulted from cotrimoxazole treatment of Q fever endocarditis, leaving the issue open to debate. Erythromycin, lincomycin, and chloramphenicol have been used with some success in the attempted treatment of chronic Q fever.[2,23,31] Mixed results have been obtained with rifampin.[24,31] The common characteristic of the overwhelming majority of antibiotics used in the treatment of persistent *C. burnetii* infection, with the possible exception of rifampin, is that they are all bacteriostatic. Along with other complicating factors, this bacteriostatic property may be a major reason why the rickettsiae are not eliminated, and, thus, treatment must continue indefinitely to prevent relapse.

An examination of reported cases shows that relapse of Q fever endocarditis followed within as many as 24 months after cessation of antibiotic therapy. McIver reported that 13 weeks of antibiotic treatment were insufficient since relapse occurred 11 months after terminating antibiotic treatment.[32] Pedoe observed that 14 weeks of tetracycline therapy did not prevent apparent relapse some 16 months subsequent to antibiotic cessation.[33] One patient suffered Q fever endocarditis relapse ultimately leading to death 24 months after receiving three 5-week courses and a continuous 4-month therapeutic regimen of tetracycline.[34] Chloramphenicol cessation has also resulted in subsequent relapse tendencies.[2,31] Still another patient's aortic valve was culture-positive for *C. burnetii*, even after 4 years of antibiotic therapy.[2]

Because of their toxic properties, some antibiotics potentially available for use in the treatment of Q fever endocarditis or chronic Q fever cannot be used on a long-term basis. Chloramphenicol (hematopoietic inhibition and Gray's syndrome in children) and streptomycin (oto- and nephrotoxicities) are primary examples of antibiotics generally contraindicated in the long-term treatment of chronic Q fever. Rifampin may be contraindicated if anticoagulants are being administered to patients with mechanical valves.[24]

A fall in *C. burnetii* complement fixation (CF) titers, reduced fever, and general patient improvement often leads physicians to reduce or terminate prematurely antibiotic treatment of chronic Q fever: relapse (or reinfection) is a relatively common result in such cases. Almost 10% of patients suffering from acute Q fever had to undergo long-term doxycycline and tetracycline therapy due to apparent relapse following 4 weeks of treatment.[35] Although CF titers have been used as an indication of infection severity,[18,19,21,22,36] it appears that changes in titers should not be the sole criterion for evaluating antibiotic success, as large increases or decreases in titer have been shown to be of little significance.[2] Because of the potential for relapse or reinfection, successful control of chronic Q fever may more accurately be termed "remission" — no evidence of true patient cure exists in the absence of long-term antibiotic therapy.

VII. FAILURE OF ANTIBIOTICS IN CONTROLLING PRIMARY CHRONIC Q FEVER

As noted, Q fever endocarditis is much more difficult to control with antibiotics than acute Q fever; the majority of cases of chronic Q fever ultimately result in patient death due to primary endocarditis, subsequent relapse or reinfection, or heart failure resulting from *C. burnetii* persistent infection. Eight weeks of tetracycline therapy failed to resolve Q fever endocarditis following 6 weeks of penicillin therapy in a 48-year-old male.[29] In a case described by Kristinsson and Bentall,[37] chloramphenicol and tetracycline administered on an intermittent basis (as suggested by Nicolau)[38] were unsuccessful in controlling Q fever endocarditis. A male patient with elevated *C. burnetii* phase I and phase II antibody levels died from congestive heart failure 23 d after initiation of tetracycline therapy.[19]

Three patients died of hemodynamic impairment caused by *C. burnetii* infection after receiving daily 2-g doses of tetracycline.[25] One of the three had been given doxycycline in place of tetracycline with no apparent benefit. Tobin and co-workers reported the deaths of four Q fever endocarditis patients treated with either tetracycline alone or tetracycline in combination with lincomycin.[21] Sulfonamides were used with no apparent success in 23 cases of suspected Q fever endocarditis, and, alternatively, *para*-aminobenzoic acid was ineffective for three other patients.[39] In a patient that could not be treated with tetracycline due to hepatic and renal failure, cotrimoxazole (initially alone) and lincomycin, subsequently changed to cotrimoxazole and rifampin, failed to suppress *C. burnetii* infection.[30] The patient died after 16 months of antibiotic therapy which ultimately consisted of cotrimoxazole, rifampin, and erythromycin. Notable in this case were the severe relapses following what appeared to be favorable initial responses with each antibiotic combination used. Penicillin in combination with gentamicin failed to resolve subacute bacterial endocarditis involving *C. burnetii* in an 86-year-old male.[40] Penicillin G and streptomycin likewise produced no beneficial effect in the management of Q fever endocarditis cases occurring in the U.S.[41] Failure of a variety of other antibiotic regimens is common in the attempted treatment of chronic Q fever or Q fever endocarditis.[24,42,43]

VIII. SURGERY IN CONJUNCTION WITH ANTIBIOTICS

Treatment of Q fever endocarditis often requires valve replacement to correct acute hemodynamic impairment, especially aortic or mitral valvular incompetence resulting in regurgitation, caused by chronic *C. burnetii* infection of heart valve leaflet tissue.[10] Such surgical procedures are generally performed under antibiotic "cover" (tetracyclines are usually administered prior to and concurrent with surgery). Although surgery may be necessary to resolve hemodynamic complications caused by Q fever endocarditis, it by no means guarantees a cure since prosthetic valves often become involved in subsequent relapse.[2,33,36,43] The source of the recolonizing rickettsiae is unknown but may include extracardiac sites such as the liver, spleen, or kidney. Although difficult to document, reinfection from environmental sources probably occurs and should not be excluded.

Until recently it was thought that patients with a history of prosthetic valve or other heart tissue compromise (rheumatic fever, etc.) were much more likely to develop Q fever endocarditis following acute Q fever than other patients who had no such heart conditions. However, analysis of more than 800 confirmed cases of Q fever in the U.K. showed that of the approximately 3 to 11% of these cases which resulted in chronic Q fever endocarditis, only one third of the patients were noted to have had any type of underlying heart lesion or impairment.[42] In contrast, a more recent review of 28 cases suggested that patient predisposition (underlying heart lesion) may have played a more important role.[3] It is clear that males are much more likely to contract Q fever and chronic Q fever than females, but this is probably occupation related.

In any case, opinions vary concerning the value of surgery in the treatment of Q fever endocarditis. Kristinsson and Bentall[37] support the view that surgery should be reserved for repair of cardiac damage caused by *C. burnetii*; others suggest that surgery may actually be an important means of removing infected tissue which would only serve as a reservoir source for subsequent relapse.[43] It is generally agreed that antibiotic therapy should be continued on a long-term basis even following apparently successful surgery.[3, 33]

IX. MODEL SYSTEMS FOR DETERMINING *C. BURNETII* ANTIBIOTIC SUSCEPTIBILITY

Three experimental model systems have been used to test the efficacies of antibiotics vs. *C. burnetii* isolates: guinea pigs, chick embryos, and, recently, cell cultures. One of the first experimental tests of an antibiotic in the control of the Q fever agent was reported by Huebner et al. in 1948,[44] when it was demonstrated that streptomycin at relatively high concentrations reduced mortality of chick embryos and guinea pigs infected with *C. burnetii* strains isolated from patients with acute Q fever. Later, Ormsbee and colleagues[27,45] used chick embryos to show for the first time that the Nine Mile and other *C. burnetii* isolates were sensitive to terramycin (oxytetracycline); they also reported that these rickettsiae were insensitive to chloramphenicol, erythromycin, and thiocymetin. More recently, Spicer et al.[46] demonstrated that the most effective antibiotics tested for protecting chick embryos from four isolates of the Q fever agent, including the Nine Mile strain, were rifampin, trimethoprim, doxycycline, and oxytetracycline; clindamycin, erythromycin, viomycin, cycloserine, and cephalothin were ineffective. A fifth isolate ("Cyprus", from an aborting sheep) was somewhat resistant to the tetracyclines. It was concluded that the effective antibiotics were rickettsiostatic because subculturing of the treated parasites resulted in their reappearance. While that may be the case, it must be kept in mind that *C. burnetii* can be found outside cells (see Reference 47) and that exposure to antibiotics might have no effect on such presumably nonmetabolizing extracellular parasites; subsequently, these extracellular *C. burnetii* may remain fully infectious.

The most recent studies testing the effect of antibiotics on *C. burnetii* have been performed in Baca's laboratory using persistently infected L929 cells.[47] *C. burnetii* isolates from animal and human sources are capable of persistently infecting cultured cell lines, including L929 mouse fibroblast cells and several macrophage cell lines such as J774 and P388D1.[48-51] These persistently infected cell populations have been maintained in continuous culture for months: L929 cells have been infected with the Nine Mile, Priscilla, and "S" isolates for over 1400, 750, and 250 d respectively. What is most interesting about these infected cells is that they do not require the addition of normal uninfected cells to maintain viable populations: the cells divide despite enormous parasite loads; the population doubling times and cell cycle progression are similar to those of normal uninfected cells.[52] This *in vitro* system has been useful in examining host-parasite interactions including the conclusive finding that the vacuole in which *C. burnetii* is sequestered is the result of fusion of lysosomes with phagocytic vacuoles and that its pH is in the acidic range.[53] This system has also proved to be a more convenient system for examining antibiotic efficacy than other experimental systems such as guinea pigs and embryonated eggs. The time required for determining susceptibility is considerably less (as early as 24 to 48 h after tests are initiated) than with other systems which take more than 1 week. Furthermore, antibiotic concentrations are more precisely controlled, whereas they can only be estimated with embryonated eggs and experimental animals. With the aid of these persistently and recently infected L929 cells, many antibiotics have been tested, including the new generation substituted quinolones. A summary of these results is presented in Table 2.

The most effective antibiotics for controlling the Nine Mile isolate were the substituted quinolone compounds and rifampin.[47] The most effective quinolone was ofloxacin followed by pefloxacin, difloxacin, ciprofloxacin, norfloxacin, and oxolinic acid.[47,55] These quinolones are very likely rickettsicidal because within 48 h after exposure to microgram amounts the number

TABLE 2
Antibiotics Tested against *Coxiella burnetii*
in Experimental Persistent Infection[a]

Antibiotic	General target	Percent reduction of infected cells
Penicillin G	Cell wall	3
Polymyxin B	Cell membrane	0
Sulfamethoxazole	Folic acid metabolism	0
Trimethoprim	39	
Streptomycin	Protein synthesis	13
Gentamicin		0
Tetracycline		0
Doxycycline		14
Chloramphenicol		38
Erythromycin		3
Rifampin	RNA synthesis	93
Novobiocin	DNA gyrase function	0
Nalidixic Acid		0
Oxolinic Acid		90
Difloxacin		97
Ciprofloxacin		96
Norfloxacin		95
Ofloxacin		98

[a] Results from tests performed on L929 mouse fibroblasts persistently infected with Nine Mile *C. burnetii* and treated for 10 d *in vitro* (see Reference 47 for experimental procedures).

of intracellular parasites declined markedly and at a rate that exceeded the generation time of the host cells (i.e., not simply a case of diluting out inhibited rickettsiae). After 10 d of continuous treatment, the number of intracellular Nine Mile *C. burnetii* was close to, if not, zero. The antibiotic concentrations used were within the physiological levels that accumulate in various tissues and fluids. Some of the other antibiotics tested which were somewhat effective in controlling the Nine Mile agent were chloramphenicol, doxycycline, and trimethoprim. With the exception of the substituted quinolones which were previously unavailable, the pattern of susceptibility observed is similar to that reported by others[27,44-46] using embryonated eggs; the effective range of antibiotic concentrations was comparable.

The Priscilla isolate, implicated in chronic disease in animals, appears to resist, to a high degree, rifampin and, compared to the Nine Mile isolate, is significantly less sensitive to the quinolones pefloxacin, ciprofloxacin, and difloxacin.[55] However, it is fairly sensitive to ofloxacin; at the highest physiologically relevant concentration tested (5 μg/ml), the number of persistently infected cells was significantly reduced after 10 d of treatment. Norfloxacin was almost as effective as ofloxacin in eliminating the Priscilla isolate from infected cells. Interestingly, norfloxacin was not significantly effective against the Nine Mile isolate. Rifampin and ciprofloxacin have been tested for their combined, possibly synergistic, effect on the Priscilla and Nine Mile isolates. The Priscilla isolate, highly resistant to either antibiotic alone, was sensitive to the rifampin/ciprofloxacin combination (0.1 + 1.0 μg/ml, respectively) resulting in almost total elimination of the parasite from within infected cells.[55]

The combined experimental data clearly support the contention that chronic isolates (e.g., Priscilla) may be more resistant to some antibiotics than isolates implicated only in acute disease (e.g., Nine Mile). These results are also in accord with the fact that chronic Q fever in humans is generally not successfully managed with antibiotics, and they point out the possible usefulness of combination antibiotic therapy which includes the substituted quinolones.

X. SUMMARY AND CONCLUSIONS

Although Q fever is at present considered a rare disease, and chronic Q fever even more so, physicians are becoming aware of the ubiquity of the rickettsial organism *C. burnetii* and are taking steps to exclude the possibility of Q fever in patients exhibiting clinical symptoms of atypical endocarditis (endocarditis which does not respond to penicillin and streptomycin). The possibility of *C. burnetii* infection should be considered in any case of blood culture-negative endocarditis which does not respond to traditional treatment so as to allow prompt antibiotic therapy in those cases which prove positive for Q fever endocarditis.

The following is a summary of observations and conclusions regarding the antibiotic susceptibility and treatment of Q fever and chronic Q fever:

1. The clinical data clearly show that antibiotic and surgical treatment of patients afflicted with chronic Q fever endocarditis, in general, has not been consistently effective.
2. With prompt diagnosis, acute Q fever is generally successfully managed using antibiotics, including the tetracyclines, erythromycin, and possibly lincomycin. The effectiveness of cotrimoxazole is still not clear. Acute Q fever patients who have heart valve prostheses or other heart tissue compromise should be carefully observed for the possibility of subsequent chronic Q fever endocarditis.
3. Early detection of chronic infection by *C. burnetii* is critical and may result in more effective management of the chronic syndrome.
4. Chronic Q fever endocarditis is generally unsuccessfully managed with antibiotic therapy and frequently requires concurrent surgical treatment.
5. Patients suffering from chronic Q fever or Q fever endocarditis should probably be continued on antibiotic therapy indefinitely as there is no evidence of complete "cure". These drugs may serve as a prophylactic measure against future episodes of chronic Q fever or Q fever endocarditis.
6. Although the tetracyclines are, at present, the most effective antibiotics in the treatment of chronic Q fever and Q fever endocarditis, clearly, a great deal of research is required to find and develop more effective drugs.

The discovery of genetically distinct *C. burnetii* isolates implicated in distinct disease syndromes coupled with *in vitro* observations that these isolates exhibit differential antibiotic susceptibilities begins to explain the problem of treatment. Use of *in vitro* model systems is beginning to facilitate the discovery of potentially effective antibiotics (such as the quinolones) and the underlying basis for the general antibiotic resistance of *C. burnetii* isolates implicated in chronic disease.

ACKNOWLEDGMENT

The authors were supported, in part, by Miles Pharmaceutical Division, West Haven, CT.

REFERENCES

1. **Derrick, E. H.**, Q fever, a new entity: clinical features, diagnosis, and laboratory investigations, *Med. J. Aust..*, 2, 281, 1937.
2. **Turck, W. P. G., Howitt, G., Turnberg, L. A., Fox, H., Longson, M., Matthews, M. B., and Das Gupta, R.,** Chronic Q fever, *Q. J. Med.*, 45, 193, 1976.

3. **Sawyer, L. A., Fishbein, D, B., and McDade, J. E.**, Q fever: current concepts, *Rev. Infect. Dis.*, 9, 935, 1987.

4. **Baca, O. G. and Paretsky, D.**, Q fever and *Coxiella burnettii:* a model for host parasite interactions. *Microbiol. Rev.*, 47, 127, 1983.

5. **S. D'Angelo, J. J., and Hetherington, R.**, Q fever treated with erythromycin, *Br. Med. J.*, 2, 305, 1979.

6. **Haldane, E. V., Marrie, T. J., Faulkner, R. S., Lee, S. H. S., Cooper, J. H., Macpherson, D. D., and Montague, T. J.**, Endocarditis due to Q fever in Nova Scotia: experience with five patients in 1981—1982, *J. Infect. Dis.*, 148, 978, 1983.

7. **Seggev, J. S., Levin, S., and Schey, G.**, Unusual radiological manifestations of Q fever, *Eur. J. Respir. Dis.*, 69, 120, 1986.

8. **Clark, W. H. and Lennette, E. H.** Treatment of Q fever with antibiotics, *Ann. N.Y. Acad. Sci.*, 55, 1004, 1952.

9. **Tiggert, W. D. and Benenson, A. S.**, Studies on Q fever in man, *Trans. Assoc. Am. Physicians*, 69, 98. 1956.

10. **Oakley, C.**, Use of antibiotics: endocarditis, *Br. Med. J.*, 2, 489, 1978.

11. **Samuel, J. E., Frazier, M. E., and Mallavia, L. P.**, Correlation of plasmid type and disease caused by *Coxiella burnetii*, *Infect. Immun.*, 49, 775, 1985.

12. **Hinrichs, D. J. and Jerrells, T. R.**, In vitro evaluation of immunity to *Coxiella burnetii*, *J. Immunol.*, 117, 996, 1976.

13. **Koster, F. T., Williams, J. C., and Goodwin, J. S.**, Cellular immunity in Q fever: specific lymphocyte unresponsiveness in Q fever endocarditis, *J. Infect. Dis.*, 152, 1283, 1985.

14. **Jerrells, T. R., Mallavia, L. P., and Hinrichs, D. J.**, Detection of long-term cellular immunity to *Coxiella burnetii* by lymphocyte transformation, *Infect. Immun.*, 11, 280, 1975.

15. **Kishimoto, R. A., Johnson, J. W., Kenyon, R. H., Ascher, M. S., Larson, E. W., and Pedersen, C. E., Jr.**, Cell–mediated immune response of guinea pigs to an inactivated phase I *Coxiella burnetii* vaccine, *Infect. Immun.*, 19, 194, 1978.

16. **Kishimoto, R. A., Rozmiarek, H., and Larson, E. W.**, Experimental Q fever infection in congenitally athymic nude mice, *Infect. Immun.*, 22, 69, 1978.

17. **Walters, J.**, Rickettsial endocarditis, *Br. Med. J.*, 4, 770, 1968.

18. **Freeman, R. and Hodson, M. E.**, Q fever endocarditis treated with trimethoprim and sulfamethoxazole, *Br. Med. J.*, 1, 419, 1972.

19. **Varma, M. P. S., Adgey, A. A. J., and Connolly, J. H.**, Chronic Q fever endocarditis, *Br. Heart J.*, 43, 695, 1980.

20. **Laufer, D., Lew, P. D., Oberhansli, I., Cox, J. N., and Longson, M.**, Chronic Q fever endocarditis with massive splenomegaly in childhood, *J. Pediatr.*, 108, 535, 1986.

21. **Tobin, M. J., Cahill, N., Gearty, G., Maurer, B., Blake, S., Daly, K., and Hone, R.**, Q fever endocarditis, *Am. J. Med.*, 72, 396, 1982.

22. **Pierce, M. A., Saag, M. S., Dismukes, W. E., and Cobbs, C. G.**, Case report: Q fever endocarditis, *Am. J. Med. Sci.*, 292, 104, 1986.

23. **Kimbrough, R. C., Ormsbee, R. A., Peacock, M., Rogers, W. R., Bennetts, R. W., Raaf, J., Krause, A., and Gardner, C.**, Q fever endocarditis in the United States, *Ann. Intern. Med.*, 91, 400, 1979.

24. **Raoult, D., Etienne, J., Massip, P., Iaocono, S., Prince, M. A., Beaurain, P., Benichou, S., Auvergnat, J. C., Mathieu, P., Bachet, Ph., and Serradimigni, A.**, Q fever in the south of France, *J. Infect. Dis.*, 155, 570, 1987.

25. **Ellis, M. E., Smith, C. C., and Moffat, M. A. J.**, Chronic or fatal Q fever infection: a review of 16 patients seen in northeast Scotland (1967-80), *Q. J. Med.*, 205, 54, 1983.

26. **Wong, J. C. and Cox, H. R.**, Action of aureomycin against rickettsial and viral infection, *Ann. N.Y. Acad. Sci.*, 51, 290, 1948.

27. **Ormsbee, R. A. and Pickens, E. G.**, A comparison by means of the complement fixation test of the relative potencies of chloramphenicol, aureomycin, and terramycin in experimental Q fever infections in embryonated eggs, *J. Immunol.*, 67, 437, 1951.

28. **Mattheis, M. S., Silverman, M., and Paretsky, D.**, Studies on the physiology of rickettsiae. IV. Folic acids of *Coxiella burnetii*, *J. Bacteriol.*, 85, 37, 1963.

29. **Ferguson, I. C., Craik, J. E., and Grist, N. R.**, Clinical, virological, and pathological findings in a fatal case of Q fever endocarditis, *J. Clin. Pathol.*, 15, 235, 1962.

30. **Subramanya, N. I., Wright, J. S., and Khan, M. A. R.**, Failure of rifampicin and co-trimoxazole in Q fever endocarditis, *Br. Med. J.*, 285, 343, 1982.

31. **Kimbrough, R. C., Ormsbee, R. A., and Peacock, M. G.**, Q fever endocarditis: a three and one-half year follow up, in *Rickettsiae and Rickettsial Diseases*, Burgdorfer, W. and Anacker, R. L., Eds., Academic Press, New York, 1981, 125.

32. **McIver, M.**, A case of Q fever endocarditis, *Med. J. Aust.*, 2, 379, 1962.

33. **Pedoe, T. H. D.**, Apparent recurrence of Q fever endocarditis following homograft replacement of aortic valve, *Br. Heart J.*, 32, 568, 1970.

34. **Meyers, D.**, Rickettsial endocarditis, *Br. Med. J.*, 4, 771, 1968.

35. **Spelman, D. W.**, Q fever: a study of 111 consecutive cases, *Med. J. Aust.*, 11, 547, 1982.
36. **Crook, B.**, Q fever endocarditis with homograft valve replacements, *Proc. R. Soc. Med.*, 65, 981. 1972.
37. **Kristinsson, A. and Bentall, H. H.**, Medical and surgical treatment of Q fever endocarditis, Lancet, 2, 693, 1967.
38. **Nicolau, P. S.**, L'etiologie rickettsienne ou pararickettsienne dans les affections cardio-vasculaires, *Bull. Soc. Pathol. P.*, 4, 690, 1963.
39. **Denlinger, R. B.,**, Clinical aspects of Q fever in southern California: a study of 80 hospitalized cases, *Ann. Intern. Med.*, 30, 510, 1949.
40. **Constantinidis, K. and Jenkins, J. P. R.**, Chronic Q fever endocarditis, *Practitioner*, 222, 533, 1979.
41. **Applefeld, M. M., Billingsley, L. M., Tucker, H. J., and Fiset, P.**, Q fever endocarditis — a case occurring in the United States, *Am. Heart J.*, 93, 669, 1977.
42. **Palmer, S. R. and Young, S. E. J.**, Q fever endocarditis in England and Wales, 1975-81, Lancet, 2, 1448, 1982.
43. **Morgans, C. M. and Cartwright, R. Y.**, Case of Q fever endocarditis at site of aortic valve prosthesis, *Br. Heart J.*, 31, 520, 1969.
44. **Huebner, R. J., Hottle, G. A., and Robinson, E. B.**, Action of streptomycin in experimental infection with Q fever, *Public Health Rep.*, 63, 357, 1948.
45. **Ormsbee, R. A., Parker, H., and Pickens, E. G.**, The comparative effectiveness of aureomycin, terramycin, chloramphenicol, erythromycin, and thiocymetin in suppressing experimental rickettsial infections in chick embryos, *J. Infect. Dis.*, 96, 162, 1955.
46. **Spicer, A. J., Peacock, M. G. and Williams, J. C.**, Effectiveness of several antibiotics in suppressing chick embryo lethality during experimental infections with *Coxiella burnetii*, in *Rickettsiae and Rickettsial Diseases*, Burgdorfer, W. and Anacker, R. L., Eds. Academic Press, New York, 1981, 375.
47. **Yeaman, M. R., Mitscher, L. A., and Baca, O. G.**, In vitro susceptibility of *Coxiella burnetii* to antibiotics, including several quinolones, *Antimicrob. Agents Chemother.*, 31, 1079, 1987.
48. **Burton, P. R., Stueckemann, J., Welsh, R. M., and Paretsky, D.**, Some ultrastructural effects of persistent infections by the rickettsia *Coxiella burnetii* in mouse L cells and green monkey kidney (Vero) cells, *Infect. Immun.*, 21, 556, 1978.
49. **Baca, O. G., Akporiaye, E. T., Aragon, A. S., Martinez, I. L., Robles, M. V., and Warner, N. L.**, Fate of phase I and phase II *Coxiella burnetii* in several macrophage-like tumor cell lines, *Infect. Immun.*, 33, 258, 1981.
50. **Baca, O. G., Aragon, A. S., Akporiaye, E. T., Martinez, I. L., and Warner, N. L.**, Interaction of *Coxiella burnetii* with macrophage-like tumor cell lines, in *Rickettsias and Rickettsial Diseases*, Burgdorfer, W. and Anacker, R. L., Eds., Academic Press, New York, 1981, 91.
51. **Roman, M. J., Coriz, P. D., and Baca, O. G.**, A proposed model to explain persistent infection of host cells with *Coxiella burnetii*, *J. Gen. Microbiol..*, 132, 1415, 1986.
52. **Baca, O. G., Scott, T. O., Akporiaye, E. T., DeBlassie, R., and Crissman, H. A.**, Cell cycle distribution patterns and generation times of L-929 fibroblast cells persistently infected with *Coxiella burnetii*, *Infect. Immun.*, 47, 366, 1985.
53. **Akporiaye, E. T., Rowatt, J. D., Aragon, A. S., and Baca, O. G.**, Lyosomal response of a murine macrophage-like cell line persistently infected with *Coxiella burnetii*, *Infect. Immun.*, 40, 1155, 1983.
54. **Yeaman, M. R. and Baca, O. G.**, manuscript in preparation.
55. **Yeaman, M. R. and Baca, O. G.**, unpublished data, 1988.

Chapter 12

PREVENTION OF *COXIELLA BURNETII* INFECTION: VACCINES AND GUIDELINES FOR THOSE AT RISK

Richard A. Ormsbee and Barrie P. Marmion

TABLE OF CONTENTS

I. BACKGROUND AND EARLY STUDIES OF VACCINE

As described in detail elsewhere in this text, acute Q fever may present not only as a debilitating fever, but also on occasion with pneumonia, hepatitis, meningoencephalitis, peripheral thrombosis, myopericarditis, bone marrow aplasia, or involvement of other body systems.[1-4] Response to tetracycline therapy may be slow.

Chronic sequelae include valvular or mitral endocarditis and chronic hepatitis.[5-11] Studies[12] currently in progress in Adelaide indicate that a relapsing, post-Q fever debility syndrome of uncertain mechanism, with features in common with the chronic fatigue syndrome, is a more frequent chronic sequel than endocarditis or hepatitis and, depending on the stringency of the diagnostic criteria, follows 20 to 40% of acute Q fever cases.

Q fever in its various forms is, therefore, not a trivial infection and is expensive. The cost and benefits of vaccine prophylaxis appear to be undeniable.

Research on Q fever in the early years (1940s) after its discovery was characterized by outbreaks of the disease in laboratory workers, sometimes at some distance from the area where the work was in progress. As a response to these problems, and given the ability to propagate the rickettsia in the chick embryo yolk sac to produce large numbers of the organism, formalin-inactivated, whole call vaccines were developed to protect laboratory workers.[13,14] These seem to have been mostly effective in preventing infection, although randomized, "blind" placebo-controlled trials of the vaccines were not undertaken at the time, perhaps because of the apparently high degree of protection that could be demonstrated in guinea pigs and the pressing need to protect the laboratory workers. Smadel and co-workers[14] prepared one of the better known early vaccines from the Henzerling strain of *Coxiella burnetii* and showed that it was highly immunogenic in man and guinea pigs. The formalin-killed, ether-extracted 10% suspension of yolk sac (Smadel vaccine) had a CF antigen titer around 1 in 8 and was probably a mixture of phase I and II coxiellae, although, of course, this was not known at the time. The overall experience with this type of vaccine suggested efficacy.[14-17]

The protective effects of Smadel-type vaccines in laboratory animals and in open trials in laboratory workers were, however, substantiated in infective challenge experiments in vaccinated volunteers. Thus, around 1959, Benenson[18] immunized 16 volunteers with three 1.0 ml doses of a Smadel vaccine containing eight complement fixation (CF) antigen units ($\sim$120 μg) of *C. burnetii* per milliliter of vaccine and made from the Henzerling strain in the 22 yolk sac pass. The volunteers were subsequently exposed to aerosols of live *C. burnetii* in a challenge dose of 150 guinea pig infective doses 50% (ID_{50}). (One to 10 guinea pig ID_{50} is sufficient to infect man.)[19] Fifteen of the 16 vaccinated subjects remained well, but 6 out of 8 unvaccinated controls developed Q fever.

The use of the Smadel and other early vaccines was accompanied by occasional severe reactions taken the form of an indolent, indurated mass at the vaccination site or the formation of a sterile abscess which pointed and discharged and sometimes formed a chronic sinus requiring excision before it healed.[20,21] The reactogenicity of the early Q fever vaccines was generally out of line with the experience of spotted fever and epidemic typhus vaccines, although the latter also produced severe reactions on occasion. For example, severe reactions were shown by recipients of Q fever vaccine in early trials in Russia.[22] On a number of occasions, adverse reactions followed a single dose of vaccine, particularly in persons living in areas of high endemicity. Similar episodes of hypersensitivity were observed in trials with Q fever vaccine in the U.S. and other countries. Thus, in 1958, vaccination of 94 employees of the Rocky Mountain Laboratory, Hamilton, MT. with a single dose of vaccine gave severe local reactions in 42 (45%) of recipients. A common perception[23] in those early years was that Q fever vaccine was effective but gave unpleasant side effects; probably this reputation was one factor in discouraging attempts to use the vaccine in circumscribed occupational groups such as abattoir workers, in which Q fever is endemic.

In recent discussions of the side effects of vaccine, a clear distinction has not always been made: on the one hand, between common, trivial reactions, e.g., erythema, transient induration, or edema, pain, and tenderness on palpation, at the vaccination site as observed with a number of vaccines; and on the other hand, the rarer, more serious effects such as a persistent mass, abscess, or fistulas. (It may be noted in passing that abscess formation is not confined to Q fever vaccine, but is sometimes provoked with alum-precipitated vaccines, e.g., diptheria toxoid. Hypersensitivity may be observed in persons repeatedly vaccinated with tentanus toxoid.)

Observations by Benenson in 1959[18] threw light on the pathogenesis of the severe, persistent reactions. He found that the incidence was related to frequent revaccination; 0.37 per 1000 persons vaccinated for the first time, progressing to 5.9 per 1000 vaccinees by the fifth and sixth booster injection, and up to 36 per 1000 vaccinees when inoculated for the ninth and tenth time. In those days, it was erroneously assumed that, as with many other killed bacterial vaccines, it was necessary to revaccinate in order to maintain immunity.

Benenson also showed that the incidence of the common, trivial reactions was closely correlated with the possession of Q fever CF antibody before inoculation.

Bell and his colleagues[20] had also noted an association between a severe persistent reaction and revaccination. From a detailed analysis of one subject with a recurrent, sterile abscess, these workers concluded that *C. burnetii* antigen persists at the inoculation site providing long-term antigenic stimulation and that the abscess is immunologically mediated as suggested by the high antibody content content in the exudate from the chronic granulomatous lesion. The chronic macrophage activation in the lesion stimulates fibrosis, and the draining track or fistulas may not collapse, but may require debridement before healing is achieved.

In the early 1960s, these observations were translated by Lackman et al.[24] into a prevaccination screening regimen to detect preexisting cellular immunity or hypersensitivity. A small dose of diluted Q fever vaccine was inoculated intradermally; sensitized subjects developed a localized indurated lesion, mostly with erythema, 5 to 7 d after the injection. Such reactors were excluded from vaccination, and adverse reactions with whole cell vaccines diminished substantially thereafter, an outcome perhaps assisted by concurrent refinements in preparation of the vaccines and identification of the protective immunogen.

After the initial phase (1945—1960) of antigenically uncharacterized, semipurified inactivated whole call vaccines to protect laboratory workers handling *C. burnetii*, vaccine development and usage evolved in several ways.

The new directions comprised (1) the identification of the phase I antigen of *C. burnetii* as the important protective immunogen for vaccine formulation; (2) the development of more effective methods to purify *C. burnetii* cells from the residual yolk sac protein; (3) exploration of fractions ("split vaccines") or solvent-extracted cells of *C. burnetii* as alternatives to inactivated whole cells; (4) a wider application of vaccines — whole cell or split vaccines — in occupationally exposed industrial populations outside the population of laboratory workers; and (5) characterization of the immune response to Q fever vaccine

These aspects are now described, and at the conclusion of the chapter, consideration is given to guidelines for vaccine use.

A. Q FEVER VACCINE AND ANTIGENIC COMPOSITION OF *C. BURNETII*

C. burnetii is unrelated antigenically to other rickettsias, and the rare reports of cross-reactions with generically unrelated organisms are of uncertain significance. There is, in effect, only one serotype, at least as demonstrated by conventional serological assays (CF or agglutination, with reciprocal absorption experiments, or by cross-protection tests in guinea pigs, hamsters, or mice with strains isolated from different hosts in many different geographic areas and not extensively passaged in the CE yolk sac).[25]

There are, however, two antigenic variations which are of significance for Q fever vaccine formulation. The first, and major one, is phase variation in the cell wall antigens of the organism, and the second is a difference in cryptic epitopes observed in certain strains from Q fever endocarditis and goats.

B. ANTIGENIC PHASE VARIATION IN *C. BURNETII*

This phenomenon was first described by Stoker and Fiset[26-29] and its biological implications are fully described in volume 1.

In brief, phase variation is now recognized to be closely similar to smooth-rough variation in the lipopolysaccharide (LPS) of enteric organisms such as *Salmonella* or *Escherichia coli*. In phase I, the state in which the organism exists in human, animal, and arthropod infections, the LPS of *C. burnetii* has its full complement of sugar units in the side chains. In phase II, after multiple passages in the chick embryo yolk sac, phase I organisms are overgrown[30] by a variant with fewer sugar units in the side chains with a concomitant, readier access of antibody to the underlying phase II membrane protein epitopes.[31-40,121]

Shortly after the discovery of phase variation, Abinanti and Marmion[41] found that antibody against phase I antigen had protective or modifying effects on infection in mice and guinea pigs, an observation later confirmed by others.[42,43] Abinanti and Marmion's observations led Fiset[44] to suggest that phase I organisms might be more virulent than their phase II counterparts and that the difference might be of significance for vaccine formulation. This was shown to be so by Ormsbee and colleagues,[45] who found that inactivated vaccines made from purified *C. burnetii* cells in phase I were 100 to 300 times more potent (w/w) in protecting guinea pigs against experimental infection than a comparable phase II antigen vaccine.

The significance of this observation was underlined by the finding[25,30] that when a late passage yolk sac culture of *C. burnetii* is freed of phase I organisms by plaque purification, so that it is a homogeneous and stable phase II strain, it will not revert to phase I on inoculation into guinea pigs and cannot be passaged serially in these animals. Thus, in three experiments, cloned phase II substrains of the Nine Mile strain, at the level of the 88 egg passage, could not be passed in normally susceptible guinea pigs and could not be recovered from the liver and spleen of guinea pigs inoculated 5 d earlier, nor could infection sufficient to elicit antibody be shown by subpassage of guinea pig tissue to fresh animals. Moreover, although guinea pigs inoculated with the plaque-purified material formed antibody against phase II antigen, they were not protected on challenge with phase I organisms.

Other investigations[30,38] established that the rate of phase antigen variation varies considerably between strains. For example, conversion of the Nine Mile strain to phase II was virtually complete after 17 yolk sac passages, whereas the Ohio and Henzerling strains still had a phase I component after 22 egg passages. Residual phase I antigen cells in yolk sac suspensions of phase II organisms may be detected by separation of cells on density gradients[46] or by immunofluorescence (IF) staining of smears with monospecific antisera.[32]

When all of these observations are taken into account, it appears that the early vaccines, prepared from different strains of *C. burnetii*, at different yolk sac passage levels, probably consisted of mixtures of phase I and phase II cells in different proportions and that their protective effect depended on the amount of the phase I component that happened to be present. With the realization that the phase I antigen is the protective immunogen, the way was clear to formulate vaccines from *C. burnetii* cells virtually entirely in the phase I antigenic state by restricting the number of serial yolk sac passages. Moreover, the amount of *C. burnetii* cells could be reduced and yet remain as protective as a greater mass of mixed phase coxiellae.

At much the same time as these observations on the importance of phase I antigen for protection were being made, it was shown that phase I antigen, like bacterial LPS, can be extracted from the coxiella with phenol-water mixtures; dimethyl sulfoxide, formamide, or trichloracetic acid; the extracts had haptenic, antigenic, or, in some instances, immunogenic activity. These observations were of importance in the development of split vaccines as possible alternatives to inactivated whole cells.

C. ANTIGENIC DIFFERENCES BETWEEN STRAINS OF *C. BURNETII* UNRELATED TO PHASE VARIATION

Despite the earlier comment that strains of *C. burnetii* are homogenous and of one serotype, as, indeed, they are by conventional serological assays, work during the last few years has nevertheless revealed some subtle differences.

Hackstadt et al.[39] extracted LPS from strains of *C. burnetii* isolated from acute Q fever or other natural sources and from isolates from chronic Q fever endocarditis. The LPS extracts were run on sodium dodecyl sulfate (SDS) gels, transferred to nitrocellulose, and developed as immunoblots with rabbit antisera to prototype strains of *C. burnetii*, such as Nine Mile, and with antisera raised against endocarditis strains. It was found that LPS from the endocarditis strains and from a strain — "Priscilla" — isolated from the products of conception of an aborting goat, gave similar immunoblot patterns and that the latter differed from the patterns given by strains from acute Q fever or prototype strains such as Nine Mile or Henzerling. It is surmised that the solvent extraction of the LPS and the immunoblot analysis reveals "cryptic" epitopes not detected by conventional serological analysis of the whole organism. Certainly, CF and IF tests with Priscilla antigen and sera from acute Q fever and chronic endocarditis show little, if any, difference in IgM, IgG, or IgA antibody patterns to those given with the Nine Mile strains.[123] Additionally, and of some importance for vaccine strategy, the Priscilla strain shows no difference as immunogen when compared in guinea pig cross-protection tests with the prototype Nine Mile strain.[47]

Last, studies[48] of the lymphoproliferative responses of the peripheral blood mononuclear cells from subjects vaccinated with the Commonwealth Serum Laboratories (CSL) Q fever vaccine (Q-vax) show equivalent reactions to Priscilla, Henzerling, and Nine Mile antigens.

It appears, therefore, that these cryptic antigenic differences between strains are unlikely to invalidate the use of a particular strain of *C. burnetii* to protect against infections with goat and other strains resembling Priscilla.

D. IMPROVED INACTIVATED WHOLE CELL *C. BURNETII* VACCINES

As a result of the identification of the phase I antigen as the determinant of the protective immunogen and of improved methods for the purification of *C. burnetii* cells from yolk sac homogenates, the current, or modern, version of Q fever vaccine evolved to replace the earlier Smadel-type vaccines.

A variety of purification methods were devised to separate the coxiella cells from yolk sac protein and lipid. These methods include ether and fluorocarbon extraction, filtration, a series of differential centrifugation cycles, fractionation on renografin-sucrose or other gradients, and sequential extraction in increasing salt concentrations and high salt-sucrose mixtures.[49-55] Residual yolk sac material in the final product may be monitored serologically with potent antisera against chicken protein.

The use of these formalin-inactivated, low dose, phase I, highly purified *C. burnetii* suspensions, together with prevaccination serotesting and skin testing to exclude subjects with existing immunity, greatly facilitated the prophylactic use of the vaccine in laboratory workers and eventually in industrial groups.

The efficacy of these newer whole vaccines was established in various ways, although formal, blind placebo-controlled trials were not undertaken at that time.

Compelling evidence of vaccine efficacy was obtained in 1966 — 1968 during vaccine trials with a purified Henzerling phase I vaccine of the improved type conducted by Richard B. Hornick of the University of Maryland School of Medicine[55,56] (Table 1).

A total of 124 subjects were given subgroup various doses of vaccine, and 57 vaccinees were challenged later with a large dose of *C. burnetii* by aerosol and the respiratory route. The upper dose of vaccine was 30 µg given subcutaneously, a dose chosen on the basis of previous assays of the protective effect of vaccine in guinea pigs.

TABLE 1
Results of Vaccination-Challenge Experiments with Q Fever Vaccine in Volunteers Conducted by the Commission of Rickettsial Diseases, U.S. Armed Forces Epidemiological Board

Vaccine dose	No. of subjects	Incidence of common or trivial reactions[a]	Antibody response[b] MA (ph I or ph II)	Antibody response[b] CF (ph II)	Duration between vaccination and challenge (months)	Response to challenge[c] sick/total	Protection %
30 μgm	64	30/60 (50%)	38/46 (82%)	12/13 (92%)	7—10	0/13	100
30 μgm	19	—	—	—	5—8	2/19	89
3 μgm	20	17/m (85%)	16/20 (80%)	3/20 (15%)	3	3/11	72
1 μgm	21	12/21 (57%)	10/21 (47%)	3/21 (14%)	3	4/14	71
NIL (controls)	6	—	—	—	—	5/6 (83%)	—

[a] Number reacting to number inoculated. All local reactions were mild, appeared within 24 to 48 h and subsided within 96 h. No systemic reactions.

[b] Microagglutination (MA) and complement fixation (CF) to phase I or phase II antigens of C. burnetii.

[c] Aerosol challenge of volunteers with AD strain of C. burnetii, phase I. 3500 GPID$_{50}$ (as determined by IP inoculation).

Overall, in the subgroup of 64 volunteers given 30 µg, 50% had common or trivial reactions, probably of the type described in detail for the trial in South Australia (below). Eighty-two percent developed antibody in the microagglutination test with phase I or II antigen, a figure matched, somewhat surprisingly, by the group which had only 3 µg of coxiellae.

Thirty-two of the subjects given 30 µg of vaccine were challenged 5 to 10 months after inoculation, and 2 (6.3%) developed Q fever. Of the 25 volunteers given doses of 1 to 3 µg and challenged 3 months after inoculation, 7 (28%) became ill. These results contrasted with Q fever in 5/6 unvaccinated controls χ^2 (= 19.8, p <0.001).

It appeared, therefore, that almost solid immunity to Q fever was given by a single dose of 30 µg, and significant protection (72%) was given by doses as low as 1 to 3 µg (represents about 3.78 to 11.3 x 10^7 organisms)[122].

The immune state of the volunteers, as judged by skin tests before vaccination, was unknown, but they were evidently negative by a sensitive serological test.

In the same decade, routine vaccination of staff began at the Rocky Mountain Laboratory, National Institute of Allergy and Infectious Diseases, Hamilton, MT, a leading center for rickettsial research. Vaccine Q58-A was made in 1960 and stored at 4°C as a saline suspension containing 22 µg/ml of coxiellae.

It was assayed in a standard guinea pig protection test in 1960, 1970, and 1980, and its protective potency did not vary significantly during the 20-year period.[30] Vaccination consisted of a single dose of 1.0 ml given subcutaneously to skin test negative individuals. The results were reviewed recently by Philip.[57] A total of 398 employees were skin tested over the 20-year period. Of these, 282 (70%) were skin test negative and vaccinated, and 79 were skin test positive and not vaccinated. Thirty-seven subjects were skin test negative but declined vaccination. A separate group of 14 subjects refused both skin test and vaccination. During the period, there were no Q fever cases in the 282 vaccinees and none in the 79 subjects with a positive skin test. Two Q fever cases occurred in the 37 skin test-negative, unvaccinated group, and 2 in the group of 14 who refused both skin test and vaccination. If the prevalence of cases is compared in the 282 vaccinees, on the one hand, and in the 51 unvaccinated persons (including the 14 without skin tests and unvaccinated), on the other hand, the difference appears to be significant (χ^2 = 20.5, p <0.001).

In other trials with a closely similar phase I vaccine in the 1960s, Luoto et al.[58] inoculated seronegative and skin test negative subjects subcutaneously with 1, 2, or 3 doses of 10 CF units (about 244 µg) of coxiellae. Some 40% of subjects reported local effects varying from tenderness or soreness on movement and, less commonly, more severe pain.

These volunteers were not challenged experimentally with living *C. burnetii* or exposed to natural infection; the immunogenicity of the inactivated Q fever phase I vaccine was measured by *in vitro* tests of antibody (CF; capillary tube agglutination [CAT]; radioimmunoprecipitation [RIP] tests) of various levels of sensitivity and by assays for cell-mediated immunity as measured by skin test conversion. A total of 141 volunteers, previously shown to be serologically and skin test negative, were inoculated with 1, 2, or 3 doses of vaccine. (Note that the dose used was about seven times higher than that used in Hornick's volunteer-challenge studies or that used in the South Austrialian trials, see below). The paper by Luoto et al.[58] shows that of 42 subjects given one dose of 10 CFU subcutaneously, 41% developed antibody by CAT and 14% by CFT. The paper also shows the number and proportion of subjects positive by serotest or skin test at varying intervals after vaccination. In the cohort of 42 individuals who had one dose of 10 CFU, 30% of subjects were seropositive, but only 19% were skin test positive at 9 weeks after vaccination. By 40 weeks after vaccination, the seropositivity rate had fallen to 13%, but 95% were now skin test positive. Last, sera from 5 subjects positive in the skin test at 40 weeks, but negative by CAT and CFT, were tested for antibody in the mouse protection test. Of these, 4 were positive, a clear demonstration of the presence of phase I antibody not detected by conventional tube tests. In a study by Bell et al.[59] in the following year, directed mainly at the assessment of the response of intracutaneous inoculation with Q fever vaccine, sera from the above trial were retested by RIPT. Of vaccinees given one dose of 10 CFU by subcutaneous inoculation, 9/20 (45%) were RIPT positive at 6 weeks after vaccination; 12/20 (60%) by 12 weeks; and 24/25 (96%) by 40 weeks.

If the volunteer-challenge studies[55,56] and those of Luoto et al.[58] are taken together, it appears that one dose of Q fever vaccine gives seroconversion in about half to three quarters of the vaccinees by antibody tests of medium sensitivity, depending on the time of sampling after vaccination, but that a high proportion — over 90% — of such vaccinees show skin test conversion and are also eventually positive by antibody tests of greater sensitivity, such as RIPT or mouse protection and are likely to be resistant to infection on challenge.

From these various experiences, it is not possible to make an exact correlation between the *in vitro* markers of immunity and protection against disease in the NIAID laboratory at Hamilton or in the volunteer-challenge experiments. About 80% of the volunteers in the latter trial developed antibody, and it may be supposed, in line with the experience of Bell et al.,[59] that a higher proportion would have been antibody positive by RIPT. Neither group was skin tested *after* vaccination, so there are no data on the cell-mediated response stimulated by vaccine. However, it is significant that there were no Q fever cases among the 79 subjects at NIAID, Hamilton, who were skin test positive.

The protective effects of inactivated, whole call Q fever vaccine just described with man have, of course, been demonstrated many times in the laboratory by vaccination and challenge of small laboratory animals such as guinea pigs or mice (for example, see Reference 18). They have also been demonstrated with larger animals. Kishimoto and colleagues[60] have confirmed the results of experiments with human volunteers in vaccination-challenge experiments with cynomolgus monkeys. The animals were given a single dose of 30 μg of inactivated whole call *C. burnetii* phase I vaccine and challenged by infective aerosol 6 to 12 months later. Protection was not absolute as some animals became ill. However, they did not develop pneumonia, and rickettsemia was of shorter duration in vaccinated animals.

Several groups of workers[61-64] have vaccinated dairy cattle with inactivated whole cell vaccines; a marked effect in natural infection, as judged by reduction in shedding of *C. burnetii* in the milk or placenta, was observed in the vaccinated animals as compared with unvaccinated animals in the same environment. Somewhat similar results were obtained by Brooks et al.[65] in sheep vaccinated with whole cell or chloroform methanol extracted coxiellae and challenged with 210,000 pfu of *C. burnetii* subcutaneously. All six unvaccinated ewes shed *C. burnetii* in the placenta or amniotic fluid and had placentitis. Two of six ewes vaccinated with whole cell vaccine and one of six given CM extracted cells shed the organism but in reduced numbers.

E. ALTERNATIVE STRATEGIES FOR Q FEVER VACCINES: SPLIT PRODUCTS FROM *C. BURNETII* AND EXTRACTED CELL RESIDUES

Efforts to separate antigenic substances from *C. burnetii* began in 1956 when Colter et al.[66] disintegrated rickettsias with ultrasonic vibration. The sonicated preparation which could not be sedimented by centrifugation at 12,000 rpm (*g* force not given) was both antigenic and stimulated protective immunity when used to immunize guinea pigs, subsequently challenged with live *C. burnetii*. No attempts were made to characterize these preparations chemically beyond nitrogen determinations. Later efforts centered around the use of organic solvents such as dimethyl sulfoxide (DMS), dimethylacetamide (DMAC), trichloracetic acid (TCA), and chloroform-methanol (CM) mixtures.

Both DMS and DMAC were effective in extracting protective immunogens from phase I *C. burnetii*, but not from phase II organisms.[67] The antigenic extracts were highly reactive in the CF test with late convalescent serum from Q fever infection and precipitated by immune serum in a capillary tube test. They showed values of 27,000 to 155,000 CFU of antigen per milligram and were proved to be almost as effective on a weight basis as intact phase I rickettsias in protecting guinea pigs against challenge with live rickettsias. A most interesting aspect was that the extracts were at least 200 times less reactive in producing delayed hypersensitivity reactions in the skins of immunized rabbits than the intact coxiella cells from which they had been prepared.

At about this time, TCA was found to be an effective solvent for the extraction of phase I antigen.[68,69] As with DMS and DMAC extracts, antigen extracted by TCA proved to be an effective immunogen and protected guinea pigs on challenge with live organisms; it was at least 100 times less reactive in producing DTH reactions in sensitized rabbits. As a result of these observations, vaccines embodying TCA extracts have been tested in field trials in Romania[70,71] and in Czechoslovakia.[72,73] The results will be described in the section on the occupational use of vaccines.

An extract with the serological properties of phase I antigen can also be obtained by phenol extraction of purified phase I *C. burnetii* cells.[74,75] It protects mice against challenge with live *C. burnetii* and provokes phase I antibody after multiple inoculations. In addition, it gives no reaction in the allergic skin test in guinea pigs previously sensitized with whole cell *C. burnetii* preparations or in the skin of human subjects previously infected with *C. burnetii*.

It appears that the difference between phenolic extracts and those made with organic solvents and TCA is that the former is lipopolysaccharide (a hapten) whereas the latter are complexes of LPS and protein;[76] phenolic extracts are unlikely to have great value as vaccines.

Despite the generally tolerable level of reactogenicity with whole call Q fever vaccine observed in the more recent trials[77] in which recipients were pretested for antibody and by skin test, the quest for a modified vaccine with high immunogenicity and minimal reactogenicity has recently gathered momentum and taken a different direction. There appear to have been two imperatives for the new directions.

First, studies[78-80] in mice with suspensions of inactivated *C. burnetii* whole cells demonstrated granulomatous changes with enlargement and necrotic foci in liver and spleen and, on histological examination, foci with lymphocytic infiltration activated macrophages, giant cell, or polykaryocyte formation. Spleen cells from the inoculated animals showed an increased blastogenic index and a diminished lymphoproliferative response to mitogens such as phytohemagglutinin (PHA).

At first sight, these observations appeared to have negative implications for the use of whole cell vaccine in man. However, the dose of whole cells given to the mice in these experiments were large on a dose/weight basis compared with the 30 μg given to a 60 to 80 kg man. With doses (3.75 to 7.5 ng) adjusted to correspond to those given to human beings, mice did not show gross or microscopic lesions in the liver; blastogenic indices were low, and the PHA response to PHA was not impaired, although mice formed antibody to *C. burnetii* as measured by immunofluorescence.[124] In the latter experiments, control mice given a large dose of whole cell vaccine, as used in the experiments of Damrow et al.,[78] showed suppression of lymphocyte mitogenesis to PHA, splenic enlargement, and granulomata on histological examination.

Nevertheless, in view of the reactions provoked, recent attempts to reduce granulomata and reactogenicity have centered on extraction of phase I *C. burnetii* cells with a chloroform-methanol mixture.[81,82] This procedure leaves a residue (CMR) which comprises 78% of the starting dry weight of coxiellae. In mice, CMR appears to be less toxic than killed whole cells. It is also immunogenic in mice and protects these animals from fever and death on challenge by i.p. route or aerosol with phase I *C. burnetii*. A strong delayed type hypersensitivity reaction without dermal granuloma was elicited by CMR in sensitized guinea pigs.

A second desirable objective in the use of CMR vaccine would be to remove the necessity for prevaccination testing for immune markers, i.e., to ensure lack of reactogenicity when vaccine is given to immune subjects.

There is no doubt that CM-extracted organisms provoke protective immunity in experimental animals to infective challenge, either in laboratory animals[82] or in domestic animals such as sheep.[65]

Recent Czechoslovakian studies[83] which compared inactivated whole cell phase I vaccine, chemovaccine (TCA extract), and CM-extracted whole cells in guinea pigs and mice, assayed for serological and cell-mediated response and challenged with living organisms, indicate that the immune response to chemovaccine or CM–extracted whole cells appears later, reaches a lower peak, and dies away more rapidly than the response to whole cell vaccine.

On the other hand, Williams and Cantrell[79] reported that CMR vaccine was more immunogenic than a whole cell vaccine as assessed by protection experiments in C57BL/10SCN endotoxin-nonresponder mice (intraperitoneal challenge with *C. burnetii*). As before, the CMR vaccine had less effect than a whole cell preparation in inhibiting responses of murine splenocytes to mitogens such as PHA.

The criteria for the validation of protective efficacy CMR vaccine in industrial, occupationally exposed human subjects are discussed under a later section.

F. ATTENUATED, LIVE STRAINS OF *C. BURNETII* AS Q FEVER VACCINES

Experience with attenuated strains of *C. burnetii* as vaccines has, for the most part, been limited to the U.S.S.R. and relates to the M-44 line of the Grita strain of *C. burnetii*.[84-86] This strain was passaged in chick embryo yolk sacs for 43 successive transfers and became known as strain M-44 at this point. When given to human subjects by the subcutaneous and percutaneous routes, reactogenicity was low and seroconversion rates acceptable. At the level of 46 egg passages, the strain produced febrile responses and lesions in guinea pigs in a manner similar to the Nine Mile strain at egg passage 88 (EP88).[87,88] The M-44 strain was used extensively in Russian field trials, but published reports of its efficacy in preventing Q fever are lacking.

Two small trials with Nine Mile EP88 were done with human volunteers in the U.S.[89] The EP88 of Nine Mile retained little toxicity for guinea pigs and mice. For example, it required intraperitoneal inoculation of about 10^6 coxiellae to cause fever in guinea pigs. In human trials of EP88,[125] groups of six volunteers were inoculated with 1.0 ml of one of a range of dilutions: 10^{-2}, 10^{-4}, 10^{-6}, and 10^{-8} of infectious yolk sac (ID_{50}) of yolk sac suspension for guinea pigs was $10^{9.6}$). The volunteers developed strong local reactions and 11 of 12 subjects developed phase II antibodies detected by CF test. On respiratory challenge by aerosol with 3000 guinea pig ID_{50} doses of phase I *C. burnetii*, those previously given 10^{-2} or 10^{-4} dilutions of infectious yolk sac resisted challenge, whereas those volunteers given 10^{-6} and 10^{-8} dilutions of yolk sac mostly became ill.

Further studies with live attenuated Q fever vaccine would seem uneconomical, given the potential complications from its use and the existence of a highly protective killed vaccine. The protective effect of the high passage yolk sac-adapted strains used presumably depends on the residual population of phase I organisms in the predominantly phase II culture (the strains were not plaque purified). The M-44 strain yields phase I organisms when passaged in guinea pigs, inducing hepatitis, splenitis, and mycarditis and surviving for a substantial period in the animal.[87,88] Infection could also be reactivated in inoculated guinea pigs by pregnancy or treatment with prednisolone, cortisone, or cyclophosphamide.[90] Consequently, mass immunization programs with live attenuated vaccine might be open to the hazards of inducing the chronic sequelae of infection (endocarditis, hepatitis, or delayed type hypersensitivity [DTH]) in vaccinees or a reactivated infection during pregnancy.

G. VACCINE PROPHYLAXIS OF Q FEVER IN OCCUPATIONALLY EXPOSED GROUPS

1. Whole Cell Vaccine

In 1979—1980, there was a rekindling of interest in vaccine prophylaxis of Q fever, arising partly from outbreaks of Q fever, at first unidentified, in research institutes and medical schools[91-95] where sheep were used for research in reproductive physiology and, in Australia,

TABLE 2
Reactions to CSL Q Fever Vaccine (Q-vax) in Volunteer
Subjects at Two Abattoirs during the Period 1981—1986

Common Reactions (N = 464)

Fever >38°C	0.2%
Headache (1 d)	9.0%
Tenderness (1—3 d)	48.0%
Erythema (1—3 d)	33.0%
Edema (1 d)	0.6%

Uncommon Reactions (N = 2682)

Abscess: drained, healed	1.0
Lump: resolved, 2 months	1.0

Note: N = number of subjects sur-
veyed either by daily monitor-
ing (common reactions) or by
inspection of medical records
and by questionnaire (uncommon reactions).

from a sharp exacerbation of the prevalence of Q fever in abattoirs following the introduction of feral goats into the slaughtering programs.[96,97]

In South Australia, the problem was addressed by arranging, with the Commonwealth Serum Laboratories, Parkville, Victoria, to produce a formalin-inactivated, *C. burnetii* vaccine with purification of the coxiellae from infected yolk sac by low-high salt extraction.[50,51] This yielded a coxiella suspension with little contaminating yolk sac protein. The *C. burnetii* strain used was Henzerling, free of adventitious agents such as avian leukosis virus and avian mycoplasmas. The seed stock was received in the first egg passage, and the vaccine, prepared from the second egg passage, was shown to be fully in the phase I antigenic state. The coxiellae were inactivated with formalin, and tests for uninactivated organisms included four serial yolk sac passages of vaccine concentrate with immunofluorescent (IF) and Gimenez staining of yolk sac smears, together with subinoculation of yolk sac suspensions into mice and guinea pigs with subsequent serological testing. The concentrate was also passed through mice before yolk sac inoculation. All methods failed to demonstrate viable *C. burnetii*.

Commencing mid-1981, the vaccine was administered to volunteers, after informed consent, at first in two and later in four abattoirs. Subjects were pretested for Q fever CF antibody at serum dilutions of 2.5, 5, 10, and 20; by IF at dilutions of 1 in 10 or higher with phase I and phase II antigens; and by an intradermal skin test with 0.02 ug of coxiella and read for reactions 5 to 7 d after inoculation.[97] Subjects without clear-cut positive immune markers in these tests were vaccinated with one 0.5 ml dose of 30 µg of organisms given subcutaneously.

In the early stages of the program, a number of subjects with equivocal antibody or skin test reactions were vaccinated.[97] Later, after a severe reaction, only those with unequivocally negative results were inoculated.

About half the abattoir population had antibody to *C. burnetii*, and another 5% had a positive skin test but no detectable antibody.

Table 2 shows the common reactions during 1981—1986 to CSL vaccine in 464 subjects; the rates were comparable to those observed in the U.S. volunteer trials[56] and by Luoto et al.[58] A detailed search of 2682 records substantiated two more uncommon reactions: one immune abscess and one indurated lump. Over the total period, 1981 — 1989 to date, in over 5000 vaccinees, there have been two immune abscesses. The first was in a master butcher of more than 30 years standing, who had low serum titers of Q fever antibody (CF 2.5, IF/IgG 10), and an

TABLE 3

Overall Figures for Vaccinees, Unvaccinated Subjects, and Q Fever Cases in Various Groups Exposed to Infection at One or Other of Four South Australian Abattoirs during the Period 1981—1986

Category	Immune markers on enrollment	Vaccine status	Subgroup (N)	Q fever cases (N)
Abattoir workers,	24—47%	Vaccinated	2716	3[a]
meat inspectors		Unvaccinated	2012	52
Supporting firms/	12 —13%	Vaccinated	292	4[a]
regular servicing		Unvaccinated	(147—150)	26
Sporadic	9%	Vaccinated	524	0
visits contact		Unvaccinated	(48—150)	19
		Total	≥5739	104

Note: ( ) Exact (N) at risk uncertain.

[a] Vaccinated in incubation period of Q fever.

equivocal skin test before vaccination. Seven days after vaccination, an acute abscess developed, pointed and discharged; it eventually required surgical excision of the track before it healed. Comparisons of the antibody content in the patient's serum and in eluates from wound swabs as measured by IF to *C. burnetii* phase I antigen and in the IgG and IgA classes, relative to albumin concentration, revealed concentrations of antibody in the eluates 139 to 479 times greater than in serum. Serum antibody titers remained static, but lymphoproliferation assays with the patient's peripheral blood mononuclear cells and *C. burnetii* antigens showed high activity.

The second reactor, antibody and skin test negative before vaccination, developed an erythematous, indurated mass, 14 d after vaccination, which eventually discharged and healed. Clear evidence of preexisting immunity has not been obtained so far, and it is possible that the reaction represents an idiosyncratic macrophage activation.

To put the matter of these two abscesses in perspective, it may be stated that vaccination of a subject with low level or borderline immune markers, though now absolutely contraindicated in the interests of safety, is not automatically followed by an indurated mass or immune abscess. In the early stages of the trial, 246 subjects at two South Australian abattoirs with such markers were vaccinated without incident.

Table 3 shows an overall summary, for the period 1981—1986, of the number of vaccinees, in three risk categories, at the four abattoirs in the trial. The regular or full-time workers — abattoir workers and meat inspectors — had the highest rate of immune markers on admission into the program, rates which contrasted with those among persons (e.g., maintenance engineers, insurance inspectors, catering school students, research workers) visiting the abattoir sporadically, but who were still at risk. It will be seen that all but 7 of the 104 Q fever cases recorded during the period were in unvaccinated persons, and further that the 7 Q fever cases in vaccinees were, in fact, not vaccine failures, but were in subjects vaccinated during the incubation period of a natural infection.

At the largest abattoir of the four in the trial (Samcor, Gepps Cross, Adelaide) a detailed search of medical records for missed cases of Q fever in vaccinees was made, together with an analysis of bias factors in the open trial, without obtaining evidence negating vaccine efficacy. Two detailed illustrations of the epidemiological pattern at Samcor may be given. Figure 1 shows the yearly prevalence of Q fever cases in vaccinated and unvaccinated persons for the total

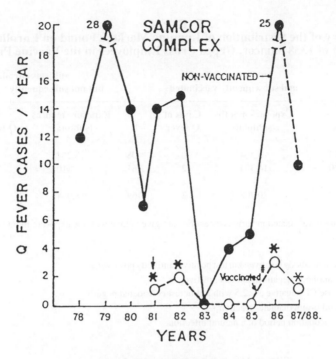

* VACCINATED IN INCUBATION PERIOD
‡ TOTAL OF 1922 VACCINATED

FIGURE 1. Yearly prevalences, 1978—1988, of Q fever cases in all
employees of and visitors to the Samcor site or complex at Gepps Cross,
Adelaide. Vaccinated and unvaccinated cases are shown separately.

period 1981 to 1988. There was a high fluctuating rate of Q fever cases in the unvaccinated workers on, or in visitors to the campus, but no Q fever cases in vaccinees who had had time (~15 d)
to develop immunity after vaccination and before exposure to infection. In 1988 there was
another Q fever case in a vaccinee, making a total of seven at Samcor, all in persons vaccinated
in the incubation period of a natural infection (Figure 1). Table 4 shows an analysis of the
incidence of Q fever and the duration of exposure ("exposure months"; a calculation necessi-
tated by the high turnover of employees), in 869 vaccinees and 469 unvaccinated subjects at
Samcor during the period 1981—1986. There were 2 Q fever cases (both vaccinated in the
incubation period of a natural infection) in 690 vaccinees who were seronegative and skin test
negative before inoculation. In contrast, there were 7 Q fever cases in 61 unvaccinated subjects
who had the same negative immune markers on enrollment into the program. When considered
in terms of Q fever cases per 1000 exposure months, the incidence ratios in 690 vaccinated vs.
the 61 unvaccinated subjects are 1:62; tests for the significance of differences in incidence ratios
by the method of Kleinbaum et al.[98] gives $p < 0.0001$.

Last, the total of eight Q fever cases in vaccinees at all four abattoirs eventually involved in
the trial and over the period 1981—1989, are analyzed in terms of the period between their
vacination and onset of illness, and compared with the cumulative total of Q fever cases in the
enrolled, but unvaccinated workers in the same abattoirs (Figure 2). It will be seen that the two
curves for cumulative totals of cases rise together over the first 10 d after vaccination or
enrollment, but that in the vaccinees reaches a plateau at 13 d after which there are no more cases.
On the other hand, the curve in the unvaccinated continues to rise over the following days and
months. In this context, it is significant that cell-mediated immunity, as judged by lymphopro-

TABLE 4
Summary of the Distribution of Immune Markers Found on Enrollment of a Group of 1338 Samcor, Gepps Cross Employees in the Vaccine Program

Immune markers on enrollment in the vaccine program	N	Subjects enrolled and subsequently vaccinated			N	Subjects enrolled but not subsequently vaccinated			
		Exposure months (subtotal)	Cases of Q fever			Exposure months (subtotal)	Cases of Q fever	Totals	
Positive[a]	179	6,364	0		408	17,254	2	587	
Negative[b]	690	14,434	2[c]		61	810	7	751	
Totals	869	20,798	2		469	18,064	9	1,338	

Note: The numbers vaccinated in various categories are given along with the exposure months in each category.

[a] Group contains those antibody positive/skin test positive, antibody positive/skin test not done, and antibody negative/skin test positive.
[b] Group contains those CF negative at <2.5 with a skin test considered negative by the physician and 0-mm diameter reaction.
[c] Vaccinated in the incubation period of a natural infection.

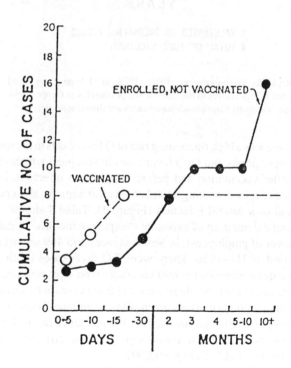

FIGURE 2. Cumulative totals of Q fever cases in vaccinees in relation to time since vaccination and in enrolled but unvaccinated subjects in relation to time from enrollment. The usual interval between enrollment and vaccination is 7 to 10 d.

liferative assays of peripheral blood mononuclear cells of vaccinees to *C. burnetii* antigens, becomes positive around 12 d after inoculation.[99]

In summary, in the South Australian trial, one dose of 30 µg of killed Q fever vaccine appeared to confer complete protection against naturally acquired Q fever by 10 to 15 d after inoculation. Cohort analysis of the vaccinees at Samcor showed that of the group vaccinated in 1981, 38% were still in the work force at the end of 1986 and had not developed Q fever. The duration of immunity appears to be at least 5 years (see also comments on cell-mediated immunity below).

The results in the South Australian trial of Q fever vaccine were confirmed in a small blind placebo-controlled trial in three Queensland abattoirs organized by Shapiro and colleagues[100] in which Q fever and influenza vaccines were administered under code. Seronegative volunteers were allocated alternately to one or other vaccine group, skin tested with the diluted vaccine, and if negative, vaccinated.

The cumulative totals of Q fever cases and the difference between the two vaccine groups were monitored in a sequential trial design (Armitage). Given the previous experience in the South Australian trial, it was possible to decide that a significant difference ($p < 0.05$) between vaccine groups would be attained if there were seven cases in one vaccine group and none in the other, when the trial could be terminated and the code broken. This point was reached after 15 months, and all Q fever cases were found in the influenza vaccine group and none in that given Q fever vaccine (Table 5).

2. Prophylaxis of Occupational Q Fever with Split Vaccines or Modified *C. burnetii* Cells

The main experience in the field with antigenic extracts of *C. burnetii* has come from eastern European countries and Czechoslovakia, in particular. As indicated above, the phase I LPS protein complex[68,69,76] can be extracted in trichloracetic acid (chemovaccine) and retains its antigenicity and protective immunogenicity in experimental animals challenged with living coxiellae. It is also antigenic in human beings and was apparently protective in laboratory staff working with the organism and preparing vaccines.[73] Some reactogenicity was still observed, nevertheless. Thus, 64% of a group of 355 workers in a cotton processing plant in South Moravia developed local reactions and 5% systemic reactions.[72,73] In a Romanian study[71] a group of 28 subjects, infected with Q fever 1 to 2 years previously, was given 1.0 ml of chemovaccine (160 complement fixing units) subcutaneously; 9 (32%) developed mild to medium level local reactions and 6 (21%) mild febrile reactions. Regional adenopathy or sterile (immune) abscesses at the inoculation site, as reported with whole cell vaccines, were not observed. Freedom from the latter is probably a significant advance, but it has to be recognized that in order to detect the infrequent severe reaction, it might be necessary to inoculate several hundred sensitized subjects before a sterile abscess would be observed (rates of 0.34 per 1000 were observed in unselected subjects in Benenson's trials,[18] and two were observed in >5000 vaccinees in the South Australialian trials.

Chloroform-methanol extracted *C. burnetii* cells, from which the substance, Lipid A, considered to be responsible for the granuloma-inducing, reactogenic effect of whole cell vaccines, has been removed,[81,82] protect laboratory animals and modify the pattern of excretion of *C. burnetii* in the milk and placentas of experimentally infected sheep.[65]

There appear to be no reports of direct tests of CM vaccine for reactogenicity in sensitized and nonsensitized human beings or of its protective efficacy against natural infection of human beings.

A recent study[101] from Bratislava compared the reactogenicity and immunogenicity of TCA extracts from *C. burnetii* whole cells, and of TCA extracts from chloroform-methanol treated coxiellas, in groups of nonsensitized human volunteers. With the TCA extract of whole cells, 62% developed local reactions and 6% general reactions (higher rates than in a study previously

TABLE 5
Summary of Prevalence of Q Fever Cases in a Randomized Blind
Placebo-Controlled Trial of Two Vaccines
in Three Queensland Abattoirs

	Vaccine given under code			
	Q fever vaccine		Influenza vaccine[a]	
Abattoir	No. vaccines	Q fever cases	No. vaccines	Q fever cases
Kilcoy	41	0	30	5
Beenleigh	21	0	28	0
Metropolitan (Brisbane)	36	0	44	2
TOTAL	98	0	102	7

Note: The vaccines were Q fever (Q-vax) and influenza vaccine (CSL Flu-vax).

[a] Placebo

Queensland vaccine trial; Shapiro, R., Siskind, V., and Marmion, B. P., unpublished data.

reported). The local and general reaction rates with TCA extract of CM-extracted cells were 40 and 4%, respectively. Antibody and skin test conversion rates were lower with the TCA extract from CM-extracted cells than with that from whole cells.

It is evident that a final judgement on the value of chemovaccine or CM-extracted cell residues must await comparative blind trials of these vaccines with a low dose, whole cell vaccine as used in the South Australian studies, preferably in high-risk groups such as abattoir workers.

H. IMMUNE MARKERS FOR Q FEVER AND VACCINE-INDUCED IMMUNITY IN MAN

Past assessments of the immune response in man to Q fever vaccine has mostly relied on measurement of antibody by CF, IF, micro-agglutination (MA) or capillary agglutination (CA) tests and more rarely, by sensitive methods such as radioimmunoassay[102] or mouse protection tests.[41] Serological tests are easy and cheap to perform in comparison with tests for cellular immunity and are frequently used to measure the immune response to many types of vaccines. There are, however, distinct limits to the information they provide on the immune response to Q fever vaccine and to their value as *in vitro* correlates of protective efficacy. CF and IF tests, although of central importance in the diagnosis of acute or chronic cases of Q fever in which the replicating organisms provide substantial antigenic mass and immune stimulation, are insensitive techniques for the measurement of the responses to an inactivated vaccine and also perhaps to those from symptomless but immunizing exposures to the organism. The point is illustrated by the marked divergence between the frequencies of CF antibody and antibody measured by radioisotope precipitation (RIP) in occupationally exposed and other populations,[102] and the lower sensitivities of the CF and capillary agglutination (CA) assays compared to the RIP or mouse protection tests in the vaccine trials of Luoto et al.[58] The divergence is particularly apparent when a single dose of vaccine is used rather than two or three doses.

The dichotomy between serological response and protection against infection is also illustrated by the findings in the human volunteer trials conducted by the Commission on

TABLE 6

Rates (%) of Antibody Detected and Distribution by Immunoglobulin Class, in Groups of Vaccinees, Serologically and Skin Test Negative before Inoculation a Few Weeks to 3 Months, and at 20—60 Months after Inoculation

Time interval from vaccination (months)	Total no.	Antibody to *C. Burnetii* antigen detected by				All Tests (CF, IF, RIA)
		Immunofluorescence				
		IgM		IgG		
		ph I	ph II	phI	ph II	
0.6—3	26	11	17	4	6	22
	(100)	(42)	(65)	(15)	(23)	(84)
20—60	47	4	9	21	6	30
	(100)	(8)	(19)	(45)	(13)	(64)

Note: ( ) = percent of total (100%) in subgroup.

Rickettsial Diseases, U.S. Armed Forces Epidemiological Board (Table 1). Eighty-two percent of subjects given 30 μg of coxiellae responded with antibody detected by the Fiset MA test;[103] the response to subsequent challenge with living organisms indicated that a number of subjects without antibody detected by this particular test were nevertheless immune. The divergence between antibody formation and resistance was even more marked in a group of 21 subjects who had one dose of 1.0 μg of vaccine; 47% of the group overall formed antibody, but only 4 (28%) of 14 challenged became ill (Table 1). The interpretation of the data is admittedly limited by the fact that more subjects were tested for antibody than were subsequently challenged, but the trend seems clear and similar to that in the earlier trials conducted by Benenson.[18]

In the Commission's trial, the subjects were not tested for skin test conversion after vaccination and before challenge, so that correlations between cellular immunity and resistance are not possible.

A similar divergence between persisting antibody and resistance to natural infection was observed in the South Australian trials. Overall, several months after vaccination, only 56 to 64% of vaccinated abattoir workers seroconverted as judged by CF or IF antibody tests.[97] Because of this seeming paradox, the kinetics of the antibody response were reexamined and reveals that in the 3 months after vaccination, 84% of vaccinees, initially serotest and skin test negative, had antibody by a variety of techniques (CF, IF, and RIA). The main response was IgM antibody, particularly to *C. burnetii* phase II. By 20 to 60 months the overall rate had fallen to 64% with a predominant response of IgG antibody to phase II antigen (Table 6).

In a low-risk population of vaccinated laboratory workers, antibody rates were generally lower, suggesting that the high-risk population of abattoir workers, although negative in the prevaccination testing, was nevertheless more often primed. Studies of lymphoproliferative responses support this conclusion (see below).

Pioneering studies by Wisseman and colleagues[104-108] of immunity to murine and epidemic typhus indicate that while antibody is probably important in clearing extracellular rickettsias by promoting phagocytosis and, perhaps, by participation in antibody-dependent cell-mediated cytotoxicity, the main component of the immune response for clearing intracellular infection and for conferring protection involves T-lymphocyte sensitization or memory and the generation of interferon (gamma-interferon) on antigen contact. In these and related studies, it was particularly significant that rickettsial vaccine did not protect experimentally infected animals unless accompanied by an adjuvant which induced a positive skin test (DTH) reaction.

Whole cell *C. burnetii* vaccines differ in that they induce a DTH reaction without added adjuvant. In other ways, they clearly depend for protective efficacy on induction of T-cell sensitization and memory rather than on antibody formation. Abinanti and Marmion[41] showed in a mouse model that antibody to phase I antigen of *C. burnetii* would neutralize the pathological effects of living *C. burnetii* in mice in the sense of reducing spleen size and the proliferation of the organism in the spleen, whereas phase II antibody had substantially less effect. This observation was subsequently confirmed by a number of workers (for example, see References 42 and 43). However, the protection by phase I antibody in the mouse has to act against a background of intact cell-mediated immunity. Congenitally athymic mice are not protected when given phase I antibody before infection, and although they develop it during infection, the pattern of the latter is not modified as in euthymic mice.[109,110] So, in this model system, antibody and cell-mediated mechanisms act together to restrict or terminate infection. Also, significantly, phase I antibody, which protects in euthymic mice, does not neutralize infection in chick embryos or cell culture, which have no immune system.[43]

Other evidence for the primacy of cell-mediated immunity in protection and resolution of infection rests, for example, on the differing patterns of infection in euthymic, athymic, and thymectomized mice; on the striking differences in the processing of the coxiella in resting macrophages, on the one hand, and in those from immune animals or those activated by immune lymphocytes, on the other hand. It also rests on correlational studies of protection to challenge and the patterns of antibody, cellular immunity, and skin reactivity in guinea pigs and mice.[109-116]

Studies of cell-mediated immunity to Q fever in man have, of necessity, been restricted to skin test reactivity and proliferation (mitogenic) assays with peripheral blood mononuclear cells to *C. burnetii* antigens. Lymphoproliferative responses have been demonstrated, sometimes in the absence of antibody, in subjects with postinfection or vaccine-induced immunity.[99,117,118] The subsets of lymphocytes involved in these reactions and the degree of similarity between lymphocyte sensitization from natural infection or from vaccine-induced immunity have not, as far as we are aware, been defined, although recent studies in Adelaide show that the lymphoproliferative response in both naturally infected and vaccinees is mainly dependent on T-lymphocytes and augmented by interleukin-2 (IL-2).[124] Nor is it known whether a natural infection or a vaccine-induced immunity both induce qualitatively similar production of cytokines, particularly, gamma interferon, that will activate macrophages and other cells to inhibit the growth of *C. burnetii*.[119]

Recent studies[99] of responses to a single 10 μg dose of Q-vax (whole cell *C. burnetii* vaccine) have shown that 80 to 90% of vaccinees in a low-risk group of subjects converted to a positive lymphoproliferative response (i.e., lymphocyte stimulation index (LSI)) to *C. burnetii* antigens after vaccination. Rates of antibody response as measured by CF or IF were lower, settling around 35% after a brief peak at 80% at 2 to 4 weeks after vaccination. Positive LSI were detected for at least 96 weeks after vaccination. In a high-risk population of abattoir workers, >95% of vaccinees had positive LSI at all intervals up to 5 years (longest period of sampling). Antibody prevalences after vaccination, measured by CF or IF, were higher than in the low-risk group and settled at about 60% overall. The differences in cellular and antibody responses between low- and high-risk groups is surmised to be due to the greater frequency of subjects in the high-risk group already primed to *C. burnetii* at a low level which was not expressed in positive CF and IF tests for antibody or in a positive skin test in the prevaccination screen tests (Table 7).

In these Australian studies, skin test conversion as a measure of vaccine efficacy was not as sensitive as assay of LSI; about 60% of the vaccinees from the high-risk group showed skin test conversion and even fewer of the low-risk group. This experience contrasts with that of Luoto et al.[58] who, however, used a larger dose of coxiellae in their vaccine than that employed in the South Australian trials. Skin test conversion has also been used successfully by Czechoslovakian workers[120] to measure the response to chemovaccine. In the latter trials, skin testing was combined with serological sampling to catch the booster effect of the skin test, thus providing an additional indicator of immunity.

TABLE 7
Patterns of Immune Response to Q Fever Vaccine (CSL Q-vax)
in Laboratory Research Staff (Outside Abattoir) with
Low Risk of Exposure to Q Fever Compared with
Those in Abattoir Workers (High Risk)

		Immune response to vaccine	
Group	N	Antibody positive[a]	Positive LSI[b]
Outside abattoir	32	11	27
	(100%)	(34%)	(84%)
Abattoir workers	81	58	77
	(100%)	(71%)	(95%)

Note: All subjects were Q fever antibody and skin test negative before vaccination.

[a] Antibody detected by CF, IF, RIA to *C. burnetii* phase I or II antigens.
[b] Positive lymphocyte stimulation index[99] *C. burnetii* phase I or II antigens.

II. GUIDELINES FOR USE OF Q FEVER VACCINE

A. OCCUPATIONAL GROUPS REQUIRING VACCINE PROPHYLAXIS

In brief, any group of workers handling cattle, sheep, or goats that are likely to be infected, and that are or have recently been pregnant require vaccine prophylaxis. The main target group is the abattoir workers, and this category includes not only the workers on the slaughter line and in the immediate processing area, but also all other persons entering the abattoir environment. The latter group of visitors include groups as diverse as electrical and mechanical operatives servicing machinery, catering school students, telephone engineers, research and diagnostic laboratory staff collecting samples, insurance office workers assessing fire risks, etc., all of whom have been infected on occasion.

The pregnant uterus and its contents, the bladder, udders, and intestine may contain the organism, and the unborn or newly born young may be infected or contaminated. The handling and processing of these by-products and the hides and fleeces may cause infection among the workers in subsidiary firms on the abattoir campus or at a distance. Drivers and other staff operating animal transport, working in the stockyards, providing veterinary services or auctioning animals have all experienced Q fever cases, although less frequently than among abattoir workers proper.

Many of the above groups have patterns of work or affiliations that enable them to be included in a vaccination program based in a medical center at the abattoir.

Q fever cases have also been detected among shearing teams, cattle and sheep farmers, and in groups raising goats, either for their wool or for their milk. It is likely that workers on the reception side of large plants handling milk, either for pasteurization and distribution or for cheese-making, may also be exposed from time to time, although, again, infection is less frequent than in the abattoir.

Geographically dispersed persons such as farmers and goat farmers are more difficult to fit into a vaccination program and must be willing to travel to a government or private clinic at which the facilities for vaccination are available.

If there is doubt about the exposure of a particular occupational group, a serological survey for antibody may resolve the matter. Among the longer term workers, 20 to 30 persons without a history of other exposure (on farms or in abattoirs) are bled and tested for Q fever CF antibody.

B. PROTOCOL FOR VACCINATION

Q fever vaccine is a powerful immunostimulant and must be handled with respect and strict adherence to pretesting protocol. It is mandatory that potential vaccinees are serotested and skin tested and that only those negative in both tests are vaccinated. The CF test has been used as the prevaccination screen test because it is widely available (at least in Australia).

From time to time "positive" CF results, with negative skin tests, are obtained which are probably due to anticomplementary effects. In these circumstances, the sera should be retested by IF or enzyme immunoassay (EIA) for Q fever antibody if these methods are available at the reference laboratory.

Antibody and skin testing together is recommended, rather than skin testing alone, so as to provide an additional element of safety in avoiding vaccine reactions. Resistance correlates best with skin test positivity, but there are occasional mistakes in performing an intradermal skin test. The inoculum may be placed too deep in the skin or in an area (e.g., over the front of the wrist) which is unsuitable. There may be dilution errors in preparing the skin test dose; again, if the diluted material is kept more than a few hours, the potency may be reduced by adsorption of the coxiella to the glass of the container. The antibody test provides a cross-check for such mishaps, and if there is a discrepancy between the antibody titer (say, CF antibody of titer 10 to 20 or more) and skin test result, then the skin test should be repeated, preferably at another site.

Finally, it should be emphasized that a negative antibody test and negative skin test does not equate completely with lack of immunity, although a high proportion of such persons will, in fact, be susceptible. The pretesting regimen is designed to avoid reactions on vaccination, not as an absolute guide to the immune state of the subject.

REFERENCES

1. **Derrick, E. H.**, Q fever, a new fever entity: clinical features, diagnosis and laboratory investigation, *Med. J. Aust.*, 2, 281, 1937.
2. **Derrick, E. H.**, The course of infection with *Coxiella burneti*, *Med. J. Aust.*, 1, 1051, 1973.
3. **Babudieri, B.**, *Advances in Veterinary Science*, Academic Press, 5, 81, 1959.
4. **Brada, M. and Bellingham, A. J.**, Bone marrow necrosis and Q fever, *Brit. Med. J.*, 281, 1108, 1980.
5. **Marmion, B. P.**, Subacute rickettsial endocarditis: an unusual complication of Q fever, *J. Hyg. Epidemiol.. Microbiol. Immunl. Prague*, 6, 79, 1962.
6. **Turck, W. P. G., Howitt, G., Turnberg, L. A., Fox, H., Longson, M., Matthews, M. B., and Das Gupta, R.**, Chronic Q fever, *Q. J. Med.*, 45, 193, 1976.
7. **Willey, R. F., Matthews, M. B., Peutherer, J. F., and Marmion, B. P.**, Chronic cryptic Q fever infection of the heart, *Lancet*, 2, 270, 1979.
8. **Peacock, M. G., Philip, R. N., Williams, J. C., and Faulkener, R. S.**, Serological evaluation of Q fever in humans: enhanced phase 1 titers of immunoglobulins G and A are diagnostic for Q fever endocarditis, *Infect. Immun.*, 41, 1089, 1983.
9. **Ellis, M. E., Smith, C. C., and Moffat, M. A.**, Chronic or fatal Q fever infection: a review of 16 patients seen in North East Scotland 1967—1980, *Q. J. Med.*, 52, 54, 1983.
10. **Weir, W. R., Bannister, B., Chambers, S., DeCock, K., and Mistry, H.**, Chronic Q fever associated with granulomatous hepatitis, *J. Infect.*, 8, 56, 1984.
11. **Palmer, S. R. and Young, S. E. J.**, Q fever endocarditis in England and Wales 1975-81, *Lancet*, 2, 1448, 1982.
12. **Marmion, B. P., Maddocks, I., Shannon, M., Izzo, A., and Oehler, S.**, unpublished, 1989.
13. **Cox, H. R. and Bell, E. J.**, The cultivation of *Rickettsia diaporica* in tissue culture and in the tissues of developing chick embryos, *Public Health Rep. Washington*, 54, 2171, 1939.
14. **Smadel, J. E., Snyder, M. J., and Robbins, F. C.**, Vaccination against Q fever, *Am. J. Hyg.*, 47, 71, 1948.
15. **Meiklejohn, G. and Lennette, E. H.**, Q fever in California 1 Observations on vaccination of human beings, *Am. J. Hyg.*, 52, 54, 1950.

16. **Vivona, S., Lowenthal, J. P., Berman, S., Benenson, A. S., and Smadel, J. E.**, Report of a field study with Q fever vaccine, *Am. J. Hyg.*, 79, 143, 1964.
17. **Marmion, B. P.**, Development of Q fever vaccines, 1937—1967, *Med. J. Aust.*, 2, 1074, 1967.
18. **Benenson, A. S.**, Q fever vaccine: efficacy and present status, in *Symposium on Q fever*, Smadel, J. E., Ed., Walter Reed Army Institute of Medical Science, Publ. No. 6, U.S. Government Printing Office, Washington, D. C., 1959, 47.
19. **Tigertt, W. D.**, Studies on Q fever in man, in *Symposium on Q fever*, Smadel, J. E., Ed., Walter Reed Army Institute of Medical Science, Publ. No. 6, U.S. Government Printing Office, Washington, D. C., 1959, 39.
20. **Bell, F. J., Lackman, D. B., Meis, A., and Hadlow, W. J.**, Recurrent reaction at site of Q fever vaccination in a sensitized person, *Milit. Med.*, 124, 591, 1964.
21. **Stoker, M. G. P.**, Q fever down the drain, *Brit. Med. J.*, 1, 425, 1957.
22. **Zdrodovskii, P. F. and Golinewich, H. M.**, *The Rickettsial Diseases*, Pergamon Press, New York, 1960, 422.
23. **Derrick, E. H.**, The Query Fever, The Elkington Oration, *Queensl. Health*, 1, 1, 1964.
24. **Lackman, D. B., Bell, E. J., Bell, J. F., and Picken, E. G.**, Intradermal sensitivity testing in man with a purified vaccine for Q fever, *Am. J. Public Health*, 52, 87, 1962.
25. **Fiset, P., Wike, D. A., Pickens, E. G., and Ormsbee, R. A.**, An antigenic comparison of strains of *Coxiella burnetii*, *Acta Virol.*, 15, 161, 1971.
26. **Stoker, M. G. P.**, Variation in complement fixing activity of *Rickettsia burneti* during egg adaptation, *J. Hyg.*, 51, 311, 1953.
27. **Stoker, M. G. P. and Fiset, P.**, Phase variation of the Nine Mile and other strains of *Rickettsia burneti*, *Can. J. Microbiol.*, 2, 310, 1956.
28. **Fiset, P.**, Antigenic Variations of Viruses and Rickettsiae with Particular Reference to *Rickettsia burneti*, Ph.D. thesis, Cambridge University, 1955.
29. **Fiset, P.**, Phase variation of *Rickettsia (Coxiella) burneti*: study of the antibody response in guinea pigs and rabbits, *Can. J. Microbiol.*, 3, 435, 1957.
30. **Ormsbee, R. A., Peacock, M. G., and Tallent, G.**, Dynamics of Phase I to Phase II antigenic shift in populations of *Coxiella burneti*, in *Rickettsiae and Rickettsial Diseases*, Kazar, J., Ed., Publishing House of the Slovak Academy of Sciences, Bratislava, Czechoslovakia, 1985.
31. **Hackstadt, T.**, Steric hindrance of antibody binding to surface proteins of *Coxiella burnetii* by phase 1 lipopolysaccharide, *Infect. Immun.*, 56, 802, 1988.
32. **Worswick, D. and Marmion, B. P.**, Antibody responses in acute and chronic Q fever and in subjects vaccinated against Q fever, *J. Med. Microbiol.*, 19, 281, 1985.
33. **Schmeer, N., Krauss, H., and Wilski, B.**, Serodiagnosis of human Q fever — demonstration of noncomplement binding of IgM antibody in the enzyme linked imunosorbent assay (ELISA), *Immun. Infekt.*, 12, 245, 1984.
34. **Schramek, S. and Mayer, H.**, Different sugar compositions of lipopolysaccharides isolated from Phase 1 and pure Phase 1 cells of *Coxiella burnetii*, *Infect. Immun.*, 38, 53, 1982.
35. **Amano, K.-I. and Williams, J. C.**, Chemical and immunological characterization of lipopolysaccharides from Phase I and Phase II *Coxiella burnetii*, *J. Bacteriol.*, 160, 994, 1984.
36. **Schramek, S., Kadziejewska-Lebrecht, J., and Mayer, H.**, 3-C-branched aldoses in lipopolysaccharide of Phase 1 *Coxiella burneti* and their role as immunodominant factors, *Eur. J. Biochem.*, 148, 455, 1985.
37. **Dahlman, O., Gareeg, P. J., Mayer, H., and Schramek, S.**, Synthesis of three 3-C-hydroxymethylpentoses with the D-ribo-, D-xylo- and L-xylo configurations. Identification of the latter with a monosaccharide derived from Phase 1 lipopolysaccharide, *Acta Chem. Scand. [B]*, 40, 15, 1986.
38. **Ormsbee, R. A., Pickens, E. G., and Lackman, D. B.**, An antigenic analysis of three strains of *Coxiella burneti*, *Am. J. Hyg.*, 79, 154, 1964.
39. **Hackstadt, T., Peacock, M. G., Hitchcock, P. J., and Cole, R. L.**, Lipopolysaccharide variation in *Coxiella burnetii*: intrastrain heterogenicity in structure and antigenicity, *Infect. Immun.*, 48, 359, 1985.
40. **Hackstadt, T.**, Antigenic variation in the Phase I lipopolysaccharide of *Coxiella burnetii* isolates, *Infect. Immun.*, 52, 337, 1986.
41. **Abinanti, F. R. and Marmion, B. P.**, Protective or neutralising antibody in Q fever, *Am. J. Hyg.*, 66, 173, 1957.
42. **Cracea, E., Dumitrescu, S., Mihancea, N., Novac, S., and Zarnea, G. H.**, Study of protective antibodies in Q fever vaccination, *Z. Immunitaetsforsch. Allerg. Klin. Immunol. (Stuttgart)*, 136, 1, 1968.
43. **Kazar, J., Brezina, R., Kovacova, E., and Urvolgyi, J.**, Testing in various systems of the neutralizing capacity of Q fever immune sera, *Acta Virol.*, 17, 79, 1973.
44. **Fiset, P.**, Serological diagnosis, strain identification and antigenic variation, in *Symposium on Q fever*, Smadel, J. E., Ed., Walter Reed Army Institute of Medical Science, Publ. No. 6, U.S. Government Printing Office, Washington, D. C., 1959, 28.
45. **Ormsbee, R. A., Bell, E. J., Lackman, D. B., and Tallent, G.**, The influence of phase on the protective potency of Q fever vaccine, *J. Immunol.*, 92, 404, 1964.
46. **Hoyer, B. H., Ormsbee, R. A., Fiset, P., and Lackman, D. B.**, Differentiation of Phase I and Phase II *C. burnetii* by equilibrium density gradient sedimentation, *Nature (London)*, 197, 573, 1963.

47. **Moos, A. and Hackstadt, T.**, Comparative virulence of intra- and interstain lipopolysaccharide variants of *Coxiella burnetii* in the guinea pig model, *Infect. Immun.*, 55(5), 1144, 1987.

48. **Izzo, A. A., Marmion, B. P., and Hackstadt, T.**, Analysis of the cells involved in the lymphoproliferative response to *C. burnetii* antigens, in preparation, 1989.

49. **Craigie, J.**, Application and control of ethyl ether-water interface effects to separation of rickettsiae from yolk sac suspensions, *Can. J. Res.*, 23, 104, 1945.

50. **Ormsbee, R. A.**, A method of purifying *Coxiella burnetii* and other pathogenic rickettsiae, *J. Immunol.*, 88, 100, 1961.

51. **Ormsbee, R., Peacock, M., Philip, R., Casper, E., Plorde, J., Gabre-Kidan, T., and Wright, L.**, Serological diagnosis of epidemic typhus fever, *Am. J. Epidemiol.*, 105, 261, 1977.

52. **Amano, K.-I., Williams, J. C., McCaul, T. F., and Peacock, M. G.**, Biochemical and immunological properties of *Coxiella burnetii* cell wall and peptidoglycan-protein complex fractions, *J. Bacteriol.*, 160, 982, 1984.

53. **Spicer, D. S. and DeSanctis, A. N.**, Preparation of Phase 1 Q fever antigen suitable for vaccine use, *Applied Environ. Microbiol.*, 32, 85, 1976.

54. **Williams, J. C., Peacock, M. G., and McCaul, T. F.**, Immunological and Biological characterisation of *Coxiella burnetii*, Phases I and II, separated from host components, *Infect. Immun.*, 32, 840, 1981.

55. **Fiset, P.**, Vaccination against Q fever, in *Proc. 1st Int. Congr. Vaccines against Viral and Rickettsial Diseases of Man*, Sci. Publ. No. 147, Pan American Health Organization, Washington, D. C., 1967, 528.

56. **Fiset, P.**, *Review of status of Q fever vaccine and vaccine studies*, Commission on Rickettsial Diseases, Armed Forces Epidemiology Board, 1970.

57. **Philip, R. N.**, personal communication.

58. **Luoto, L., Bell, J. F., Casey, M., and Lackman, D. B.**, Q fever vaccination of human volunteers. I. The serologic and skin test response following subcutaneous injections, *Am. J. Hyg.*, 78, 1, 1963.

59. **Bell, J. F., Luoto, L., Casey, M., and Lackman, D. B.**, Serologic and skin test response after Q fever vaccination by the intracutaneous route, *J. Immunol.*, 93, 403, 1964.

60. **Kishimoto, R. A., Gonder, J. C., Johnson, J. W., Reynolds, J. A., and Larson, E. W.**, Evaluation of a killed Phase I *Coxiella burnetii* vaccine in cynomologous monkeys (Macacca faskicularis), *Lab. Anim. Sci.*, 31, 48, 1981.

61. **Luoto, L., Winn, J. F., and Huebner, R. J.**, Q fever studies in Southern California. XIII. Vaccination of Dairy Cattle against Q fever, *Am. J. Hyg.*, 55, 90, 1951.

62. **Sadecky, E., Brezina, R., Kazar, J., Schramek, S., and Urvolgyi, J.**, Immunisation against Q fever of naturally infected dairy cows, *Acta Virol.*, 19, 486, 1975.

63. **Sadecky, E., Brezina, R., and Michalovic, M.**, Vaccination of cattle against Q fever, *J. Hyg. Epidemiol. Microbiol. Immunol.*, 19, 200, 1975.

64. **Biberstein, E. L., Reimann, H. P., Franti, C. E., Behymer, D. E., Ruppanner, R., Bushnell, R., and Crenshaw, G.**, Vaccination of dairy cattle against Q fever (*Coxiella burnetii*) results of field trial, *Am. J. Vet. Res.*, 38, 189, 1977.

65. **Brooks, D. L., Ermel, R. W., Franti, C. E., Ruppanner, R., Behymer, D. E., Williams, J. C., and Stephenson, J. C.**, Q fever vaccination of sheep: challenge of immunity in ewes, *Am. J. Vet. Res.*, 47, 1235, 1986.

66. **Colter, J. S., Brown, R. A., Bird, H. H., and Cox, H. R.**, The preparation of a soluble immunizing antigen from Q fever rickettsiae, *J. Immunol.*, 76, 270, 1956.

67. **Ormsbee, R. A., Bell, E. J., and Lackman, D. B.**, Antigens of *Coxiella burnetii*. I. Extraction of antigens with nonaqueous organic solvents, *J. Immunol.*, 88, 741, 1962.

68. **Anacker, R.L., Lackman, D. B., Pickens, E. G., and Ribi, E.**, Antigenic and skin reactive properties of fractions of *Coxiella burnetii*, *J. Immunol.*, 89, 145, 1962.

69. **Brezina, R. and Urvolgyi, J.**, Extraction of *Coxiella burneti* phase 1 antigen by means of trichloracetic acid, *Acta Virol.*, 5, 193, 1961.

70. **Cracea, E., Dumitrescu, S., Botez, D., Toma, E., Bandu, C., Sabin, S., Ioanid, L., and Chirescu, N.**, Immunisation in man with a soluble Q fever vaccine, *Arch. Roum. Pathol. Exp. Microbiol.*, 32, 45, 1973.

71. **Cracea, E., Dumitrescu-Constantinescu, S., Botez, D., and Ioanid, L.**, Q fever soluble vaccine effects in *Coxiella burneti* sensitized humans, *Zentralbl. Bakteriol. Parasitenkd. Infektionskr.*, 238, 413, 1977.

72. **Brezina, R., Schramek, S., Kazar, J., and Urvolgyi, J.**, Q fever chemovaccine for human use, *Acta Virol.*, 18, 269, 1974.

73. **Kazar, J., Brezina, R., Palanova, A., Tvrda, B., and Schramek, S.**, Immunogenicity and reactogenicity of a Q fever chemovaccine in persons professionally exposed to Q fever in Czechoslovakia, *Bull. H. O.*, 60, 389, 1982.

74. **Brezina, R., Pospisil, V., and Schramek, S.**, Study of the antigenic structure of *Coxiella burnetii*. VII. Properties of phenol extracted Phase 1 antigenic component, *Acta Virol.*, 14, 295, 1970.

75. **Brezina, R. and Pospisil, V.**, Study of the antigenic structure of *C. burnetii*. VIII. Immunogenicity of phenol extracted Phase I antigenic component, *Acta Virol.*, 14, 302, 1970.

76. **Schramek, S.**, Rickettsial endolipopolysaccharides, in *Rickettsiae and Rickettsial Diseases*, Kazar, J., Ormsbee, R. A., and Tarasevich, I. V., Eds., Veda Slovak Academy of Sciences, Bratislava, Czechslovakia, 1978, 79.

77. **Ascher, M. S., Berman, M. A., and Ruppanner, R.**, Initial clinical and immunologic evaluation of a New Phase 1 Q fever vaccine and skin test in humans, *J. Infect. Dis.*, 148, 214, 1983.

78. **Damrow, T. A., Williams, J. C., and Waag, D. M.**, Suppression of *in vitro* lymphocyte proliferation in C57BL/ 10 ScN mice vaccinated with Phase I *Coxiella burneti*, *Infect. Immun.*, 47, 149, 1985.

79. **Williams, J. C. and Cantrell, J. L.**, Biological and immunological properties of *Coxiella burnetii* vaccines in C57BL/10ScN endotoxin-nonresponder mice, *Infect. Immun.*, 35, 1091, 1982.

80. **Kokorin, I. N., Pushkareva, V. I., Kazar, J., and Schramek, S.**, Histological changes in mouse liver and spleen caused by different *Coxiella burneti* antigenic preparations, *Acta Virol. (Praha)*, 5, 410, 1985.

81. **Ascher, M. S., Williams, J. C., and Berman, M. A.**, Dermal granulomatous hypersensitivity in Q fever: comparative studies of the granulomatous potential of whole cells of *Coxiella burnetii* phase 1 and subfractions, *Infect. Immun.*, 42, 887, 1983.

82. **Williams, J. C., Damrow, T. A., Waag, D. M., and Amano, K.-I.**, Characterization of a Phase 1 *Coxiella burnetii* chloroform-methanol residue vaccine that induces active immunity against Q fever in C57BL/10ScN mice, *Infect. Immun.*, 51, 851, 1986.

83. **Kazar, J., Votruba, D., Propper, P., and Schramek, S.**, Onset and duration of immunity in guinea pigs and mice immunised with different types of Q fever vaccine, *Acta Virol.*, 30, 499, 1986.

84. **Genig, V. A.**, Attenuated variant "M" of *Rickettsia burneti* as a possible live vaccine against Q fever, *Vestn. Akad. Med. Nauk S.S.S.R.*, 15, 46, 1960 (in Russian).

85. **Genig, V. A.**, Experiences of the mass imunization of man with the live vaccine M-44 against Q fever. II. Cutaneous and oral methods of immunization (translation), *Vopr. Virusol.*, 10, 703, 1965.

86. **Genig. V. A.**, A live vaccine 1/M-44 against Q fever for oral use, *J. Hyg. Epidemiol. Microbiol. Immunol.*, 12, 265, 1968.

87. **Johnson, J. W., Eddy, G. A., and Pederson, C. D.**, Biological properties of the M44 strain of *Coxiella burnetii*, *J. Infect. Dis.*, 133, 334, 1976.

88. **Johnson, J. W., McLeod, C. G., Stookey, J. L., Higbee, G. A., and Pedersen, C. E.**, Lesions in guinea pigs infected with *Coxiella burnetii* strain M-44, *J. Infect. Dis.*, 135, 995, 1977.

89. **Wisseman, C. L., Jr. and Fiset, P.**, Annual progress report, AFEB Commission on Rickettsial Diseases, unpublished data, 1971.

90. **Sidwell, R. W., Thorpe, B. D., and Gebhardt, L. P.**, Studies of latent Q fever infections in effect of multiple cortisone injections, *Am. J. Hyg.*, 79, 320, 1964.

91. **Bayer, R. A.**, Q fever as an occupational illness at the National Institutes of Health, *Public Health Rep. Washington*, 97, 58, 1982.

92. **Hall, C. J., Richmond, S. J., Caul, E. O., Pearce, N. H., and Silver, I. A.**, Laboratory outbreak of Q fever acquired from sheep, *Lancet*, 1, 1004, 1982.

93. **Curet, L. B. and Paust, J. C.**, Transmission of Q fever from experimental sheep to laboratory personnel, *Am. J. Obstet. Gynecol.*, 114, 566, 1972.

94. **Schachter, J., Sung, M., and Myer, K. F.**, Potential danger of Q fever in a university hospital environment, *J. Infect. Dis.*, 123, 301, 1971.

95. **Meiklejohn, G., Reimer, L. G., Graves, P. S., and Helmick, C.**, Cryptic epidemic of Q fever in a medical school, *J. Infect. Dis.*, 144, 107, 1981.

96. **Buckley, B.**, Q fever in Victorian general practice, *Med. J. Aust.*, 1, 593, 1980.

97. **Marmion, B. P., Ormsbee, R. A.,. Kyrkou, M., Wright, J., Worswick, D., Cameron, S., Esterman, A., Feery, B., and Collins, W.**, Vaccine prophylaxis of abattoir-associated Q fever, *Lancet*, 2, 1411, 1984.

98. **Kleinbaum, D. G., Kupper, L. L., and Morgenstern, H.**, *Epidemiologic Research Principles and Quantitative Methods*, Wadsworth, CA, 1982.

99. **Izzo, A. A., Marmion, B. P., and Worswick, D. A.**, Markers of cell-mediated immunity after vaccination with an inactivated whole cell Q fever vaccine, *J. Infect. Dis.*, 157(4), 781, 1988.

100. **Shapiro, R., Siskind, V., and Marmion, B. P.**, unpublished data.

101. **Kazar, J., Schramek, S., Lisak, V., and Brezina, R.**, Antigenicity of chloroform methanol treated *Coxiella burnetii* preparations, *Acta Virol.*, 31, 158, 1987.

102. **Tabert, G. G. and Lackman, D. B.**, The radioisotope precipitation test for study of Q fever antibodies in human and animal sera, *J. Immunol.*, 94, 959, 1985.

103. **Fiset, P., Ormsbee, R. A., Silberman, R., Peacock, M., and Spielman, S. H.**, A microagglutination technique for the detection and measurement of rickettsial antibodies, *Acta Virol.*, 13, 60, 1969.

104. **Murphy, J. R., Wisseman, C. L., and Fiset, P.**, Mechanisms of immunity in typhus infection: some characteristics of *Rickettsia mooseri* infection in guinea pigs, *Infect. Immun.*, 21, 417, 1978.

105. **Murphy, J. R., Wisseman, C. L., and Fiset, P.**, Mechanisms of immunity in typhus infection: adoptive transfer of immunity in *Rickettsia mooseri*, *Infect. Immun.*, 24, 387, 1979.

106. **Murphy, J. R., Wisseman, C. L., and Fiset, P.**, Mechanisms of immunity in typhus infection: analysis of immunity to *Rickettsia mooseri* infection of guinea pig, *Infect. Immun.*, 27, 730, 1980.
107. **Crist, A. E., Wisseman, C. L., and Murphy, J. R.**, Characteristics of lymphoid cells that adoptively transfer immunity to *Rickettsia mooseri* infection in mice, *Infect. Immun.*, 44, 55, 1984.
108. **Wisseman, C. L. and Waddel, A.**, Interferon-like factors from antigen and mitogen stimulated human leucocytes with anti rickettsial and cytolytic actions on *Rickettsia prowazekii* infected human endothelial cells, fibroblasts and macrophages, *J. Exp Med.*, 157, 1780, 1983.
109. **Kishimoto, R. A., Rozmiarek, H., and Larson, E. W.**, Experimental Q fever infection in congenitally athymic nude mice, *Infect. Immun.*, 22, 69, 1978.
110. **Humphres, R. C. and Hinrichs, D. J.**, Role for antibody in *Coxiella burnetii* infection, *Infect. Immun.*, 31, 641, 1981.
111. **Kishimoto, R. A., Veltri, B. J., Shirey, F. G., Canonico, P. G., and Walker, J. S.**, Fate of *Coxiella burnetii* in macrophages from immune guinea pigs, *Infect. Immun.*, 15, 601, 1977.
112. **Kishimoto, R. A. and Walker, J. S.**, Interaction between *Coxiella burneti* and guinea pig peritoneal macrophages, *Infect. Immun.*, 14, 416, 1976.
113. **Kishimoto, R. A., Veltri, B. J., Canonico, P. G., Shirey, F. G., and Walker, J. S.**, Electron microscopic studies on the interaction between normal guinea pig macrophages and *Coxiella burneti*, *Infect. Immun.*, 14, 1087, 1976.
114. **Kishimoto, R. A. and Burger, G. T.**, Appearance of cellular and humoral immunity in guinea pigs after infection with *Coxiella burneti* administered in small particle aerosols, *Infect. Immun.*, 16, 518, 1977.
115. **Kishimoto, R. A., Johnson, J. W., Kenyon, R. H., Ascher, M. S., Larson, E. W., and Pedersen, C. E.**, Cell mediated immune responses of guinea pigs to an inactivated Phase I *Coxiella burneti* vaccine, *Infect. Immun.*, 19, 194, 1978.
116. **Kazar, J. and Schramek, K.**, Relationship between delayed antigenic hypersensitivity reaction and resistance to virulent challenge in mice immunised with different *Coxiella burneti* antigen preparations, in *Proc. 3rd Int. Symp. Rickettsiae and Rickettsial Diseases*, Kazar, J., Ed., Veda, Bratislava, 1985.
117. **Jerrells, T. R., Mallavia, L. P., and Hinrichs, D. J.**, Detection of long term cellular imunity to *Coxiella burneti* as assayed by lymphocyte transformation, *Infect. Immun.*, 11, 280, 1975.
118. **Hinrichs, D. J. and Jerrells, T. R.**, *In vitro* evaluation of immunity to *Coxiella burneti*, *J. Immunol.*, 117, 996, 1976.
119. **Turco, J., Thompson, H. A., and Winkler, H. H.**, Interferon-gamma inhibits growth of *Coxiella burnetii* in mouse fibroblasts, *Infect. Immun.*, 45, 781, 1984.
120. **Kazar, J., Schramek, S., and Brezina, R.**, The value of skin test in Q fever convalescents and vaccinees as indicator of antigen exposure and inducer of antibody recall, *Acta Virol.*, 28, 134, 1984.
121. **Amano, K.-I., Williams, J. C., Missler, S. R., and Reinhold, V. N.**, Structure and biological relationships of *Coxiella burnetii* lipopolysaccharides, *J. Biol. Chem.*, 262, 4740, 1987.
122. **Ormsbee, R., Peacock, M., Gerloff, R., Tallent, G., and Wike, D.**, Limits of rickettsial infectivity, *Infect. Immun.*, 19, 239, 1978.
123. **Worswick, D. and Marmion, B. P.**, unpublished, 1989.
124. **Izzo, A. A. and Marmion, B. P.**, unpublished, 1989.
125. **Hornick, R. B.**, unpublished, 1967.

INDEX

T - #0263 - 101024 - C0 - 254/178/14 [16] - CB - 9780849359842 - Gloss Lamination